Science and Emotions after 1945

Science and Emotions after 1945

A Transatlantic Perspective

EDITED BY FRANK BIESS
AND DANIEL M. GROSS

The University of Chicago Press   Chicago and London

FRANK BIESS is professor of history at the University of California, San Diego, and the author of *Homecomings: Returning POWs and the Legacies of Defeat in Postwar Germany.*
DANIEL M. GROSS is associate professor of English at the University of California, Irvine. He is the author of *The Secret History of Emotion: From Aristotle's "Rhetoric" to Modern Brain Science.*

The University of Chicago Press, Chicago 60637
The University of Chicago Press, Ltd., London

Printed in the United States of America

23 22 21 20 19 18 17 16 15 14   1 2 3 4 5

ISBN-13: 978-0-226-12634-0 (cloth)
ISBN-13: 978-0-226-12648-7 (paper)
ISBN-13: 978-0-226-12651-7 (e-book)
DOI: 10.7208/chicago/9780226126517.001.0001

Library of Congress Cataloging-in-Publication Data

Science and emotions after 1945 : a transatlantic perspective / edited by Frank Biess and Daniel M. Gross.
pages cm
Includes bibliographical references and index.
ISBN 978-0-226-12634-0 (hardcover : alkaline paper)—ISBN 978-0-226-12648-7 (paperback : alkaline paper)—ISBN 978-0-226-12651-7 (e-book) 1. Emotions—Psychological aspects. 2. Affective neuroscience—History—20th century. 3. Psychology—Germany—History—20th century. 4. Psychology—United States—History—20th century. I. Biess, Frank, 1966– editor of compilation. II. Gross, Daniel M., 1965– editor of compilation.
BF531.S38 2014
152.409′045—dc23

2013032851

♾ This paper meets the requirements of ANSI/NISO Z39.48-1992 (Permanence of Paper).

Contents

INTRODUCTION

Emotional Returns

FRANK BIESS
DANIEL M. GROSS

Writing in 1940, the doyen of American sociology, Talcott Parsons, outlined what he considered to be the main differences between American democracy and German fascism. Among these differences was an emphasis on the "rationalistic character of [American] culture," which he then contrasted to "fundamentalism, not only in religion, but in any field," as one characteristic feature of National Socialism, and which he saw as challenging "the status of critical rationality in our culture."[1] This diagnosis was symptomatic of a widespread association of German fascism with political irrationalism and excessive emotionality. In the contemporary American academic discourse of the 1930s and 1940s, Nazism was often understood in terms of a collective psychopathology, and this diagnosis was later extended to communism.[2]

Postwar Rationalism and the Marginalization of Emotions after 1945

Leading social scientists' association of Nazism with excessive emotionalism further enhanced the marginalization and deep suspicion of emotions in Western social science that predated the post–World War II period but assumed a new meaning after 1945. Emotions, at least as far as they were visible and operative in social and political life, appeared largely as symptoms or causes of political and social pathologies and hence did not assume a central place in outlining prescriptive guidelines for a democratic, antitotalitarian society. While Parsons's work is a case in point, he did not discount emotions entirely. In fact, the contrast between *affectivity* and *affective neutrality* constituted the first pattern variable in Parsons's explanation of social action.[3] Affective neutrality enabled the renunciation of instant gratification and hence was instrumental to the stability and the functioning of the social system. Still, Parsons's theory offered, in the words of Jack Barbalet, "a paradigm case of a sophisticated discounting of the significance of emotion for understanding social processes."[4] On this account emotions flourished in the private sphere of the family and friendships but became increasingly less relevant in higher-ranked "secondary institutions" of the modern state. If anything, unfettered emotions threatened to destabilize the social order and hence needed to be contained by mechanisms of social control. In this respect, emotions remained present in postwar social theory yet increasingly represented the irrational, premodern and potentially totalitarian "other" to modern "rational" liberal democracy.[5]

Dominant social science paradigms of the postwar period such as modernization theory and behaviorism reinforced this marginalization of emotions. Modernization theory rejected the "emotionality and spiritualism of romanticism" and propagated instead the "ideals of the Enlightenment: the power of science, the importance of control, and the possibility of achieving progress through application of human will and instrumental reason."[6] Behaviorism did not develop a theory of emotions but tended to see emotions as conditioned responses to external stimuli.[7] If behaviorists discussed emotions, they portrayed them—in line with the dominating irrationalism model—as politically damaging. The political scientist Harold Laswell's psychopathological model, for example, defined emotions as the displacement of unconscious impulses onto the field of politics.[8] In the context of the highly ideological conflicts of the Cold War, one of the goals of postwar social science was the containment and management of emotions. As Hunter Heyck has argued, Cold War science aimed at the "production of reason" by developing models of decisionmaking that shifted the focus away from the potentially irrational "chooser" to the process of rational choice or decisionmaking. Emotions were relegated in this process either to preexisting "givens (like values and preferences)," or they were "intrusions that short-circuited the normal processes of decision."[9] Even psychoanalysis in its Americanized version of

Table 1: Postwar Rationalism

Philosophy	Economics	Political Science	Sociology	Psychology/ Psychoanalysis
Logical positivism	Rational choice	Habermas[a]	Weber (US)	Skinner
	Samuelson textbook[b]	Rawls	Parsons	Hartmann
		Laswell		

[a]A view from the US perspective ca. 1978 is summarized in Ronald Rogowski, "Rationalist Theories of Politics: A Midterm Report," *World Politics* 30.2 (1978): 296–323. German terms for the debate are famously established in the *Positivismusstreit* 1961–69, which set Karl Popper's critical rationalism against the critical theory of the Frankfurt school, which included Habermas, whose development of the lifeworld sets the cognitive horizon against a background of practices and competencies, including affective. Hence, Habermas is a rationalist only from a limited theoretical perspective.
[b]Paul A. Samuelson, *Economics: An Introductory Analysis* (New York: McGraw-Hill, 1948). For a discussion, see Frevert, chapter 12 in this volume.

ego psychology supported the larger project of postwar rationalization and normalization. Psychodynamic therapy was supposed to promote the self-adaption of the individual to the cultural norms of postwar society, including heterosexuality and the nuclear family. It treated the individual as, in the words of Ely Zaretsky, a "rational, self-regulating actor whose maturation would be facilitated by forms of intervention that refrained from external direction."[10]

Postwar "rationalism" thus extended across a variety of disciplines in the social sciences and beyond. In the spirit of heuristics, not completism, we offer in table 1 a sketch of this trend marked by a few key figures and schools of thought. This relative marginalization and pathologization of emotions in midcentury Western social science constituted a distinctly postwar phenomenon, illustrated by a formative moment in the intellectual development of Jürgen Habermas, whom we could classify under sociology and philosophy as well as political science.

In 1953 his fellow student Karl-Otto Apel handed Habermas the reissue of Martin Heidegger's 1935 lectures *Introduction to Metaphysics*, in which Habermas encountered antirationalism, including "idolatry of the nationalist spirit," Platonist devaluation of "intelligence," and the rejection of Enlightenment principles of equality and universalism.[11] Upset that this postwar reissue appeared with no meaningful qualification of Heidegger's avowed National Socialism, the twenty-four-year-old Habermas published a critique in the *Frankfurter Allgemeine Zeitung* (July 25, 1953): "Mit Heidegger gegen Heidegger denken: Zur Veröffentlichung von Vorlesungen aus dem Jahre 1935" (Thinking with Heidegger against Heidegger: On the publication of lectures from 1935). Soon thereafter, while working as Theodor W. Adorno's research assistant and reading American social scientists, including Parsons preeminently, Habermas started developing a broad, social modernization framework for understanding the relationship between the political failures of German philosophers—Heidegger, Schmitt, Jünger, Gehlen—and democracy movements that had taken root elsewhere in the West but not yet securely in 1950s Germany. He drew important analytic distinctions between rational forms of social organization and "pathological" forms, with the root word *pathos* suggesting diagnosis and cure. Under what conditions, he asked, do public spaces and the political public sphere either "betray the pathological traits of anomie or repression" or, alternatively, provide the space for a complex society to cohere normatively, according to "abstract, legally mediated forms of solidarity among citizens?"[12] Habermas explains in *Zwischen Naturalismus und Religion*, his 2005 intellectual autobiography, that this question motivated his abiding and influential pursuit of "communicative rationality" as a model for reconstructing the norms and procedures by which agreement might be reached by very different sorts of people. Rationalism in this context functioned almost as a form of "magical thinking," that is, both a means and an end for containing and controlling perceived irrational and populist impulses that were deemed to have formed the emotional basis of fascism and might still pose a threat to postwar liberal democracy in the present.[13] In a similar vein, the eminent West German historian Hans Ulrich Wehler attributed the neglect and marginalization of emotions in postwar history writing to violent aftereffects of war and fascism. "For those generations who had experienced the Second World War, flight, and the postwar period," he argued, "the control of affects was an indispensible precondition of physical and psychological survival." Emotions were "given in to only among family and close friends" and entered "academic and political engagement only in a mediated fashion." This suspicion of emotions distinguished the postwar generation, according to Wehler, from the later generation, for whom emotions became both more important and less problematic.[14] What these examples indicate to the historian of emotions is how National Socialism, the Second World War, and their lingering aftereffects served as a point of departure, in Germany and Europe more broadly, for postwar projects in rational reconstruction particularly applicable to the social sciences.

The turn to rationalism in postwar social thought, however, followed on the heels of, and partly disavowed, the discovery and broad thematization of emotions across a wide range of academic disciplines during the "long" turn of the century from the 1890s to the 1930s.[15] As Daniel Morat and Uffa Jensen's collection *Die Rationalisierung des Gefühls* (The rationalization of feelings) has persuasively demonstrated, emotions came increasingly into the purview of the humanities and the sciences during this period. This was also the period when the scientific view of emotions as purely physiological, nonintentional, and noncognitive processes took shape in its postromantic form.[16] William James's 1884 essay "What Is an Emotion" marked an important culmination point in this development. Yet, as Thomas Dixon has shown, while James's view remains influential up to this day, it was already subject to much contemporary criticism, and later treatments of emotions in other disciplines did not necessary follow his view of emotions as purely physiological.[17]

The late nineteenth century and the early twentieth also saw an intense engagement with emotions across a variety of disciplines, such as sociology, philosophy, and history writing. For the German sociologist Georg Simmel, emotions formed an important social kit in society.[18] Emotions also featured prominently in the grand narratives of modernity as they were developed in the early twentieth century. Recent research has highlighted the long-forgotten centrality of emotions in Max Weber's work. The "Protestant ethic," for example, was centrally driven by a "fear of condemnation."[19] Norbert Elias's *On the Process of Civilization*, written in the 1930s but not widely received until the 1960s and 1970s, identified the increasing control of affects as central in the process of modernization. And the historian John Huizinga highlighted the intense and "childish" emotionality of the Middle Ages in contrast to a more mature and restraint modernity.[20] All these narratives, including Freud's civilization theory, shared what Barbara Rosenwein has called a "hydraulic" model of emotions, namely, a model holding that emotions are "like great liquids within each person, heaving and frothing, eager to be let out."[21] The narrative of modernity became identical with the development of increasingly refined practices of individual and collective emotional restraint and (self-)control. The perceived failure of these mechanisms of emotional control then informed Lucien Febvre's seminal call for a history of the "emotional life of man and all its manifestations," including "the history of hate, the history of fear, the history of cruelty, the history of love," at the height of the Nazi Empire in 1941.[22] The post-1945 marginalization of emotions in social theory and their association with political irrationalism thus did not constitute an entirely new development. They rather hardened and intensified the identification of emotions as modern rationality's other that had already emerged in the earlier part of the twentieth century.

Historicizing the Emotional Turn

Our volume seeks to explain this reconfiguration in the historical relationship between science and emotions after 1945. What happened to the pre-1945 academic preoccupation with emotions? Why and by what disciplinary means were emotions as a category increasingly marginalized and delegitimated in the sciences and humanities? And what prompted a gradual rediscovery of emotions as an analytic category and as an object of research from the 1960s on? In answering these questions, this volume seeks to provide greater historical depth to the current fascination with emotions across a wide range of academic disciplines. It seeks to historicize what some observers in the humanities and social sciences have already called the *affective* or *emotional turn*.[23]

The difference in terminology between *affect* and *emotion* requires some additional commentary. Considered historically, and occasionally in this volume, *affect* can simply be a synonym for *emotion*. But, from a transdisciplinary perspective, this terminology can be a point of inadvertent but also sometimes sharp disagreement about priority. Historically, *affect* can designate a general physiological disposition that precedes emotion theoretically, temporally, phylogenetically, and/or ontogenetically. In Shakespeare's *Love's Labour's Lost* (1.1.149) we read: "For euery man with his affectes is borne." But we would not expect to hear that we are born angry, let alone jealous, which are emotions in Shakespeare and beyond. From one polemic perspective with historical roots, affect comes first, and then we grow into emotions as individuals and as a species. This formulation in more recent scientific psychology considers affect the experience of a feeling or an emotion, and hence it would be something (unlike emotion) that we share with lower organisms.[24] According to Ruth Leys, who has written pointedly about this conceptual distinction, the definition of *affect* has recently consolidated across some neuroscience and also some high theory as "independent of, and in an important sense prior to, ideology—that is, prior to intentions, meanings, reasons, and beliefs—because they are nonsignifying, autonomic processes that take place below the threshold of conscious awareness and meaning."[25] Her analysis is based in part on a critical response to Brian Massumi's "The Autonomy of Affect,"[26] which appears in her reading as the inadvertent return of Cartesian mind/body dualism, where affect is associated with body and emotion with the mind. Alternatively, from the perspective of social construction, where the appropriate level of analysis is phenomenological or rhetorical, emotion is in fact fundamental, so there is no affect prior to, underneath, or behind the nameable emotions such as jealousy or love or anger.[27] Finally, *emotion* designates a familiar semantic field discernible as we historicize key terms like *fear* or *anger* or *love*—hence, *emotion* is the organizing term we use in this volume, which is ultimately housed in the discipline of history, where the "priority of affect" would take us beyond the historical object of study (for discussion, see also Reid, chapter 15 in this volume).

By analyzing the changing relationship between science and emotion after 1945, the volume thus seeks to provide a genealogy of the current fascination with emotions across many disciplines. Presently, emotions are no longer seen as an irrational province restricted to and best expressed in the fine arts or treated by psychoanalysis. Instead, they have become the subject of proliferating interdisciplinary research in the social and natural sciences, for reasons we will explore in this book. The editors of the recently founded and already significant periodical *Emotion Review* celebrate emotions as now constituting "one of the great hubs in the scientific study of the human condition,"[28] and social scientists and humanities scholars write about the emotional turn.

In several disciplines, the rationalism of the postwar period has been replaced by approaches and theories that provide more extensive space for emotions. Once again in the spirit of heuristics, not completism, we offer a sketch of this trend marked by a few key figures and schools of thought (see table 2). In this volume you will read about situated cognition (Gross and Preston, chapter 3), behavioral economics (Frevert, chapter 12), the Goffman-Hochschild "Berkeley" school of sociology (Flam, chapter 13),[29] and Lazarus et al.'s appraisal theory in psychology (Reddy, chapter 1) because each maintains a strong claim on social science. Generally, however, the emotional turn in the social sciences necessitates some new kind of humanistic inquiry: in each of these discipline categories, schools and figures after the emotional turn invite critical, historical, interpretive, and rhetorical work that would qualify the stricter social science of Parsons and his contemporaries outlined above.

But which historical factors have made it once again safe for serious scientific work on emotion next to—and sometimes including—humanistic work on emotion that no longer appears sympathetic to

Table 2: The Emotional Turn

Philosophy	Economics	Politics	Sociology	Psychology
Situated cognition	Behavioral economics	Neorepublicanism	Hochschild	Lazarus
Nussbaum		Communitarianism		Averill[a]
		Critical theory[b]		

[a]James R. Averill, "An Analysis of Psychophysiological Symbolism and Its Influence on Theories of Emotion," *Journal for the Theory of Social Behavior* 4 (1974): 147–90; Richard S. Lazarus, "Thoughts on the Relations between Emotion and Cognition," *American Psychologist* 37 (1982): 1019–24.
[b]A view from the US perspective ca. 2000 is summarized in George Marcus, "Emotions in Politics," *Annual Review of Political Science* 3 (2000): 221–50. Neorepublican work on emotion has roots in Leo Strauss, *The Political Philosophy of Hobbes: Its Basis and Its Genesis* (1935; Chicago: University of Chicago Press, 1952). Examples of what is sometimes called communitarian work on emotion include Jean-Luc Nancy drawing on Heidegger, the American school around the Berkeley sociologist Robert Bellah, and Michael Walzer, notably *Politics and Passion: Toward a More Egalitarian Liberalism* (New Haven, CT: Yale University Press, 2004). Critical theory of emotion appears, for example, in the work of political philosopher Wendy Brown, and also Sara Ahmed in the UK. Raymond Williams, associated with the New Left, introduced "structures of feeling" tied to the analytics of class and class conflict; see his *Marxism and Literature* (Oxford: Oxford University Press, 1977), 128–35.

fascism or other forms of irrationalism? The answer to this question is complex and will occupy most of the chapters included in this volume. While increased attention to emotions across several disciplines and social movements dates back to the 1960s, the emotional turn is closely associated with the more recent rise of the neuroscientific study of emotion. Our claim is that the relatively recent (last decade of the twentieth century and first decade of the twenty-first) neuroscientific study of emotion did not initiate but instead consolidated the emotional turn by clearing the ground for all sorts of work on the emotions, now unencumbered by the postwar pathologies introduced above. Emotion studies in the social sciences and even in the humanities now can work around postwar models of rationality versus irrationalism while invoking biological and physiological models of emotion offered by neuroscience. Sometimes, as in the case of Martha Nussbaum, this invocation of neuroscience in the social sciences and humanities is explicit. Sometimes it remains only in the expert and popular background cultures that make a variety of emotion studies seem worthwhile beyond the stigma of irrationalism, even if the particular study is disconnected from neuroscience (e.g. Walzer, *Politics and Passion*). Even humanities work that is critical of neuroscience finds its opportunity largely within the vast expanse of legitimation that such neuroscience provides (e.g., Leys, chapter 2 in this volume; Gross and Preston, chapter 3).[30]

From this perspective, the emotional turn proper could be dated to

1994, when the neuroscientist Antonio Damasio popularized his expert work in *Descartes' Error*, followed by a further popularized version in the psychologist Daniel Goleman's 1995 international bestseller, *Emotional Intelligence*.[31] In simplest terms, integrating laboratory work on the relationship between emotion and cognition with anti-Cartesian philosophies of mind (Churchland, Lakoff, Dennett), Damasio overturned the midcentury paradigm that prioritized rationality and pathologized emotion. In fact, as Damasio summarized: "The action of biological drives, body states, and emotions may be an indispensable foundation for rationality. The lower levels in the neural edifice of reason are the same that regulate the processing of emotions and feelings, along with global functions of the body proper such that the organism can survive. These lower levels maintain direct and mutual relationships with the body proper, thus placing the body within the chain of operations that permit the highest reaches of reason and creativity. Rationality is probably shaped and modulated by body signals, even as it performs the most sublime distinctions and acts accordingly."[32] Although the cognitive function of emotion had been argued already in philosophy (e.g., Martha Nussbaum, Robert Solomon, Amélie O. Rorty),[33] in cognitive psychology (e.g., Lazarus), and in the literary humanities (e.g., Nussbaum, Adela Pinch),[34] Damasio and related neuroscientific research had once again made it safe across the disciplines to prioritize emotion as a positive and at least unavoidable human phenomenon and therefore as a legitimate object of study now disassociated from its postwar critique. A symptomatic shift in this mode of argumentation can be tracked from Martha Nussbaum's 1990 *Love's Knowledge*, which relies on stoicism and narrative theory to demonstrate the emotional foundation of reason, to her influential 2001 *Upheavals of Thought*, which also relies on cognitive science research. In this way emotion research has reappeared in traditionally humanistic fields now producing subfields called, for instance, *neuroaesthetics*, *neurotheology*, and the *cognitive approach to literature*, each of which draws from neuroscientific work demonstrating how higher-order cognition is grounded in bodily states, including emotional, that orient thinking and provide motivation that is often linked to the explanations of evolutionary psychology and biology.

This emotional turn is reinforced by persistent nonrationalist frameworks with deeper nonscientific histories, including religious passion and its theology, the rhetoric of emotion, Heideggerian phenomenology,[35] studies of everyday life (e.g., Elias and Febvre), and psychologies of motivation (e.g., Ludwik Fleck). In other words, the emotional turn has a complex history that we will explore in this book, especially as it applies to the natural and the social sciences. As a consequence of this historical work, we hope that the emotional turn can be understood beyond the postwar paradigm and beyond the varied demonstrations of a recent neuroscientific discovery: emotional phenomena need not be reduced to what can be explained by the latest science. In fact, several contributors point to the 1960s and 1970s—hence prior to the more recent predominance of neuroscience—as a period when emotions again became an important research domain at certain locations within the humanities and social sciences, where social movements of the time, including feminism and black power, provided some significant motivation. Hence a summary of our historical framework: the 1960s and 1970s help compose the relevant prehistory of the emotional turn; the turn proper we see consolidated in the last decade of the twentieth century and the first decade of the twenty-first, when disparate projects are newly legitimated and offered opportunities in the broader context of neuroscientific research and its popularization. Indeed, this context is powerful enough to account even for some projects that would have nothing to do with neuroscience per se or, like this one, seek to historicize neuroscience.

Historical analysis can help us identify this emotional turn and explain how emotion studies have taken their current shape. Historians too have taken an emotional turn by developing a reconceptualized history of emotions that updates but also partly transcends older historical approaches to emotions dating back to the 1940s. Lucien Febvre's seminal 1941 essay, as mentioned above, did not really inspire mainstream historiography.[36] And, while emotions did feature prominently in the works of Jean Delumeau and Theodore Zeldin, these approaches did not inspire a strong historiographic tradition immediately discernible by way of intellectual history (although we will pick up this thread below, by way of Foucault).[37] Equally short-lived were the psychohistorical approaches of the 1970s, which were directly inspired by psychoanalysis. Second-wave feminism and the rise of women's and gender history certainly sensitized historians to emotions as a potential subject of historical research.[38] The rise of poststructuralism influenced fields beyond literary theory from the 1970s onward, showing how emotions are structurally constructed and culturally contingent rather than universal and hardwired.[39] Yet it was only with the work of Peter and Carol Stearns that the history of emotion became a self-described and growing subfield within the academic discipline.[40] "Emotionology," however, remained wedded to a civilizational narrative that posited an increased repression or "dampening" of emotions with the advent of modernity. More recent approaches by historians such as Barbara Rosenwein and William Reddy, by contrast, are strongly influenced by cognitive psychology and seek to transcend the Cartesian dualism between reason and emotion.[41] These new histories tend to focus on "emotional communities" (Rosenwein) or other cultural formations in political, institutional, and everyday life that have an essential emotional component.

What has largely been missing in this newly emerging subfield of the history of emotions—certainly for the twentieth century—is a distinct history-of-science perspective. Despite some important work primarily for earlier periods, the changing ways in which emotions have been constructed as a scientific category in the postwar period have not received much attention.[42] The history of emotions and the history of postwar science have thus evolved largely separately and in isolation. Yet, because of the normative force that scientific conceptions of emotions exert in modern society, Jan Plamper has recently argued, the "history of emotions in the modern period will have to be written to a large extent as the history of science."[43]

Science and Emotions after 1945

This volume responds by analyzing the changing relationship between science and emotions after 1945 in a transatlantic and comparative perspective. It is based on the assumption that the humanities still offer the best methodological and theoretical tools to study emotions in the history of postwar science. Historical and related interpretive approaches—including critical, rhetorical, and narratological—are not only sensitive to changing scientific perspectives on emotions; they also are best equipped to place the sciences of emotion into larger social and cultural contexts.

The contributions conceptualize the changing relationship between science and emotions in four distinct ways. First, they ponder the role of emotions in the scientific process itself. While recent histories of science have outlined the rise of the ideal of objectivity in the postwar period, historians of science have also begun to unearth the significance of emotions within lager "thought styles" (Ludwig Fleck) in the twentieth century. As Lorraine Daston and Peter Galison have argued, changing scientific epistemologies were linked to shifting (and, we would add, always emotional) scientific selves.[44] Along similar lines, the sociologist

Jack Barbalet has called for an investigation of the ways in which conscious and unconscious emotional motivations inform the process of scientific research and discovery.[45] The significance of scientists' emotional subjectivity becomes apparent in several contributions to this volume. Frank Biess (chapter 7) and Rebecca Plant (chapter 8) demonstrate how military psychiatrists' emotional attitudes and their membership in distinct "emotional communities" significantly influenced their diagnoses and conceptualizations of soldiers' psychic responses to war and violence. Uffa Jensen (chapter 10), in turn, discerns a shared "epistemtic fear" among scholars in the sciences and humanities, while Nayan Shah (chapter 9) documents how the responses of medical and psychiatric experts to political hunger strikes were conditioned by the expectation of "objectivity" devoid of political passions.

Second, the volume investigates the changing status of emotions as an object of scientific study. The contributions underline the ways in which science has constituted and transformed the category of emotion over time. As Paul White writes: "The role that sciences have played in transforming not only the study, but also the nature, of emotions has a long and complex cultural history, and this history has only just begun to be written."[46] By examining a wide variety of disciplines—from medicine and psychiatry to sociology, economics, and historiography—the volume traces the changing fortunes of emotions as a scientific object in the postwar period.

The complex internal histories of several disciplines cannot be fully integrated into a general framework; this volume focuses on the fate of one important topic (emotion and science), and even this focus is perforce at times aporiatic and deeply ambiguous. Sometimes these ambiguities and aporias are central to the story we wish to tell about how the topic is newly consolidated, for instance, when feminist theory of the 1970s and 1980s runs into normal science (Lutz, chapter 14). Sometimes they provide a foothold for future research. For example, several contributions confirm our hypothesis that emotions were relatively marginalized during the first two postwar decades. As the contributions by Ute Frevert (chapter 12) on economics and Helena Flam (chapter 13) on sociology demonstrate, Parsonian structural functionalism and the neoclassical synthesis in postwar economics marginalized or ignored emotions. In other cases, emotions assumed a distinctly negative valence, such as the "fear of fear" in the treatment of cancer patients (Hitzer, chapter 6), in the pathologization of panic (Biess, chapter 7), or in the aversion to the emotions articulated in Holocaust survivor testimonies (Dean, coda). Otniel Dror's contribution, by contrast, demonstrates how the academic status of emotions was not just the product of a distinct postwar moment but also derived from alternative, long-term genealogies, especially from physiological enactments of emotions in the laboratory, that is, from the study of adrenaline and "emotional excitement," which then directly inspired the rise of cognitivist approaches and appraisal theory in psychology from the 1960s on.

What is equally apparent as the repression and/or marginalization of emotions across various disciplines is the (re)discovery or reconceptualization of emotions as an object of study from the 1960s on: be it in the form of a distinct sociology of emotions (Flam, chapter 13 in this volume), the rise of behavioral economics (Frevert, chapter 12), the emergence of appraisal theory in cognitive science (Reddy, chapter 1; Dror, chapter 4), or the attack on pathologizing Cold War conceptions of emotions (Biess, chapter 7; Hitzer, chapter 6; Gere, chapter 5), emotions came back with a vengeance as an object of scientific inquiry in the 1960s and 1970s. Rather than focusing on individual disciplines, the volume's multidisciplinary approach allows us to identify this pattern across the disciplines. At the same time, individual contributions offer a finite set of causal explanations for this transformation. For Helena Flam (chapter 13), the discourse of social inequality in the United States, exemplified by the civil rights and the feminist movements, prompted the emergence of a distinct sociology of emotions. This explanation is further elaborated by Catherine Lutz (chapter 14), who elucidates the precise impact of feminism on the conceptualization of emotions across a variety of academic disciplines. Ute Frevert (chapter 12) attributes the rise of behavioral economics and concepts of bounded rationality to flaws within the ontologies on which neoclassical economics was based. Shifting conceptions of subjectivity, or, as William Reddy (chapter 1) calls it, *personhood,* underlie some of these transformation as well. The decline of *homo oeconomicus* inaugurated a more complex self that was not purely driven by economic self-interest, while the rise of the doctrine of informed consent in medical ethics (Gere, chapter 5; Hitzer, chapter 6) takes emotions more seriously and makes individuals managers of their own emotions. Conversely, as Nayan Shah (chapter 9) shows, advanced psychiatric and medical research again questioned the notion of the autonomous individual underlying the doctrine of informed consent, at least in the case of severely deprived and malnourished hunger strikers. In general, these contributions show how shifting relationships between science and emotions were closely related to what the German sociologist Andreas Reckwitz has described as changing "subject cultures." New conceptions of personhood, or what Reckwitz calls the emergence of a "postmodern consumption-oriented creative subject," from the 1960s on enabled a renewed attention to and a reevaluation of emotions across several academic disciplines.[47]

The volume also underlines the transatlantic context as a crucial factor in explaining the shifting status of emotions as an object of scientific inquiry. It inserts emotions as an object of study into the longer history of the transatlantic migration of ideas dating back at least to the late nineteenth century.[48] While European influences on the American academy tended to dominate from the late nineteenth century to the mid-twentieth, the rise of the United States as a global superpower after 1945 and the extensive US influence on the reconstruction of postwar (Western) Europe established the predominance of American science and reversed the direction of transatlantic exchanges.[49] Supported by the funding power of such philanthropic institutions as the Rockefeller and Ford Foundations, American ideas and approaches now exerted a decisive influence on the European academy, especially the social sciences. Transatlantic scientific links were strongest between the United States and Germany: up until 1914, German social science, especially economics, exerted a decisive influence in the United States. Two world wars as well as the rise of fascism not only led to the emigration of many German academics to the United States; the Nazi period also cut off the German academy from international developments and hence made it more receptive to American influences after 1945. Several contributions to the volume thus reflect the particular depth and intensity of transnational linkages between Germany and the United States (Gere, chapter 5; Hitzer, chapter 6; Biess, chapter 7; Jensen, chapter 10; Frevert, chapter 12; Flam, chapter 13).[50]

Yet transatlantic science did not simply entail the imposition of an American model on German or European traditions. Instead, the transfer of approaches and ideas often involved multilayered processes of exportation and reimportation. In the 1940s and 1950s, for example, a group of German émigré scholars working together with such American intellectuals as Carl Schorske and Stuart Hughes sought to invent a new kind of intellectual history, one that sought to account for the irrationalism of European politics and, by drawing on psychology and psychoanalysis, to counter the then dominant behaviorism in the United States.[51] As Uffa Jensen (chapter 10 in this volume) demonstrates, the reception of Max Weber's sociology in the United States was decisively shaped by Talcott Parsons, who emphasized the rational aspects of Weber's work and tended to neglect the important role of emotions. This "rationalized" Weber was, then, in the 1950s and 1960s reimported to Europe, where the important role of emotions in his work (as well as its more pessimistic, Nietzschean strands) was only gradually rediscovered. A similar process defined the transatlantic history of psychoanalysis, which rose to prominence in the United States in the 1920s and was then reimported to Germany in its particularly Americanized version with the considerable support of the Rockefeller Foundation.[52] These examples notwithstanding, it would also be misleading to posit a strict dichotomy between a European tradition that was receptive to emotions and an American academic context that was not. As several contributions highlight, American influences were crucial to transforming traditional conceptions of emotions in Europe or to directing attention to emotions as an object of study and analytic category in the first place. Such transatlantic exchanges became operative in the rise of psychosomatic explanations of cancer (Hitzer, chapter 6), in the transformations of West German conceptions of panic (Biess, chapter 7), and in the evolution of a distinct sociology of emotions (Flam, chapter 13), to name just a few examples.

Third, the volume considers the extent to which scientific treatments of emotions both shaped and were shaped by larger emotional cultures.[53] In so doing, it contributes not only to the history of science but also to the history of emotions more broadly. The professionalization of science and, with it, the increasing cultural authority of scientific expertise achieved a new high point in the post-1945 period. In an age of the "scientification of the social," the social sciences, in particular, were invested with considerable authority in engineering the postwar social order and in constructing the postwar state.[54] With the rise of a therapeutic society in the United States in the 1950s and, two decades later, in Western Europe as well, the psy-sciences (psychiatry, psychology) enjoyed a similar cultural authority, only to be supplanted in recent years by the neurosciences.[55] As a result, the definition and the status of emotions in a variety of scientific disciplines entailed the potential to determine and shape emotional norms and emotional practices in culture and society at large.

Scientific treatments of emotions, moreover, defined not only what William Reddy calls "emotional regimes," that is, normative assumptions regarding the expression and experience of emotions, but also the ways in which they inform actual emotional practices, what Reddy calls "emotives."[56] These academic/scientific discussions of emotions also responded to larger social and cultural contexts. For example, in his classic study *American Cool*, Peter Stearns argues that scientific devaluation of strong emotions followed rather than shaped a larger cultural emphasis on emotional anti-intensity, which he sees as a dominant trend in the emotional history of the United States in the twentieth century.[57] Among the contributors to this volume, Cathy Gere (chapter 5) reveals the link between an emerging Holocaust memory and the decline of utilitarianism in US neurology. Likewise, Jordanna Bailkin (chapter 11) highlights how decolonization and the expansion of the postwar welfare state inaugurated new emotional ties that sought to bind together young elites in Britain with those in the former colonies, thus creating new types of social bonds now understood within an evolving international human rights framework. The impact of the Cold War on scientific conceptions of emotions becomes apparent in the discussion offered by Biess (chapter 7) of shifting conceptions of panic in postwar Germany, and it factors prominently in the discussion offered by Gere (chapter 5) of the delayed implementation of the doctrine of informed consent in medical ethics in the United States.

Fourth, the volume seeks to offer a historical perspective on how and why emotion studies more generally have taken their current shape. It especially engages the rise of the neurosciences as the paradigmatic discipline for the study of emotions, and it probes the potential and pitfalls of neuroscientific understandings of emotions from a humanities perspective. How are the preferred objects of emotion studies, like empathy and terror, shaped by our postwar cultures? How are the focal points for emotion science, like the face and the brain, situated in a larger narrative that has previously sought emotions elsewhere, for example, in the traumatized body or beyond the individual organism altogether? What are the presuppositions that have produced emotion as the consilient object par excellence shared among the humanities, the human sciences, and the natural sciences?[58] From the perspective of the humanities, the contributions by William Reddy (chapter 1), Ruth Leys (chapter 2), and Daniel M. Gross and Stephanie Preston (chapter 3) offer a critique of basic emotion theory that considers emotion universal, primarily biological, and "legible" via facial expressions and the underlying facial muscle movements. Yet they also highlight the diversity of the discipline of neuroscience, and they demonstrate how certain approaches in the neurosciences that situate emotion in broader cognitive and environmental contexts—such as appraisal theories and situated cognition theories, respectively—are compatible with humanistic inquiry.

The Scope of the Volume

This volume brings together leading representatives of the history of emotions with specialists from several other disciplines (literature, rhetoric, sociology, neuroscience) to trace the changing relationship between science and emotions after 1945. It is organized according to different scientific disciplines in which emotions have featured prominently, including medicine, psychiatry, neuroscience, and the social sciences, in each case bringing a humanities perspective to bears. We are aware of the fact that the boundaries between these disciplines have always been fluid and that individual disciplines within these larger categories—for example, economics and sociology within the social sciences—did not always follow the same rhythms of historical change. Still, this division allows us to discern larger patterns in the relationship between science and emotions across several fields. Moreover, the editors recognize that the status of emotions in historical writing itself is of course also subject to historical change. The volume therefore also entails a self-reflexive coda by Carolyn Dean that subjects postwar historiography, and especially the emotionally charged historiography of the Holocaust, to the same interpretive methodologies. The contributions combine overviews of entire disciplines with more precise case studies of specific emotions or historical episodes in which emotions feature prominently.

In light of the paradigmatic significance of neuroscience for the recent emotional turn, part 1 analyzes (and problematizes) recent neuroscientific research on emotions from the perspective of the humanities. While pointing out the conceptual pitfalls of experimental design in the neurosciences, the chapters nevertheless also probe the potential for collaboration on emotions as a topic of shared concern.

As William Reddy argues in chapter 1, humanists are often dismayed when they turn to current research in experimental psychology and in the cognitive and affective neurosciences because they witness experimentalists relying on traditional and predominantly Western commonsense notions of the person and the faculties that make up the person. This reliance, moreover, can lead to a performative contradiction that Foucault warned us against, where, for instance, human scientists themselves are not subject to the same behavioral laws they find in the humans they study. However, Reddy argues, humanists with the epistemological commitments and training associated with interpretive method and ethnographic, cultural, and literary readings should recognize that their own research and critical reflection on their own methods in fact align them closely with versions of appraisal theory,[59] emotion regulation theory, and nonmodular understandings of neural functioning. Hence this chapter establishes three important themes of this volume. First is a historiographic observation: the history of emotion is in some important way a corollary to a history of personhood. Second is a methodological injunction: when studying emotion, one should be able to characterize the relationship between the subject and the object of analysis. Third is a critique of "two cultures": the work of humanists is not incommensurable with the work of scientists; humanists and scientists share significant interests in emotion studies. Indeed, Reddy concludes, we run the risk of placing ourselves in performative contradiction if we step back and view the modern science of emotion strictly as historians.

In chapter 2, Ruth Leys elaborates on the critique of basic emotion theory that is already implicit in chapter 1. Her contribution displays the critical potential that is inherent in the interpretive method of the humanities when applied to certain approaches in neuroscientific research on emotions. It offers a comprehensive critique of the theoretical assumptions and laboratory practices of two experiments designed to prove mirror-neuron theory and the neural basis of empathy—one 2003 experiment by Wicker et al. and a follow-up experiment from 2007 by Jabbi, Swart, and Keysers.[60] Both experiments, Leys argues, uncritically adopt the central assumptions of basic emotion theory as developed by Sylvan Tomkins and Paul Ekman, who conceptualize emotions as noncognitive, hard-wired, and pan-cultural entities that can be truthfully represented in a subject's facial expression. Notwithstanding the fact that both experiments asked professional actors to perform the emotion of disgust, both remain oblivious to the "performative-transactional nature of facial and other displays" (p. 78), that is, the possibility that facial expression does not necessarily align with inner feelings. The first experiment ignored the subjects' own reports of their feelings, and, when the second experiment tried to address them, it necessarily undermined its own assumptions regarding the noncognitive and nonintentional nature of emotions. Like Reddy, Leys thus articulates a clear preference for cognitive or appraisal theories of emotions within psychology and neuroscience that reject central assumptions of basic emotion theory. Yet she also points to unresolved theoretical and philosophical problems in the research on empathy that ultimately require the assistance of the humanities' interpretive method.

Written jointly by a rhetorician, Daniel M. Gross, and a neuroscientist, Stephanie Preston, chapter 3 offers a concrete example of the analytic benefits entailed in interdisciplinary collaboration between the two academic cultures. By first outlining various limitations and constraints of laboratory science on emotions, Gross and Preston highlight the hardening of the two cultures divide in recent research on emotions. In particular, a recent injunction against "reverse inference" in neuroscientific practice has rendered the establishment of casual connection between brain activation and emotional states more difficult. Still, Gross and Preston discuss two sets of experiments that show how laboratory research on emotions can be modified with the interpretive method of the humanities. Recent studies of fear, for example, document the persistence of unconscious racial bias while, at the same time, neglecting larger historical and social contexts. More promising from a humanities perspective appear to be experiments that draw on a "situated emotions" approach, one that explicitly incorporates language and audience considerations into experimental design and interpretation. Recent experiments on situated empathy with terminally ill patients by Stephanie Preston and her associates reveal surprising similarities to Carolyn Dean's discussion of emotional reactions to Holocaust survivors.[61] Empathy with the suffering of others is conditioned by historically specific and culturally normative assumptions regarding victimhood that also include consideration of gender. Rather than a naive endorsement of interdisciplinary collaboration, the chapter remains sensitive to significant barriers between the two cultures yet nevertheless outlines necessary concessions from each side: a realization among humanists that laboratory research requires the isolation of one key variable over others and a recognition among scientists of the historically and socially specific environment in which emotions take shape.

Part 2 focuses on medicine and medical ethics in the postwar period. In presenting a preliminary genealogy of excitement in terms of the *operational* history of emotion inside the laboratory, Otniel Dror (chapter 4) challenges the contemporary dominant emphasis on the intellectual history of post–World War II emotion. This intellectual perspective often presents postwar emotion in terms of the legacies of nineteenth-century emotions, that is, in terms of a history of Darwin's and James's (and Freud's) emotions. Dror's argument is that important aspects of postwar emotion and of postwar constructs of the relationships between emotion, excitement, intensity, activation, and *physiological* arousal emerged from early twentieth-century laboratory enactments of emotion—that is, from the study of adrenaline and emotional excitement. Hence, this chapter provides an altogether new prehistory of noncognitive emotion science ranging from the work of Zajonc in the 1980s to mirror-neuron research today. It thus forms an essential prehistory to contemporary neuroscientific research as discussed by Leys (chapter 2) and Gross and Preston (chapter 3).

In her discussion of postwar neurology, Catherine Gere (chapter 5) argues that the trial of Adolf Eichmann in 1961 and Stanley Milgram's "obedience to authority" experiment that was inspired by it represent a turning point in the history of electrical stimulation research on human emotions and thus a turning point in the broader history of emotion where utilitarian calculus still operates. Like numerous medical research projects now considered unethical, Robert Heath's electrical stimulation research was justified according to a utilitarian calculus in which the rightness or wrongness of a particular action was to be judged by a cost-benefit analysis of its consequences. This utilitarian justification had a disturbing similarity to Nazi doctors' justifications, and partially for that reason the structure of medical practice and clinical research was reformed in 1973 around the principle of informed consent. Gere demonstrates, however, that this reform has left us a problem with roots in conflicting models of emotional personhood. While the sovereign individual of the informed-consent regime—a rational, autonomous, and almost disembodied being—dominates contemporary medical ethics, the utilitarian self—the suffering human animal in need of deliverance from the ultimate indignity of bodily anguish—has not ceased to resonate with the aims, methods, and realities of medical practice and research.

Tracking emotional regimes around cancer demonstrates how historically specific institutional frameworks help produce particular kinds of embodied personhood. Focusing on cancer as an example of broader postwar phenomena, Bettina Hitzer (chapter 6) identifies two conflicting emotional regimes in the late 1950s and early 1960s. While one regime perceived the balanced and optimistic attitude of cancer patients as something basically positive, the other identified this attitude as a fundamentally harmful repression of one's inner feelings. This paradox of fear, moreover, is at the same time evident in educating children, in the rhetoric of the 1950s peace movements, and in civil defense brochures: fear registers as omnipresent as it is branded negative, rational, pathological, or even harmful. Hitzer then identifies two factors that contributed to a gradual shift in emotional regimes around cancer: the emerging concept of informed consent and the rise of psychosomatic medicine, which changed the therapeutic encounter drastically. Finally, 1960s cancer treatment, especially the practice of radical mastectomy in the United States, triggered an angry protest movement against established medicine, which was branded as paternalistic and devoid of emotions.

In part 3, on psychiatry, the contributions by Frank Biess (chapter 7) and Rebecca Plant (chapter 8) compare and contrast medical psychiatrists' responses to war trauma and the emotional challenges of actual or anticipated violence during the Second World War and the Cold War. Biess documents the persistent influence of traditional concepts of panic dating back to the Second World War in post-1945 West Germany. By examining the genesis, production, and reception of military training manuals on panic at a moment of escalation Cold War tensions in the early 1960s, he offers a case study of the ways in which a specific scientific perspective shaped emotional norms in the institutional context of the West German military. He then traces the gradual unraveling of this particular Cold War nexus of science and emotions. Transatlantic influences of US psychiatry and an emerging psychiatric reform movement within West Germany ultimately transformed Cold War conceptions of panic and inaugurated a more complex and less pathologizing definition of this emotional phenomenon.

Rebecca Plant's article draws on Barbara Rosenwein's concept of "emotional communities" to analyze competing views on neuropsychiatric casualties among U.S. Army troops during World War II. The war is generally understood to have paved the way for not only a massive expansion of the psychiatric profession but also the rise of the psychodynamic and psychoanalytic approaches that dominated postwar psychiatry. Plant argues that a closer look at the experiences of psychiatrists and medical officers who actually treated men at the front complicates this picture. Her chapter focuses on the psychiatrist and psychoanalyst John W. Appel, who in 1944 traveled to the Italian theater at the behest of the Surgeon General's Office to study why so many American soldiers were being evacuated as neuropsychiatric casualties. Thereafter, he authored what is arguably the most important psychiatric paper to come out of World War II, in which he advocated the need for limited tours of duty—a concept that would ultimately play a major role in shaping manpower policies in Korea and Vietnam. Drawing on an unpublished diary, Plant shows how Appel's emotionally charged experiences profoundly altered his views of war trauma in ways that led him to implicitly reject psychodynamic and psychoanalytic thinking. On the one hand, he came to see the environment—the extraordinary stress of combat itself—as the most important etiological factor that caused men to break down. On the other hand, he increasingly embraced the moralistic views of military officers who believed that all men should be able demonstrate a certain level of emotional endurance. According to Plant, this case study suggests that the militarization of American psychiatry during World War II exerted pressure on the profession and its practitioners in ways that have yet to be fully appreciated. At the same, it reveals how published psychiatric studies tended to obscure the subjective and emotional experiences that undergirded their authors' conclusions.

Nayan Shah (chapter 9) addresses the (self-)management of emotions among hunger strikers in the twentieth century as well as the medical and psychiatric responses to the self-starving body. Casting his net widely to include Gandhi's hunger strikes in colonial India in the 1930s, Cesar Chavez's in the US West in the 1960s, and those by detainees in apartheid South Africa in the 1980s, Shah offers a case study of the ways in which intense emotions were kept at bay in passionate and existential political conflicts. Whereas Gandhi and Chavez transmuted their anger, frustration, and despair over social and political injustice into fearlessness, medical and psychiatric experts channeled their emotive responses to hunger strikers into the concept of scientific objectivity and neutrality. Shah also underlines the persistent efficacy of religious emotional communities for the popular resonance of hunger strikes. Like Gere (chapter 5) and Hitzer (chapter 6), he highlights the significance of the informed-consent doctrine in defining medical ethics in the postwar period. At the same time, he also discusses more recent research regarding the psychological impact of starvation that ultimately calls into question the model of the autonomous, rational self capable of making informed decisions about his or her own well-being. In this sense, he also links the history of emotions to the history of changing conceptions of the self.

Part 4 discusses the shifting status of emotions in the transnational social sciences of the postwar period. Invested with tremendous prestige after 1945, the social sciences were directly relied on by postwar governments to construct the postwar state and maintain social order.

Uffa Jensen (chapter 10) brings together the analysis of a specific emotional thought style as developed by the biologist and physician Ludwik Fleck with the changing status of emotions in the works of Max Weber and Sigmund Freud. In all three cases, he is able to demonstrate how the transatlantic transfer from Europe to the United States led to a significant downplaying or even erasure of emotions. This was true for Thomas Kuhn's *The Structure of Scientific Revolutions*,[62] which drew heavily on Fleck, as well as for the Parsonian Weber reception and American ego psychology. By providing strong evidence for the causal nexus between the transatlantic migration of ideas and postwar rationalism, Jensen confirms one of the central working hypotheses of this volume. Moreover, he also points to historically specific conditions that determine the relative presence/absence of emotions across different disciplines, such as the need for scientific rationality during the passionate ideological conflicts of the Cold War. By invoking Fleck's concept of emotional thought styles, he points to analytic possibilities for transcending the gap between the two academic cultures—the sciences and the humanities—since they may have both participated in the same "thought collectives" and may have shared "epistemic fear."

Shifting the focus to Europe's relationship with its former colonies, Jordanna Bailkin (chapter 11) focuses on British efforts to manage emotion during the era of decolonization. She highlights the distinctiveness of postcolonial regimes of emotion: welfare and decolonization were both expected to bring about dramatic changes in material and political conditions but also to inaugurate new affective experiences and new types of social bonds. These issues are explored by looking at the founding of Voluntary Service Overseas (VSO), which served as the British predecessor to the Peace Corps. Bailkin takes VSO as a case study of (1) the centrality of emotion in the intersecting histories of welfare and decolonization and (2) the diverse and sometimes conflicting ways in which emotion was understood by social scientists, their critics, and the decolonizing state. The reconstruction of personal relationships was, in the VSO view, crucial to the West's success in the Cold War as well as Britain's peaceful management of decolonization and the new challenges to personal character brought by the rise of the welfare state—which, in some sense, transformed all Britons for the first time into both givers and receivers of aid. The VSO example thus illustrates how a postwar regime of democratized and universal well-being was in part established on the uneven terrain of decolonization.

The discussion by Ute Frevert (chapter 12) of emotions and economics charts the rise of postwar rationalism and the notion of *homo oeconomicus* as a rational actor in the field of economics. Similarly to developments in sociology, the neoclassical synthesis in the United States displaced earlier traditions that had been more receptive to emotional and psychological influences, such as the German historical school or Keynes's notion of animal spirits as a central element of modern subjectivity (with antecedents dating back to the early modern period). While the neoclassical synthesis owed its postwar triumph both to the dominance of US science and to disciplinary pressures for exact mathematical modeling, it was increasingly subjected to criticism from the 1970s on. Frevert shows how attention to emotions is inextricably intertwined with shifting conceptions of personhood: the orthodox *homo oeconomicus* gave way to a more complex figure operating under conditions of bounded rationality and motivated by more complex factors than the pursuit of mere self-interest. The new field of behavioral economics thus constitutes an essential element of the emotional turn as it draws on neuroscience and cognitive psychology to identify the cognitive dimension of emotions.

Along similar lines, Helena Flam (chapter 13) charts the genealogy of the sociology of emotions as it emerged in the 1960s and 1970s in the United States. In line with the general thesis of this volume, she shows how the dominant structural functionalism in postwar US sociology did not provide a language for emotions and largely relegated them to the irrational and deviant. Impulses for a sociology of emotions came from within the discipline—mainly Erving Goffman's symbolic interactionism—but also derived from the impact of larger social and political contexts, especially the preoccupation with social, racial, and gender inequalities. Arlie Hochschild's classic study *The Managed Heart* constituted the prime example for the emerging constructivist perspective on emotions,[63] which then radiated outward from the University of California, Berkeley, and began to influence researchers in Western Europe. As an American import, the sociology of emotions initially displaced a long-standing European tradition in the field (Simmel, Elias) but has meanwhile evolved into a multidirectional and genuinely transatlantic discourse.

Drawing substantially from her home discipline of anthropology, Catherine Lutz (chapter 14), whose 1988 *Unnatural Emotions* more than any other work exemplified the social constructionist approach to emotions,[64] demonstrates from another angle how the emotional turn comes at the vexed intersection of 1960s/1970s social movements and the rise of brain science—a central thesis of this volume. She contrasts feminist and normative approaches to the emotions. Feminist approaches are distinguished by their attention to the material, institutional, and cultural capillaries of power through which discourses of emotion operate. Normative approaches restrict their questions to the limited power that emotion—as culturally and conventionally defined in Western academic circles—has to shape individual behavior. Finally, Lutz speculates about the place of feminist approaches to emotion in the broader context of late twentieth-century knowledge production about emotions. It can be no coincidence, she notes, that intense interest in emotion per se in the academy developed in the 1980s with the maturation of the women's movement and the influx of women and, to a lesser extent, racial minorities into research positions. While these social changes have helped create a feminist literature on emotion, normative science also revalorized emotion as an object worthy of scientific study in this same period. This latter work, Lutz concludes, may be at least partially motivated by the attempt to co-opt and evacuate the emotional of its association with the female and irrationality.

Focusing on the contemporary context, Roddey Reid (chapter 15) analyzes the emotional implications of market economies' promise of universal well-being and corresponding regulation against hostile environments, especially in the workplace. As he demonstrates, market economies depend on humiliation as a way of life. Demonstration proceeds as he analyzes US and European legal and medical responses to bullying and psychological intimidation, especially in the workplace. Reid argues that the different arenas of the school, the workplace, media, and politics operate as multiple sites of bullying so as to produce a public culture of intimidation that links the most subjective, individual experiences with those of collective life itself. Historical explanation has one important anchor in the United States in the 1960s and 1970s. In the aftermath of the civil rights and women's movements, defeat in Vietnam, and early globalization's massive deindustrialization of the economy, Reid argues, a new macho white populism arose in the 1970s in the United States that forged strong links between deep-seated feelings of individual insecurity, collective economic uncertainty, and distress at the diminishing fortunes of a superpower on the verge of decline. However, the subjective experience of contemporary suffering in its connections to collective life has been the subject of research beyond the United States, in Europe, and the phenomena studied have been global. Reid thus links his work on bullying to studies that highlight how the interplay of economic and noneconomic humiliations and indignities, encountered in daily life under market economies and the War on Terror, undermines personhood and identity.

In a coda on postwar historiography, Carolyn Dean highlights the epistemological benefits of applying a history-of-emotions perspective to the already very well developed historiography of the Holocaust. In particular, she examines "emotions invested in the reception of Holocaust survivors' testimony" (p. 387) as well as the extent to which "historians' own emotions intrude into the interpretation and use of such memoirs" (ibid.). Contexts, she argues, that determined the reception of Holocaust memoirs after 1945 were also affective contexts that were based on cultural assumptions about the nature of the ideal victim. As she documents on the basis of numerous examples, Holocaust victims experienced aversion and rejection if they did not correspond to the cultural norm of the innocent, modest, and reticent victim. Their "anxiety of transmission" was thus also an anxiety of not living up to cultural expectations. Similarly to other contributors to this volume, Dean rejects the conceptualization of emotional responses to victim testimony as merely natural and biological and instead pleads for a historically specific reconstruction of the "social and rhetorical organization of affect" (p. 394) in the postwar period. Her analysis of the emotional dimension of Holocaust memoirs moves beyond the traditional dichotomy between history and memory and ultimately makes it possible to conceptualize the existence of not one but multiple truths in victims' testimony.

A Methodology for Emotion Studies: Suggestions for Further Research

Taken together, the essays in this volume also point to a more general methodology for emotion studies, which, in the interest of stimulating further research, we outline here. At any given historical juncture, and in the purview of a particular region or locality, a researcher might consider the following four points, exemplified here by way of happiness because it is an important postwar emotional phenomenon touched on only briefly by Frevert (chapter 12) and therefore provides a good test case for the methodology as a generative and not just a descriptive resource for emotion studies.

1. *Which emotions are explicitly referenced and appear dominant in a particular field?* Most immediately this is text-based, historical work that maps semantics fields in their explicit references. For example, emotion studies and the history of emotion now should be interested in happiness, which has recently emerged as a significant phenomenon across psychology, neuroscience, economics, politics, and public policy as well as literary studies. In one instance the regional purview might span from the tiny Himalayan country of Bhutan, which in 1972 started indexing "gross national happiness" instead of gross domestic product,[65] to the Anglo-American context where happiness indexes have increasingly gained the attention of politicians and policymakers in Canada, Great Britain, and the United States. So, when US Federal

Reserve Chairman Ben Bernanke endorses the movement and defines *happiness* as a "short-term state of awareness that depends on a person's perceptions of one's immediate reality, as well as on immediate external circumstances and outcomes,"[66] the emotion researcher might begin by assessing the reach of this definition across seemingly disparate expert discourses and populations. In this volume along with *happiness* (Frevert, chapter 12), the other emotional phenomena that appear particularly significant in the semantics fields of post-1945 science include *fear* (Hitzer, chapter 6; Biess, chapter 7; Plant, chapter 8; Gross and Preston, chapter 3; Reid, chapter 15) and *empathy* (Gross and Preston, chapter 3; Leys, chapter 2; Dean, coda). Also, the index to this volume provides a rough survey of which emotions are salient in our period of inquiry.

2. *Which emotions are superseded, excluded, or otherwise rendered unavailable or beyond the range of immediate experiential or discourse-analytic possibilities?* This point in the methodology might include theoretical and comparative work, where the researcher analyzes the "shadow" of a semantic field to identify what is obfuscated by a particular emotional regime—for example, the systematic suffering that can be obfuscated by hedonic psychology, state-sponsored happiness indexes, and the broader culture of "happyism" recently critiqued by Deirdre McCloskey.[67] Clinically speaking, happiness is contrasted with depression. Behavioral economics and latter-day utilitarianism, however, would not contrast it with sadness or depression; instead, in these fields a zero point anchors a scale that measures only the positive emotion while relegating negative emotions like resentment to systematic or local distortions in what is measured as normal and normative, baseline happiness. Emotion researchers studying happiness in the post-1945 context, therefore, should be able to document and then analyze at least these two conflicting models contouring this emotional regime differently, with different but not unrelated ranges of presence and absence.[68]

In this volume we work on the principle that histories of emotion are not just about the positive object per se—that is, an emotion like fear—but also about the historical conditions that make certain emotions appear more readily than others and that make for certain significant kinds of emotional absence, including, for instance, medical patient reticence or resilience (Gere, chapter 5; Gross and Preston, chapter 3; Dean, coda) and fearlessness (Shah, chapter 9). Hence, the concepts of emotional regime and emotional culture help only so far. We also have to think in terms of political economy and economies of scarcity in particular (see no. 4 below).

3. *What are the trajectories of salient emotional arcs over a relevant period of time?* Once a positive semantic field has been identified and its shadows analyzed comparatively, the historian of emotion has the opportunity to explain trends and trajectories over time. For instance, McCloskey, an economist, locates happyism in the long history of utilitarianism. The literary scholar Vivasvan Soni maps the long history of "trial narratives" (like Samuel Richardson's *Pamela; or, Virtue Rewarded*) that index happiness according to the person and not according to civic virtues, as had been the case previously and paradigmatically in Solon's Greek antiquity.[69] Both McCloskey and Soni critique and historicize the current understanding that reduces happiness to pleasure.

An emotional arc introduced in this volume (but treated extensively only in the middle period) is *anxiety–fear and panic–terror,* where anxiety is most readily documented in 1920s phenomenology and psychiatry, fear and panic are evident in the Hot and Cold War cultures of the 1940s and 1950s, and terror characterizes the post-9/11 years. What research questions are marked by this trajectory? The historian of emotion might first address the implication that the trajectory is upward, with anxiety as the moderate emotion and terror as the extreme. However, this very broad stroke is interesting only insofar as it qualifies the equally broad thesis that modernization increasingly tempers strong emotion with institutions like the Securities and Exchange Commission and the insurance industry.[70] In other words, a trajectory like anxiety–fear and panic–terror can be documented up to a point, though, in terms of methodology, its primary function is to generate research projects. For instance, how does the 1929 panic and depression figure into this trajectory? What might it mean that anxiety is an existential condition, Cold War fear a function of the state (Biess, chapter 7), and terror in many ways post- or antistatist? These two questions partially fall outside the historical purview of this volume, but Gross and Preston (chapter 3) touch on the following: What can we learn from the observation that a vintage formation of political fear (1940s and 1950s) is now for scientists the quintessential basic emotion exemplified in the reflexive response to a coiled snake?

4. *What is the political economy of emotion?* Once a dynamic emotional field has been mapped in terms of its positive and negative elements, emotion studies are in a good position to analyze causality as well as opportunism in the Foucauldian sense in which the (emotional) phenomenon is constituted by the ways it is taken up, used, incited, and manipulated. Causal explanations that are strictly internal do not suffice: "Happiness has become important because Google Trends upward

2004–10." Nor are causal explanations that beg the question by relying on the topos of technological revelation: "Happiness has become important because we now have the tools to study it adequately." Yes, happiness is an obviously important, transnational phenomenon of interest. Why now, and why in the particular form it has taken? What political economies of emotion are at play, foregrounding certain emotional phenomena and not others? These are, methodologically, Foucauldian questions that help emotion studies explain historical trends by documenting confluent interests beyond intentioned and explicit policy.[71] So, extending our example of happiness research, we might observe that certain kinds of hedonic psychology and neuroscience reinforce the liberal-individualistic model that locates happiness in the person—not in a social situation—despite the fact that particular psychological or scientific study may have nothing to do with liberal individualism and might even be the product of a researcher who would reject the relevant politics. With McCloskey, then, we might note that happiness indexes have a hard time with the long-standing ethical expectation that villains like Adolf Hitler should not, by definition, qualify as happy regardless of how they might describe themselves.

In certain important respects, we see, emotion studies pick up where Michel Foucault left off in 1984, exploiting opportunities that his work provided, though they were never pursued systematically by Foucault himself.[72] Foucault studied phenomena that set up some important aspects of emotion studies without actually doing emotion studies: the emergence of institutionalized irrationalisms around the Enlightenment (*Madness and Civilization*), semantics fields that have ontological consequences (*Archaeology of Knowledge*), regimes of incorporation (*Discipline and Punish*), and the "care of the self," including most obviously the detailed analysis of Seneca's *On Anger.*[73] But, for Foucault, Seneca's *On Anger* demonstrates how Hellenistic philosophy developed important practices of self-administration that provide a contrast with Christian examinations of conscience or Freudian deep psychology. *Care of the Self* is not a study of anger per se, nor does Foucault pursue any part of the four-step methodology outlined above, which owes to him a significant debt: (1) identify the positive semantic field, (2) analyze structurally the shadow, or negative valence, of a semantic field, (3) analyze temporally the emotional arc, and (4) analyze genealogically the emotional economy as it is structured by regimes and practices of power. One might even say that emotion studies represent one of the missed opportunities of Foucault's academic career. Nevertheless, emotion studies have taken shape, in some measure, thanks to the questions, concerns, procedures, and critical ontologies that Foucault helped establish. Indeed, recent histories of emotions have been made possible, in part, by a relatively unacknowledged and even discontinuous set of historical research practices and insights from the French school broadly conceived, including the unlikely pairing of Febvre and Foucault. On this reading, Febvre's analysis of a negative field implicated by a feeling structure[74] and Foucault's analysis of bodily sentiments deployed by power[75] are in fact converging tributaries when viewed from the perspective of current history of emotion research.

With Foucault and William Reddy (chapter 1 in this volume), the emotion researcher is informed by the questions: What kind of person are we talking about? What are the ideologies and formative structures of personhood at any particular time and place? Focusing on the postwar context, we see a subject of human rights, personal welfare, and well-being (Bailkin, chapter 11). We are offered a progressive version of personhood where the positive emotions of happiness and empathy would naturally increase or at least democratize as they become increasingly available to all human beings. At the same time, the postwar self, at least in Europe, was also tortured by memories of a past violence and loss as well as the anticipation of future catastrophes in the Cold War, giving rise to a distinctly postwar discourse of uncertainty and fear. More recently, radical subjection is sold in terms of free market logics and sensibilities (Reid, chapter 15). Hence, humiliation is in some ways the shadow of happiness and empathy when universal emotion is the lens through which subject formation is analyzed. Antipathy or hatred figures in hate crimes, hate speech, and genocide projects—shadows of universal emotion like empathy that would mitigate against antisocial or divisive impulses. Finally, anger is a post-1960s response of the subjectified, including at least women (Flam, chapter 13; Lutz, chapter 14), blacks and Latinos (Shah, chapter 9), gays, then (angry) white men (Reid, chapter 15).

As the editors of this volume we believe that the history of science and emotion after 1945 is best written through a carefully selected variety of academic perspectives that can—only in concert—sketch the relevant historiography and the factors contributing to the current shape of our lived inconsistencies. It would be impossible to do this topic justice from the perspective of any one of our contributing scholars, including ourselves, each of whom is lodged in emotional communities and disciplinary habits that allow only certain objects of study to appear and only modest perspective. Taken together, however, the chapters in this volume do provide a rich texture we think particularly appropriate to this new field in which we work.

NOTES

1. Talcott Parsons, "Memorandum: The Development of Groups and Organizations Amenable to Use against American Institutions and Foreign Policy and Possible Measures of Prevention," in *Talcott Parsons on National Socialism*, ed. Uta Gerhardt (New York: De Gruyter, 1993), 101–30, 110–11, 124. For an analysis of this document, see Uta Gerhardt, *Talcott Parsons: An Intellectual Biography* (New York: Cambridge University Press, 2002), 76–77.
2. Ron Robin, *The Making of the Cold War Enemy: Culture and Politics in the Military Intellectual Complex* (Princeton, NJ: Princeton University Press, 2001); Nils Gilman, *Mandarins of the Future: Modernization Theory in Cold War America* (Baltimore: Johns Hopkins University Press, 2003); Benjamin Alpers, *Dictators, Democracy, and American Public Culture: Envisioning the Totalitarian Enemy, 1920s–1950s* (Chapel Hill: University of North Carolina Press, 2002).
3. Talcott Parsons, *The Social System* (New York: Free Press, 1951), 58–67.
4. Jack M. Barbalet, *Emotion, Social Theory, and Social Structure: A Macrosociological Approach* (New York: Cambridge University Press, 2001), 16–17. See also Jan Plamper, *Geschichte und Gefühl: Grundlagen der Emotionsgeschichte* (Munich: Siedler, 2012), 139.
5. Barbalet, *Emotion, Social Theory, and Social Structure*, 16–18, 139; Helena Flam, *Soziologie der Emotionen: Eine Einführung* (Konstanz: UVK, 2002), 106–10, 14.
6. Gilman, *Mandarins of the Future*, 7–8.
7. See B. F. Skinner, *Science and Human Behavior* (New York: Macmillan, 1953), 160–70 (chap. 10, "Emotion"). According to Jan Plamper, the master narrative in psychology textbooks has it that, with the rise of behaviorism, the "lights went out" regarding research of emotions. However, Plamper emphasizes an ongoing continuity of laboratory experiments on emotions, which is also discussed in Dror, chapter 4; see Plamper, *Geschichte und Gefühl*, 214–15.
8. Robin, *Making of the Cold War Enemy*, 64–69.
9. Hunter Heyck, "Producing Reason," in *Cold War Science: Knowledge Production, Liberal Democracy, and Human Nature*, ed. Mark Solovey and Hamilton Cravens (New York: Palgrave Macmillan, 2012), 99–116, 111.
10. Ely Zaretsky, *Secrets of the Soul: A Social and Cultural History of Psychoanalysis* (New York: Random House, 2004), 281.
11. Jürgen Habermas, *Zwischen Naturalismus und Religion* (2005), translated by C. Cronin as *Between Naturalism and Religion* (Cambridge: Polity, 2008).
12. Ibid., 21–22.

13. On this notion, see Till von Rahden, "Clumsy Democrats: Moral Passions in the Federal Republic," *German History* 29.3 (2011): 485–504, 487.
14. Hans Ulrich Wehler, "Königsweg zu neuen Ufern oder Irrgarten der Ilusionen: Die westdeutsche Alltagsgeschichte 'von innen' und 'von unten,'" in *Geschichte von unten, Geschichte von innen: Kontroversen um Alltagsgeschichte*, ed. Jürgen Kocka and Franz Josef Brüggemeier (Hagen: Fernuniversität Hagen, 1985), 17–47, 30.
15. Daniel Morat and Uffa Jensen, eds., *Die Rationalisierung des Gefühls: Zum Verhältnis von Wissenschaft und Emotionen* (Munich: Wilhelm Fink, 2008).
16. For this prehistory, see, e.g., G. J. Barker-Benfield, *The Culture of Sensibility: Sex and Society in Eighteenth-Century Britain* (Chicago: University of Chicago Press, 1992); and Jessica Riskin, *Science in the Age of Sensibility: The Sentimental Empiricists of the French Enlightenment* (Chicago: University of Chicago Press, 2002). See also Dror, chapter 4 in this volume.
17. Thomas Dixon, *From Passion to Emotion: The Creation of a Secular Psychological Category* (Cambridge: Cambridge University Press, 2003), 204–30.
18. Flam, *Soziologie der Emotionen*, 16–42.
19. Ibid., 44–51; Plamper, *Geschichte und Gefühl*, 60–61.
20. Plamper, *Geschichte und Gefühl*, 61–64.
21. Barbara H. Rosenwein, "Worrying about Emotions in History," *American Historical Review* 107.3 (2002): 834.
22. Lucien Febvre, "Sensibility and History: How to Reconstitute the Emotional Life of the Past," in *A New Kind of History: From the Writings of Febvre*, ed. Peter Burke (New York: Harper & Row, 1973), 12–26, 13, 26. On other approaches to emotions in the historiography of the interwar period, see Jakob Tanner, "Unfassbare Gefühle: Emotionen in der Geschichtswissenschaft vom *Fin de Siècle* bis in die Zwischenkriegszeit," in Jensen and Morat, ed., *Rationalisierung des Gefühls*, 35–59; and Plamper, *Geschichte und Gefühl*, 55.
23. Two examples among many: Patricia Clough and Jean Halley, eds., *The Affective Turn: Theorizing the Social* (Durham, NC: Duke University Press, 2007); and the proceedings of the conference "The Emotional Turn in the Social Sciences," University of California, Los Angeles, November 2011, the program of which is available at http://lifeconference2011.wordpress.com/panels-papers/the-emotional-turn/.
24. See Robert B. Zajonc's influential "Feelings and Thinking: Preferences Need No Inferences," *American Psychologist* 35.2 (1980): 151–75.
25. See Ruth Leys, "The Turn to Affect: A Critique," *Critical Inquiry* 37.3 (2011): 434–72, 437.
26. Brian Massumi, "The Autonomy of Affect," *Cultural Critique* 31 (1995): 83–109.
27. The social construction of emotion is articulated in Catherine A. Lutz, *Unnatural Emotions: Everyday Sentiments on a Micronesian Atoll and Their Challenge to Western Theory* (Chicago: University of Chicago Press, 1998).

28. James A. Russell and Lisa Feldman Barrett, "Editorial," *Emotion Review* 1.1 (2009): 2.
29. See also James M. Jasper, "Emotions and Social Movements: Twenty Years of Theory and Research," *Annual Review of Sociology* 37 (2011): 285–303.
30. Along these lines, see also Suparna Choudhury and Jan Slaby, *Critical Neuroscience: A Handbook of the Social and Cultural Contexts of Neuroscience* (Chichester: Wiley-Blackwell, 2012).
31. Antonio R. Damasio, *Descartes' Error: Emotion, Reason and the Human Brain* (New York: Avon Books, 1994); Daniel Goleman, *Emotional Intelligence* (New York: Bantam, 1995). William Reddy (chapter 1 in this volume) fills out this picture in the expert domain with the earlier rise of appraisal theory (Richard S. Lazarus's "Thoughts on the Relations between Emotion and Cognition," *American Psychologist* 37 [1982]: 1019–24, following Joseph J. Campos and C. Richard Sternberg, "Perception, Appraisal, and Emotion: The Onset of Social Referencing," in *Infant Social Cognition*, ed. Michael E. Lamb and Lonnie R. Sherrod [Hillsdale, NJ: Erlbaum, 1981], 273–314), and then, later, emotion regulation theory and nonmodular understandings of neural functioning.
32. Damasio, *Descartes' Error*, 200.
33. Martha Nussbaum, *Love's Knowledge: Essays on Philosophy and Literature* (New York: Oxford University Press, 1990), and *Upheavals of Thought: The Intelligence of Emotions* (New York: Cambridge University Press, 2001); Robert Solomon, *The Passions* (New York: Doubleday, 1976); Amélie O. Rorty, ed., *Explaining Emotions* (Berkeley and Los Angeles: University of California Press, 1980).
34. Nussbaum, *Love's Knowledge*, and *Upheavals of Thought*; Adela Pinch, *Strange Fits of Passion: Epistemologies of Emotion, Hume to Austen* (Stanford, CA: Stanford University Press, 1996).
35. See Daniel M. Gross, *The Secret History of Emotion: From Aristotle's "Rhetoric" to Modern Brain Science* (Chicago: University of Chicago Press, 2006), and "Being-Moved: The Pathos of Heidegger's Rhetorical Ontology," in *Heidegger and Rhetoric*, ed. Daniel M. Gross and Ansgar Kemmann (Albany: State University of New York Press, 2005), 1–45. An early and influential work on the deep history of passion's reasonableness is Albert O. Hirschman, *The Passions and Interests: Political Arguments for Capitalism before Its Triumph* (Princeton, NJ: Princeton University Press, 1977). Hirschman quotes, e.g., d'Holbach: "Reason . . . is nothing but the act of choosing those passions which we must follow for the sake of our happiness" (ibid., 27).
36. Febvre, "Sensibility and History."
37. Jean Delumeau, *Sin and Fear: The Emergence of a Western Guilt Culture, 13th–18th Centuries* (New York: St. Martin's, 1990); Theodore Zeldin, *A History of French Passions, 1848–1945* (New York: Oxford University Press, 1977–81).

38. See Lutz, chapter 14 in this volume.
39. Plamper, *Geschichte und Gefühl*, 135.
40. Peter N. Stearns and Coral Z. Stearns, "Emotionology: Clarifying the History of Emotions and Emotional Standards," *American Historical Review* 90.4 (1985): 813–36. On the evolution of the history of emotions as a field, see also Plamper, *Geschichte und Gefühl*, 62–72; and Bettina Hitzer, "Emotionsgeschichte—ein Anfang mit Folgen," *H-Soz-u-Kult*, posted November 23, 2011, http://hsozkult.geschichte.hu-berlin.de/forum/11-11-001.pdf.
41. Rosenwein, "Worrying about Emotions in History"; William Reddy, *The Navigation of Feeling: A Framework for the History of Emotions* (Cambridge: Cambridge University Press, 2001).
42. For the nineteenth century, see esp. Dixon, *From Passion to Emotion*. For the later period, see Morat and Jensen, eds., *Die Rationalisierung des Gefühls*; as well as the work of Otniel Dror. For the post-1945 period, the work of Ruth Leys is pathbreaking. See, e.g., Ruth Leys, "How Did Fear Become a Scientific Object and What Kind of Object Is It?" *Representations* 110.1 (Spring 2010): 66–104. See also "The Emotional Economy of Science," special issue, *Isis* 100.4 (2009). A focus on emotions appears to become more popular in the history of science. See, e.g., "The Varieties of Empathy in Sciences, Art, and History," special issue, *Science in Context* 25.3 (2012).
43. Plamper, *Geschichte und Gefühl*, 85.
44. Lorraine Daston and Peter Galison, *Objectivity* (Cambridge, MA: MIT Press, 2007).
45. Jack Barbalet, "Science and Emotions," in *Emotions and Sociology*, ed. Jack Barbalet (Oxford: Blackwell, 2002), 132–50.
46. Paul White, introduction to "The Emotional Economy of Science," special issue, *Isis* 100.4 (2009): 792–97, 794.
47. Andreas Reckwitz, *Das hybride Subjekt: Eine Theorie der Subjekulturen von der bürgerlichen Moderne zur Postmoderne* (Göttingen: Velbrück, 2006).
48. Daniel Rodgers, *Atlantic Crossing: Social Politics in a Progressive Age* (Cambridge, MA: Harvard University Press, 1998).
49. Mark Solovey, "Cold War Science: Specter, Reality, or Useful Concept," in Solovey and Cravens, eds., *Cold War Social Science*, 3–22.
50. The volume's focus on the German-US transatlantic relationship is rooted in the specific histories of Germany and the United States, where 1945 marks a motivated turn toward rationalism that had repercussions beyond these particular national and nation-specific academic contexts. Alternatively, Shah (chapter 6 in this volume) and Bailkin (chapter 11 in this volume) provide a transnational focus around the British postcolonial period. More work on science and emotion after 1945 in which other important national and transnational perspectives might be brought to bear would be welcome. For instance, in his review of the manuscript of this collection, Jan Plamper reminded us of the Russian/Soviet schools of

Pavlov and Vygotsky, each of which played important transnational roles in the natural and social sciences of emotion, Vygotsky particularly in education theory. See also, e.g., Anne DiPardo and Christine Potter, "Beyond Cognition: A Vygotskian Perspective on Emotionality and Teachers' Professional Lives," in *Vygotsky's Educational Theory in Cultural Context*, ed. Alex Kozulin (Cambridge: Cambridge University Press, 2003), 317–48.

51. Tim B. Müller, *Krieger und Gelehrte: Herbert Marcuse und die Denksysteme im Kalten Krieg* (Hamburg: Hamburger Edition, 2010).
52. On the reintroduction of psychoanalysis in postwar Germany, see Tobias Freimüller, *Alexander Mitscherlich: Gesellschaftsdiagnose und Psychoanalyse nach Hitler* (Göttingen: Wallstein, 2007).
53. We are aware of the problematic opposition between "internalist" and "externalist" explanations in the history of science. While our volume does not intend to contribute to this debate, it nevertheless seeks to distinguish between emotions as an object of study within a variety of academic disciplines and the ways in which these scientific practices shaped and were shaped by larger cultural norms. On the internalism/externalism debate, see Steven Shapin, "Discipline and Bounding: The History and Sociology of Science as Seen through the Externalism and Internalism Debate," *History of Science* 30 (1992): 333–69.
54. This is most apparent in the formation of economic advisory boards. See Michael A. Bernstein, *A Perilous Progress: Economists and Public Purpose in Twentieth Century America* (Princeton, NJ: Princeton University Press, 2001); and Alexander Nützenadel, *Stunde der Ökonomen: Wissenschaft, Politik und Expertenkultur in der Bundesrepublik, 1949–1974* (Göttingen: Vandenhoek & Ruprecht, 2005). On the social sciences more broadly, see Gilman, *Mandarins of the Future*; Robin, *Making of the Cold War Enemy*; and Uta Gerhardt, *Denken der Demokratie: Die Soziologie im atlantischen Transfer des Besatzungsregimes* (Stuttgart: Fritz Steiner, 2007). See also, in general, Lutz Raphael, "Die Verwissenschaftlichung des Sozialen als methodische und konzeptionelle Herausforderung für eine Sozialgeschichte des 20. Jahrhunderts," *Geschichte und Gesellschaft* 22.2 (1996): 165–93.
55. On the therapeutic society in the United States, see Ellen Herman, *The Romance of American Psychology: Political Culture in the Age of Experts, 1940–1970* (Berkeley and Los Angeles: University of California Press, 1995). On the West German context, see Sabine Maasen, Jess Elberfeld, Pascal Eitler, and Maik Tändler, eds., *Das beratene Selbst: Zur Genealogie der Therapeutisierung in den "langen" Siebzigern* (Bielefeld: Transcript, 2011).
56. See Reddy, *Navigation of Feeling.*
57. Peter Stearns, *American Cool: Constructing a Twentieth-Century Emotional Style* (New York: New York University Press, 1994), 55, 99.
58. *Consilience* is a term coined in 1840 by William Whewell and revived by Edmund O. Wilson in *Consilience: The Unity of Knowledge* (New York: Knopf, 1998). Consilience is literally a "jumping together" of knowledge

by "the linking of facts and fact based theory across disciplines to create a common groundwork of explanation." As described by Wilson, it is a scientific methodology rooted in "a conviction, far deeper than a mere working proposition, that the world is orderly and can be explained by a small number of natural laws." Ibid., 8, 4. For discussion, see Gross and Preston, chapter 3 in this volume.

59. Lazarus's "Thoughts on the Relations between Emotion and Cognition," following Campos and Sternberg's "Perception, Appraisal, and Emotion."
60. Bruno Wicker et al., "Both of Us Disgusted in *My* Insula: The Common Neural Basis of Seeing and Feeling Disgust," *Neuron* 40 (2003): 655–64; Mbemba Jabbi, Marte Swart, and Christian Keysers, "Empathy for Positive and Negative Emotions in the Gustatory Cortex," *Neuroimage* 34.4 (2007): 1744–53.
61. Stephanie Preston and Alicia J. Hofelich, "The Many Faces of Empathy: Parsing Empathic Phenomena through a Proximate, Dynamic-Systems View of Representing the Other in the Self," *Emotion Review* 4.1 (2012): 24–33.
62. Thomas S. Kuhn, *The Structure of Scientific Revolutions*, 2nd ed. (Chicago: University of Chicago Press, 1970).
63. Arlie Hochschild, *The Managed Heart: Commercialization of Human Feeling* (Berkeley: University of California Press, 1983).
64. Catherine A. Lutz, *Unnatural Emotions: Everyday Sentiments on a Micronesian Atoll and Their Challenge to Western Theory* (Chicago: University of Chicago Press, 1998).
65. See http://www.grossnationalhappiness.com.
66. See, e.g., Ben S. Bernanke, "Economic Measurement," speech delivered at the Thirty-second General Conference of the International Association for Research in Income and Wealth, Cambridge, MA, August 6, 2012, http://www.federalreserve.gov/newsevents/speech/bernanke20120806a.htm: "Following the growing literature, I define 'happiness' as a short-term state of awareness that depends on a person's perceptions of one's immediate reality, as well as on immediate external circumstances and outcomes. By 'life satisfaction' I mean a longer-term state of contentment and well-being that results from a person's experiences over time. Surveys and experimental studies have made progress in identifying the determinants of happiness and life satisfaction. Interestingly, income and wealth do contribute to self-reported happiness, but the relationship is more complex and context-dependent than standard utility theory would suggest. Other important contributors to individuals' life satisfaction are a strong sense of support from belonging to a family or core group and a broader community, a sense of control over one's life, a feeling of confidence or optimism about the future, and an ability to adapt to changing circumstances. Indeed, an interesting finding in the literature is that the overwhelming majority of people in the United States and in many other

countries report being very happy or pretty happy on a daily basis—a finding that researchers link to people's intrinsic abilities to adapt and find satisfaction in their lives even in very difficult circumstances." See also Ben S. Bernanke, "The Economics of Happiness," speech delivered at the University of South Carolina commencement ceremony, Columbia, SC, May 8, 2010, www.federalreserve.gov/newsevents/speech/bernanke 20100508a.htm.

67. Deirdre N. McCloskey, "Happyism: The Creepy New Economics of Pleasure," *New Republic*, June 8, 2012, 16–23.
68. See also, e.g., Ute Frevert's *Emotions in History: Lost and Found* (Budapest: Central European University Press, 2011), a study of the decline of honor and the rise of empathy.
69. Vivasvan Soni, *Mourning Happiness: Narrative and the Politics of Modernity* (Ithaca, NY: Cornell University Press, 2010).
70. A version of this argument can be found in Stearns, *American Cool*.
71. It should be noted, however, that the selected causal place of Foucault in the evolution of the history of emotions is contested. For a more critical view, one that sees the history of emotions more as a reaction against Foucault and certain strands of poststructuralism (rather than as emerging from it), see Plamper, *Geschichte und Gefühl*, 349. See also the controversial discussion of this point in Nicole Eustace, Eugenia Lean, Julie Livingston, Jan Plamper, William M. Reddy, and Barbara H. Rosenwein, "*AHR* Conversation: The Historical Study of Emotions," *American Historical Review* 117.5 (December 2012): 1487–1531. See also the remarks of Uffa Jensen and Lyndal Roper in "Discussion Forum: The History of Emotions," *German History* 28.1 (2010): 67–80.
72. "We believe that feelings are immutable, but every sentiment, particularly the noblest and most disinterested, has a history." Michel Foucault, "Nietzsche, la genealogie, l'histoire," in *Hommage à Jean Hyppolite*, by S. Bachelard et al. (Paris: Presses Universitaire de France, 1971), 145–72, translated by Donald F. Bouchard and Sherry Simon as "Nietzsche, Genealogy, History," in *Language, Counter-Memory, Practice*, ed. Donald F. Bouchard (Ithaca, NY: Cornell University Press, 1977), 139–64, 153. This passage is quoted as an epigraph to an early historiographic contribution: Adela Pinch, "Emotion and History: A Review Article," *Comparative Studies in Society and History* 53 (1995): 100–109.
73. Michel Foucault, *Madness and Civilization* (1961, New York: Vintage, 1988), *The Archaeology of Knowledge* (1969, New York: Pantheon, 1972), *Discipline and Punish* (1975, New York: Vintage, 1979), and *Care of the Self*, vol. 3 of *The History of Sexuality* (1984, New York: Vintage, 1988), esp. 61 (on Seneca).
74. Roland Barthes and Fernand Braudel recognized that Foucault owed more to the Annalists—and in particular Lucien Febvre—than might be expected and that, in the case of Febvre, the "history of mentalities" finds meaningful uptake in Foucault despite his statements to the contrary (see

n. 75 below). For instance, Braudel writes the following about Febvre: "In the first two parts of the work [*Le problême de l'incroyance au XVIe siêcle: La religion de Rabelais* (1942)], he treated Rabelais in the traditional way, using evidence from his life and his works, but he devoted the third part to the 'mental apparatus' of the period—the words, the feelings, the concepts that are the infrastructure of the thought of the century, the basis on which everything was constructed or could be constructed, and which may have prevented certain things from being constructed. Although the book was much praised, its originality limited its appreciation by historians. Febvre's work was ahead of its time, and it is only recently that the structuralists of the new literary criticism (for example, Michel Foucault, in his *Les mots et les choses: Une archéologie des sciences humaines*, 1966) have done similar research into the culture of a society." Fernand Braudel, "Lucien Febvre," in *International Encyclopedia of the Social Sciences* (17 vols.), ed. David L. Sills and Robert K. Merton (New York: Macmillan, 1968), 5:348–50, 349.

75. In *The History of Sexuality*, vol. 1, *Introduction*, trans. Robert Hurley (1976, New York: Pantheon, 1978), 52, Michel Foucault distinguishes "history of bodies" from the Annales school and its "history of mentalities": "The purpose of the present study is in fact to show how deployments of power are directly connected to the body—to bodies, functions, physiological processes, sensations, and pleasures. . . . Hence I do not envisage a 'history of mentalities' that would take account of bodies only through the manner in which they have been perceived and given meaning and value; but a 'history of bodies' and the manner in which what is most material and most vital in them has been invested."

PART ONE

Neuroscience

ONE

Humanists and the Experimental Study of Emotion

WILLIAM M. REDDY

How should humanists approach experimental research in cognitive and affective neuroscience? By *humanist* I mean a scholar whose epistemological commitments dictate reliance on interpretive method. By *interpretive method* I mean any method that attempts to explore meaning, intention, or the meaningful dimensions of language, text, or action. It is appropriate to call such commitments *humanist* because their purpose is to grasp the personal. Humanists, in this sense, for the most part, assume that there are persons and attempt to acquire knowledge about persons as persons, where a person is an entity with a measure of self-consciousness, an entity that acquires and deploys linguistic and behavioral repertoires while engaged in formulating, weighing, and pursuing goals. Humanists do not pursue explanation of personal phenomena as mechanisms, cause-effect chains, or algorithms; the assumption is that persons are flexible, to a very significant extent, and that their flexibility is achieved by reflection and manifest in intentional behavior called *action*.

By *cognitive and affective neuroscience* I mean the lines of research deriving from experimental psychology, on the one hand, and neurophysiology, on the other, whose union has been made possible by brain-imaging technologies and other methodological breakthroughs. In these closely related fields it is now possible to study such classic phenomena of experimental psychology as priming effects, the Stroop effect, cognitive load, subliminal perception, and automaticity while mapping the brain activations of participants. Affective responses, whose close integration with cognition was already being recognized prior to the imaging revolution, have also been subjected to new kinds of scrutiny. Entirely new concepts have been devised to make sense of these findings, such as "top-down processing," "mirror neurons," "embodied cognition," and "emotion regulation." At present, researchers are obliged to consider the neurological dimension of any explanatory scheme they devise for cognitive or affective phenomena and to gather evidence on it if possible.

These developments have coincided with a remarkable convergence of interests and approaches around the question of emotions across a number of disciplines, including philosophy, anthropology, history, and literary studies. Do these two trends have anything in common? Can their proponents learn from each other? Some humanists have sounded a note of caution, asking why anthropologists, historians, and other scholars who rely on interpretive method should concern themselves at all with developments in neuroscience. Jan Plamper recently raised this question, warning historians of emotions: "At bottom, both Barbara [Rosenwein] and [William Reddy] are saying that scientific hypotheses confirm their interpretation of historical data. Yet others could have interpreted these data differently or adduced different data in the first place. They could have cited competing scientific hypotheses in support. All of which suggests that it is problematic to invoke universalizing natural science to bolster contingent humanities claims."[1] The epistemological framework of neuroscientific research may be too different from that which prevails among scholars in "interpretive" disciplines such as history and anthropology. As a result, convergent or parallel findings may be intriguing, but they do not add up; that is, findings about emotional striving or effort on one side of the epistemological divide cannot be used to confirm similar findings on the other side. In addition to this problem of *incommensurability*, as some have called it, Plamper notes that, in the large interdisciplinary space of neuroscientific research, there are many models and many trends.[2] How is the scholar who turns to such research to be sure that she is not just picking out trends that happen to resonate with her own interests?

The discussion that follows will (1) examine the historical development of theories about the concept of person; (2) consider whether experimentalists who treat personhood or consciousness as strictly a mechanical matter are engaged in performative contradiction, a form of contradiction that, it has been shown, has plagued social scientific and philosophical thinking about human nature in recent centuries; (3) examine a trend in emotion research among experimentalists and neuroscientists that sees the brain as the support of an interpreter or agent; and (4) suggest that the line of incommensurability should be drawn, not *between* experimental neuroscience and the humanist disciplines, but *within* neuroscience.

The Concept of the Person

That historians, literary scholars, and others have become interested in the emotions recently is partly a reflection of a renewed interest in the history of the person or self. This renewal of interest followed what might be termed a *poststructuralist interlude* that effectively put into question every aspect of personhood that had previously been accepted as commonsensical among Western-influenced humanists. That a person possessed rationality and intentions and used meaningful language—these assumptions were forcefully challenged. (Interestingly, poststructuralists had little to say about emotions, at least at first.) The claim was that these assumptions (that persons were rational, intentional, meaningful) were the inadvertent side effects of certain arbitrary ways of using language, such as the illusion of "presence" denounced by Derrida or the idea of an "author" questioned by Foucault.[3] This renewal of interest in the history of the person also follows on the development of a powerful critique of gendered, racial, ethnic, and Orientalist practices that had wrongly justified the preeminence of the adult white European male. This preeminence had, for a long time, been explained in terms of a superior rationality and self-discipline, conceptualized in terms of a now questionable understanding of the person.

While its excesses have led to its partial eclipse, the "linguistic turn" has nonetheless accomplished a salutary clearing of the decks. What a person is and how a person may come to be are no longer regarded as self-evident or as matters that historians can afford to leave to the care of philosophers or psychologists. Recent Western-influenced cultural anthropology even before the linguistic turn, insofar as it has focused on cultural features of personhood, self-understanding, gender, and identity, had also brought forward lessons about the great variability in what counts as persons in various places, how persons are imagined to be structured, and how they interact.[4]

To summarize one thread among these lessons, cultural historians and other humanists have come to recognize that in Western-influenced contexts for the last twenty-five hundred years Western thinkers, moralists, and ideologues relied on various commonsense schemas to describe persons, schemas that are actually the parochial constructs of a particular tradition, with no special claim to universality. Since the Greeks, schemas of personhood have generally included what are sometimes called *faculties*—each author or school drawing on a variable list of such faculties that included the will, reason, memory, imagination, emotion, senses, and appetites. Persons were, in Western-influenced contexts, believed to be frequently drawn into internal tugs-of-war in which, for example, reason contended with appetite for influence over the will or emotion dictated goals to reason or else overwhelmed reason. In works of literature, one sometimes finds these struggles staged as allegorical internal debates in which personified Reason and Desire argue with each other, each attempting to win Will over to its own side. (A prominent example is the thirteenth-century *Roman de la Rose*.)[5] Christian theologians believed that original sin had resulted in a weakening of reason and the will and a strengthening of the passions, especially the passion of sexual desire, but also including avarice, gluttony, anger, and others.[6] The historian John Martin has argued that, well into the early modern period, the European self was viewed as "porous," that is, open to all kinds of outside influences, including grace, temptation, and magic spells.[7]

Descartes's rethinking of this tradition aimed at situating a self or mind within an entirely mechanical universe. Descartes did this by rigidly separating the world of the *res extensa* (things that take up space) and that of the *res cogitans* (things that think). Their sole point of interface, he believed, was in the brain, which mediated between sensory input and mind. He also argued that appetites and passions communicated with, influenced, or shook the mind through the pineal gland.[8] Controversy soon arose as to whether personal behavior was strictly mechanical and predictable, as, for example, Thomas Hobbes argued, relying on a model derived from geometry, or included a spiritual component, as Descartes believed. Locke avoided the issue; for him, the mind was possessed of reason and received all inputs from the senses.[9] From such seventeenth-century attempts at an explanation of human nature arose a very new vision of the person. Compatibly with the worldview of the scientific revolution, the person was imagined to be free of outside influences except for sensation. But the new thinking about the person continued to rely on the old vocabulary of faculties. In particular, the relation between reason and passion was imagined differently, but the two were quite consistently distinguished, as were other faculties such as memory and imagination. With the rise of a late modern secular outlook, the self has been reconceived as isolated, autonomous, but still containing faculties. One can see the old faculty style of thinking at work in the Freudian notions of the id, ego, and superego.

Such Western-influenced approaches to the person can be contrasted with, for example, that of the medieval South Asian sect known as the Kaula, as recently described by David Gordon White.[10] Kaula ritual focused on the propitiation of yoginīs, who were petulant, powerful goddesses first mentioned in sixth-century CE Sanskrit manuscripts. Kaula initiates sought to win the yoginīs' favor by offerings of a number of prohibited substances, including blood, meat, fish, wine, and semen.[11] After eating these offerings, the dangerous yoginīs revealed themselves as ravishing young women willing to confer various supernatural powers on their devotees.[12] Just as humans offered yoginīs substances that pleased them, such as blood and semen, so yoginīs conferred their favors through the sexual discharge of their vulvas. In Kaula ritual a yoginī would take possession of a woman, whose female discharge during coitus (*maithuna*) was believed to be from the yoginī herself. This precious fluid was consumed by practitioners along with the transformed offerings of other unclean substances.[13] Drinking female discharge was not intended to be arousing, according to White, and has nothing to do with "the 'bliss' and 'fun' offered by the modern-day Tantric sex trade." Instead: "The 'happy ending' of these rituals is described time and again in the Tantras as well as the adventure and fantasy literature of the medieval period . . . : both Yoginī and Vira [virile hero] fly up into the sky, to sport there together for eons of time."[14] The exchange of reproductive substances also transformed the Kaula initiate into a member of a divine clan or lineage able to pass its secret knowledge down through the ages.

In this cult's thinking, as was common more generally in South Asia, no distinction is made between reason and passion; the self is not divided between a spiritual part and a material part.[15] Spiritual transformations involve the whole self, including what in Western thinking would have been called the *body* and its *appetites*. In the South Asian

Sanskrit tradition, pleasure was the fruit of careful training. According to Daud Ali, the sex act entailed no pleasure unless it was carried out with "a vast array of accoutrements, material, verbal, and gestural, which were thought to be integral to its enjoyment."[16]

Among Christian theologians, by contrast, pleasure was little more than an inborn incentive to satisfy the physical needs of the flesh. So Peter Lombard explained it in 1154: "And we say that concupiscence is always evil, because it is a burden and a penalty of sin, but it is not always sinful. For often we see that the Savior took pleasure according to the flesh in one or another thing, such as resting after labors, eating when hungry. Such enjoyment is no sin unless it be immoderate."[17] Examples of cultural configurations that do not rely on anything resembling faculty theory or the spiritual-material divide implicit in the distinction between reason and passion could be multiplied at will.

The anthropologist Unni Wikan found that the Balinese also do not distinguish between reason and emotion, collapsing both into *keneh* (the "feeling-mind"). In Bali, when Wikan did fieldwork there in the 1980s, shaping the feeling-mind was an urgent collective concern. Many people strove to avoid even the slightest expression or experience of a negative emotion such as anger, fear, or grief. Such feelings rendered one vulnerable to black magic, which was believed to be responsible for approximately half the deaths on the island. Negative emotions meant a breach in the boundaries, or defenses, of the person, a draining away of strength. They were also regarded as contagious. To display grief after the death of a loved one was to endanger public health. Wikan's informants did not merely strive to avoid *expressing* negative emotions. By displaying the "bright face" they believed that they could gradually transform the feeling-mind. These two parts of the self, face and feeling-mind, were thus in constant, dynamic interaction.[18]

In the languages of Southeast Asia, generally, no distinction is made between emotion and thought, just as none can be found in Homeric Greek or ancient Hebrew.[19] In the Sanskrit tradition, there is no word corresponding to *emotion*. In the Kaula world, sexual arousal is not conceived of as a mere function of the material body. In all these examples, one can readily recognize that there are persons, that is, entities with intentions, capable of meaningful utterances and of action. In addition, in many non-Western contexts, as in many past Western-influenced contexts, personhood is not confined to human beings. More often, humans are just one type of person, and fundamental distinctions are often made among humans as well.

The Public Sphere and Performative Contradiction

Just as many scholars in humanist disciplines have learned that they must not take the construction of the person for granted, so they have learned to scrutinize with great care the relationships between authors and publics. In Western-influenced contexts, such relationships are practical realizations of a specific, local model of the person. With reference to authors and publics, one of the salient issues that arose during the poststructuralist interlude was the problem of "performative contradiction." *Performative contradiction* means acting in a way that conflicts with what one says. For example, it would be a performative contradiction to work hard to demonstrate to an audience that there is no such thing as a person. Such a goal, the meaningful practices it requires, and the assumptions those practices entail about the audience imply a strong belief in the existence of persons. One of the grounds on which Habermas critiqued Foucault's work was that Foucault fell into such performative contradiction.[20] Hilary Putnam has similarly criticized Richard Rorty's pragmatism for failing to accord with Rorty's own intellectual practices.[21]

But one way of elucidating Foucault's critique of the modern social sciences would be to say that social scientists systematically fall into performative contradiction themselves.[22] The *activities* of social scientists are grounded on the assumption that their research is shared among, and evaluated on its merits by, rational persons, that is, their fellow social science experts organized into scientific publics. Yet the *content* of much social scientific work is grounded on the assumption that individual humans and human societies behave mechanically, operating on the basis of chains of cause and effect that can be uncovered through empirical research similar to that carried out by astronomers, physicists, and chemists. Such mechanical causation of behavior can be conceptualized as not rational, for example, when social psychologists explore "how stereotypes of social groups become activated automatically on the mere perception of the distinguishing features of a group member."[23] Or the rationality of actors can be conceptualized as their so rigidly pursuing their own advantage that their behavior can be summarized by algorithms, as in rational choice theory.[24]

The only way to avoid performative contradiction in such circumstances is for social scientists to imagine themselves to be, or to treat themselves in practice as, exceptions to the rule. Participants in an expert public must presume that they and their colleagues are prepared to be persuaded by what Habermas calls "the force of the better argument."[25] They must presume that colleagues will bow to such force even if it is not in their own rational self-interest and even if the argument activates unconscious, irrational aversive responses.

But, in Foucault's view, the presumption that purposive rationality (rather than algorithmic pursuit of a predefined goal) is a shared attribute of the expert public—but not of human beings in general—is a most undemocratic maneuver. By implication this presumption seems to empower sociologists, economists, psychiatrists, criminologists, and other experts with inordinate and unwarranted authority—an authority similar to that arrogated by priesthoods and imperial bureaucracies in other times and places.

In short, in the last two or three decades, many humanist scholars have been sensitized to the great relativity and variability in the ways persons are imagined to be constituted. They have been sensitized, as well, to the dangers of performative contradiction and the dangerous political implications that may follow when social scientists avoid performative contradiction by assuming, implicitly or explicitly, a privileged status as persons capable of intentional action and meaningful communication while their subjects are presumed to be phenomena of the natural world, controlled by the mechanical laws of cause and effect. This privileged social scientific outlook has been evident in various configurations, some more egregious than others, in the thinking of Orientalists and Bolsheviks, Social Darwinists and Weberians, Charcot and Krafft-Ebing, racial theorists and eugenicists. That is, arrogation of social scientific insight by an in-crowd bent on exercising or accumulating power, or even more modestly on reshaping public policy, can easily be shown to have a dark history.[26]

Scholars who have learned to be alert to this kind of error, and who spend a lot of their time and effort attempting to remedy its effects in their own disciplines, are often dismayed when they turn to current research in experimental psychology and in the cognitive and affective neurosciences because they find plentiful indications that experimentalists continue to rely on very traditional commonsense notions of the person and the faculties that make up the person and that experimentalists also appear to fall frequently into that form of performative contradiction Foucault warned us against.

The persistence of such styles of thinking in these fields is dismaying. But such styles of thinking are not inherent in experimental research on human beings. On the contrary, (1) not every experimenter embraces such faulty thinking, and (2) the embrace of such faulty thinking is, in any case, not incompatible with carrying out interesting, even fascinating research. In fact, it seems plausible that, as research continues to advance, the raw findings are likely to force experimentalists to abandon traditional commonsense ideas about the person and recognize the traces, in psychological mechanisms and in brain structures, of the paramount status of meanings, intentions, and goals in human nature—the kinds of things that motivate researchers themselves.

The approach to experimental psychology and the neurosciences advocated here is not that of the history of science. Post-Kuhnian history of science concerns itself with the validity of scientific findings at best only secondarily. As Peter Dear has put it, in the history of science, understanding does not arise from the idea that Copernicus was right. It arises from appreciating all the reasons he came to think as he did when he did.[27] By contrast, a scholar with the humanist commitments outlined above seeks to enter into dialogue with experimentalists as the advocate of a point of view and may ally with certain experimentalists and certain research trends and against others. The humanist may legitimately do so on the grounds that, when experimentalists assume that human beings, as the objects of research, are machines while treating their fellow experimentalists as persons, their behavior may contradict their assertions.

When Daniel Wegner, for example, encourages his fellow psychologists to treat human beings more consistently as robots, he is subject to the same criticism that Habermas aims at Foucault.[28] Like Foucault's claim to engage in genealogy, Wegner's call for "robot analysis" stands only so long as he fails to apply this method to his own thinking. That is, the question whether a robot's thinking (i.e., Wegner's thinking if Wegner is a robot) is valid cannot be answered except by a self-conscious person capable of critical reflection. Is Wegner himself robot or person? If his audience is made up of robots, what do we make of their susceptibility to persuasion based on the force of the better argument? How can we know whether these robots are good scientists? As Habermas once complained of Nietzsche: "If, however, all proper claims to validity are devalued and if the underlying value-judgments are mere expressions of claims to power rather than validity, according to what standards should critique then differentiate?"[29] One only need substitute *mechanically realized algorithms* for *claims to power* in this complaint to apply it to Wegner. Self-consistency is just as respectable a criterion for judging the rationality of psychological research as is the adequacy of method for answering the question posed, or the coherence of the procedure with the questions, or the statistical significance of the findings. That peer reviewers too often neglect it is no reason to treat self-consistency as extraneous; it is instead all the more reason to raise the issue with insistence. A person addressing an expert public sphere made up of persons all of whom share the presuppositions of disenchanted "modern" selves must, to be consistent, treat the brain as a site associated somehow with the realization of personhood, not as a piece of machinery for realizing robot-like response patterns.

The Rise and Fall and Rise of Appraisal Theory

Wegner might reasonably respond that his call is only for a stance that has heuristic value in the short term. Before that knotty problem can be addressed, one must become aware that a difficulty exists and what its contours are. There is, in any case, a wealth of research in experimental psychology and cognitive neuroscience published over the last twenty years or so that seems to offer support for a view of the person in sharp contrast to Wegner's robots. Neuroscience researchers are quoting Wittgenstein and starting to talk about meaning and interpretation.

During the 1980s, a debate developed among experimental psychologists over the relation between affect and cognition. R. B. Zajonc launched the debate in a 1980 article in which he claimed: "Affective judgments may be fairly independent of, and precede in time, the sorts of perceptual and cognitive operations commonly assumed to be the basis of these affective judgments."[30] This was a kind of updated version of Hume's famous dictum that reason is the slave of passion.[31] By itself, according to this theory, cognition can only discover relationships and structures; it cannot attribute value. It deals with means, not ends. Affect, on its own, automatically, rapidly, and unconsciously delivers evaluations to the cognitive system, rather than evaluating input derived from cognitive processing. This view was compatible with the dominant place that basic emotion theory had come to hold in the minds of psychologists by the 1980s. Basic emotion theory treated emotions as hardwired, genetically preprogrammed response patterns. Various versions of the theory named different lists of basic emotions; but the most influential, that of Paul Ekman, named six basic emotions on the grounds that there were six distinct, naturally occurring facial expressions that could be recognized readily by any human being of any cultural background: happiness, sadness, anger, fear, disgust, and surprise. Ekman's tests were carried out in many parts of the world and appeared to confirm his claims; his views were routinely reported in basic psychology textbooks by the 1980s.[32]

Responding to the challenge of Zajonc's 1980 article, Richard S. Lazarus in 1982 insisted that no affect could occur until some cognitive processing had been accomplished and that there was no need to suppose that all cognitive processing was "deliberate, rational, and conscious," as Zajonc and others appeared to believe.[33] A fearful response to the sight of a bear required that cognitive processing first identify the stimulus as a bear, for example. By this view, cognitions that gave rise to emotions included an evaluation of the relevance of the stimulus to the self and the self's goals. Such special cognitions, although learned, could be rapid and peripheral to consciousness; Lazarus, following Campos and Sternberg and others, called them *appraisals*, to distinguish them from other cognitions.[34] And, in the course of the 1980s, the debate continued, pitting basic emotions theorists against appraisal theorists.

By the early 1990s, the appraisal theorists had scored a number of points. (And here, in part, I am traversing territory recently explored by Ruth Leys.) In a 1990 review article, Ortony and Turner compared the main basic emotions theories, noting the high degree of inconsistency and arbitrariness in various proposed lists of basic emotions. Ekman's insistence that facial expression was the key to affect seemed to leave out obvious candidates that other theorists included, such as love, pride, shame, and guilt. His only reason for doing so appeared to be that it was difficult to identify a distinctive facial expression that corresponded to each of them. The need to measure was dominating the theoretical formulation—the tail was wagging the dog—in the view of Ortony and Turner.[35]

In addition, basic emotions theory could not account for the great range of things that emotion words apply to. For example, it could not account for the difference between fear of a bear that one sees right now and fear of cancer that one might get someday. The stark contrast between an emotional *episode* that includes an immediate conscious response and a background emotional *orientation* that may not be activated for days, months, or years is, in fact, an issue that many theories of emotion have simply failed to address, as Ortony and Turner realized. They argued for a component theory of emotions. Some *components* of emotions, they believed, *were* basic, that is, hardwired response patterns, including things such as the furrowed brow, the curled lip, the heightened heart rate, the fight-or-flight response, and tears. Other components might be learned, such as the ability to read a yellow sign with the word *danger* on it. Each emotional response began with the specific appraisal that gave rise to it, and this appraisal then assembled an appropriate array of other components, including autonomic arousal, facial expressions, attentional focus, etc. Therefore, a key component of every emotion was an appraisal—a cognition related to the needs or goals of the self that could be either conscious and controlled or learned and automatic (or nearly automatic).

In 1994, James Russell published a detailed review of the Ekman-inspired research and identified some profound weaknesses in it. Ekman's standard procedure involved "forced choice"; that is, subjects were shown photographs of six emotional faces and forced to label them by choosing from six emotion terms. In addition, they were given training in how to do this prior to the administration of the test. Even under these favorable circumstances, however, Russell pointed out, agreement was high cross-culturally only for happy faces; for the other five of Ekman's basic emotions, it dropped off quickly in some populations. If subjects were permitted to choose their own words, the results failed utterly to support the idea of basic emotions.[36]

In response to these challenges, basic emotions theorists revised some of their views. But basic emotions theory remained dominant among experimentalists.[37]

From the early 1990s, as new brain-imaging technology began to be used in research on affect, PET and fMRI scans seemed at first to confirm basic emotions theory by, for example, identifying the amygdala as a region of the brain that seemed to respond rapidly and automatically to fearful stimuli, in particular to photographs of fearful or angry faces.[38] Because the amygdala was subcortical, it was considered to be part of the subcortical limbic system, so-called, a set of regions thought by many neurophysiologists to be responsible for emotions. Because the responses of the amygdala seemed to occur without attentional or voluntary input, this region was soon being heralded as the center of affective, or at least of aversive or fear-related, responses, based probably on hardwired patterns. The evolutionary advantage of being able to recognize fearful or angry faces seemed enough explanation to many researchers to make it plausible that the amygdala was the site of such a hardwired recognition system. The amygdala's responses seemed to operate independently of, and more rapidly than, the brain regions responsible for cognition. That the amygdala has a role in such rapid and automatic responses has by now been confirmed by scores of studies and is a widely recognized finding, frequently mentioned in the popular press.

But, as brain imaging became more sophisticated, persistent problems emerged with these results. For example, as Ochsner and Gross recently noted: "Amygdala activation in response to emotional stimuli was found inconsistently . . . , and prefrontal systems not important in animal work were often activated in human studies."[39] In addition, the amygdala was found to be involved in a variety of processes not necessarily linked with affect such as Pavlovian learning and the maintenance of attention. In animal studies, direct electronic stimulation of the amygdala resulted, not in a fear response, but in heightened attention to surroundings. Thus, the fear response caused by viewing emotional faces might simply be a special case of a more general function of the amygdala: the arousal of, and maintenance of, heightened attention.[40]

To account for these problematic findings, some neuroscientists turned to appraisal theory. Cognitive processes such as appraisal, according to Ochsner and Gross, "have been associated with regions of lateral and medial prefrontal cortex (PFC) thought to implement processes important for regulatory control, and regions of dorsal anterior cingulate cortex (ACC) thought to monitor the extent to which control processes are achieving their desired goals." Involvements of such systems might explain the inconsistencies in early imaging results: "If participants are controlling their attention to, and appraisal of, emotionally evocative stimuli, that could explain at least some instances of PFC activity, and potentially failures to observe amygdala activity as well."[41]

This hypothesis helped give rise to "emotion regulation" theory, which has attracted a growing number of researchers in the last ten years. In describing its features here, I follow Ochsner and Gross. Emotion regulation theory affirms the existence of bottom-up processing of emotional stimuli in the amygdala and basal ganglia coupled with top-down regulation of that processing by regions of the PFC and ACC. The top-down processing may be intentional, or it may be an unintended side effect of attending to a specific task.

For example, some studies show that activity of the amygdala decreases if participants are directed to pay attention to the emotional features of a facial expression but increases if they are asked to identify the face's gender. This might be explained if the amygdala were involved in background monitoring of input for affective cues—a function not needed when these cues are themselves the focus of higher-level directed processing. Amygdala response may also be reduced by having participants accomplish distracting tasks, purposefully flooding their attention with thoughts irrelevant to the stimulus. "Regions of orbitofrontal cortex (OFC), medial PFC (MPFC), ACC, and dorsolateral PFC (dlPFC) may be more active during distraction," note Ochsner and Gross.[42]

Amygdala response can also be "up-regulated" or "down-regulated" by instructing participants to interpret images pessimistically or optimistically. Such reappraisal "activates dorsal ACC and PFC systems that presumably support the working memory, linguistic, and long-term memory processes used to select and apply reappraisal strategies," report Ochsner and Gross. Further, "instrumental avoidance of aversive stimuli" as well as "reversal of stimulus-reward associations . . . depend on interactions between the amygdala, Nacc [nucleus accumbens], and ventral PFC, OFC, and/or ACC." Ochsner and Gross hypothesize that the cognitive control of emotion works in one of two ways: there is a "top-down description-based appraisal system" that activates a "bottom-up perceptual appraisal system," and there is also a "top-down outcome-based appraisal system" that works with a "bottom-up affective appraisal system." In the former of these two, the individual redescribes to herself an emotional stimulus or a feeling state "in a symbolic format that often is verbalizable."[43] In doing so, she draws attention to certain aspects or implications of the stimulus and ignores or sets aside others, with the result that the bottom-up perceptual appraisal system actually perceives or appraises the stimulus differently.[44]

An intriguing development in emotion regulation research was presented recently by Urry et al.[45] They used a combination of fMRI scans, electrodermal signals, heart rate monitoring, and pupil dilation data to examine what happened when participants were asked to alter their response to emotionally arousing photographs. Four seconds after display of the photographs began, participants were instructed either to "increase," "decrease," or "maintain" their emotional reactions. The researchers found that activation rose in certain regions of the PFC in response to instructions to increase response. Simultaneously, amygdala activations went up. When instructed to decrease their response, certain other regions in the subjects' PFC were activated, and amygdala activation decreased. Thus, the PFC regions appeared to be regulating the activation level of centers in the amygdala. Urry et al. also found that electrodermal signals increased or decreased along with amygdala activations, while pupil dilations and heart rate increased with activations of PFC regions. This finding confirmed the results of other studies, suggesting that pupil dilation and heart rate increases reflected increased cognitive demand in general and increased attention to a stimulus. Only the electrodermal signals correlated directly with increased or decreased amygdala activation, reflecting presumably an amygdala-directed arousal of the autonomic nervous system.

The Urry et al. study as well as neuroimaging studies of emotion regulation reflect growing interest in what is generally called *top-down* processing. In top-down processing, regions of the PFC are found to have a role in the very early stages of detection and identification of stimuli. Sketchy information is apparently rushed from early stage sensory-processing areas to the PFC, which responds with guesses about the nature of the stimulus; these guesses are then rushed back down to speed full and accurate identification at intermediate levels of sensory processing. Top-down processing has been uncovered in both visual object identification and speech recognition, although the exact nature of the upward and downward linkages is still under debate. It is significant that early PFC processing can be multimodal; that is, the guesses formulated in the PFC rely on inputs from more than one sensory modality and coordinate attention to multiple sensory inputs. This involves a prodigious and extremely rapid translation of sensory data into abstract representations that can yield expectations about coordinated data in other senses. The mutual influence of facial and auditory input on speech recognition is a well-known and much-studied example of such rapid multimodal coordination. Comparison of auditory signals and visual signals within the PFC can be carried out, completely unconsciously, within a time frame of 150 milliseconds after stimulus onset.[46]

In this context, emotion regulation involving linkages between PFC regions and the amygdala appears to represent simply the affective dimension of the PFC's general involvement in top-down processing. By these models, perception involves an ongoing process of "matching," in which estimates or control signals that are rushed from frontal regions via top-down processing are matched with inputs processed by sensory cortical regions.

The discovery of mirror neurons can be understood within the same general context. Visual input from the posterior superior temporal sulcus (STS) is passed to the motor cortex and from there to the ventral premotor cortex, which responds with simulated movement commands that represent approximate matches of the visualized movements. This system is important for affective processing because it is involved in the recognition of the affective meaning of facial expressions. Numerous studies have detected unconscious and very slight movements of facial muscles when subjects look at an emotional facial expression.[47] These are believed to be effects of the activation of mirror neurons. In one recent study, researchers found that botox, which temporarily paralyzes muscle tissue near the skin, "selectively slowed the reading of sentences that described situations that normally require the paralyzed muscle for expressing the emotions evoked by the sentences." Presumably, this was because mirror neurons were less readily able to run simulations of the appropriate facial expressions to speed in the comprehension of the sentences that the participants read.[48]

But the precise role of mirror neurons is subject to as much uncertainty as are the roles of other brain regions of recent interest. In a 2009 opinion piece, Scott, McGettigan, and Eisner, for example, argue that mirror neurons are of only limited use in speech perception. Lesions of the motor cortex, they note, do not cause deficits in understanding speech. Instead, they suggest, the decoding of the stream of speech occurs near the auditory cortex in the superior temporal gyrus, while mirror neurons serve only to track the rhythm and rate of other speakers and thus anticipate when one can begin speaking oneself.[49]

Top-down processing, mirror neurons, distributed networks, paired coupling of neural regions—these are some of the top buzzwords of recent neuroscience research, insofar as it has implications for affect. Another buzzword is *recruitment*, referring to the way in which specific regions are recruited into different specialized response networks depending on what kind of task or stimulus, or mix of tasks and stimuli, is present.

Salzman and Fusi, in a recent review proposed an interesting theoretical explanation of the overlapping, multifunctional character of distributed processing networks that are being revealed in neuroimaging research, note:

> If neurons represent only one mental state variable at a time, like stimulus identity, or stimulus valence, then "binding" the information about different variables becomes a substantial challenge. In this case, there must be an additional mechanism that links the activation of the neuron representing pleasantness to the activation of the neuron representing the first stimulus. One simple and efficient way to solve this problem is to introduce neurons with mixed selectivity to conjunctions of events, such as a neuron that responds only when the first stimulus is pleasant. In this scheme, the representations of pleasantness and stimulus identity would be entangled and more difficult to decode, but the number of situations that could be

> represented would be significantly larger. . . . [D]ifferent brain areas may contain representations with different degrees of entanglement.

In their view, the regions of the PFC constitute "a site of multimodal convergence of information about the external environment." These same regions of the PFC also handle information about motivations and emotions. This makes the PFC a "potential anatomical substrate for the representation of mental states." Its massive connections with the amygdala indicate that "the neural circuits mediating cognitive, emotional, physiological, and behavioral responses may not truly be separable and instead are inextricably linked." The inextricable overlapping of functions is therefore a necessary consequence of the way that neural architecture solves the binding problem. The authors conclude with a bow to Wittgenstein: "Much in the way that Wittgenstein argued that philosophical controversies dissolve once one carefully disentangles the different ways in which language is being used . . . , we argue that the debate between scientists about the origin of emotional feelings—whether visceral processes precede or follow emotional feeling—dissolves. Instead, all these parameters may be linked and together form the representation of our mental state."[50]

The June 2012 issue of *Behavioral and Brain Sciences*, for example, contained an article that powerfully reinforced the leanings of Salzman and Fusi.[51] Lindquist et al. reanalyzed data from 656 PET or fMRI scans, drawn from 234 studies, to test whether specific emotions can consistently be assigned to specific structures in the brain, that is, the "locationism" theory or "basic emotions view."[52] They concluded that the evidence decisively refutes this "locationist" approach.

Following the article are comments from twenty-eight different authors or teams of authors.[53] Not one of these commentators attempted to defend the locationist view of emotions. In other words, not one defended the idea that fear is a response located in or triggered by the amygdala, or that disgust is located in or triggered by the insula, or that anger is located in or triggered by the orbitofrontal cortex. A handful of commentators defended revised forms of basic emotions theory. Ross W. Buck, for example, asserted that genetically preprogrammed emotional responses could be based on distinctive "neurochemical cocktails" that could pervade various brain regions.[54] But even his comments break decisively with the doctrine, long cherished among experimentalists, of independent emotional and cognitive wiring.

None of these authors would deny the involvement of subcortical regions such as the amygdala in emotional responses. What they do deny is that subcortical regions constitute a separate system. Luis Pessoa, in his comment, remarked that brain regions should not be regarded as either cognitive or emotional. Traditionally, "regions whose function involves homeostatic processes and/or bodily representations have been frequently viewed as 'emotional,' whereas regions whose function is less aligned with such processes have been viewed as 'cognitive.'" But Pessoa urges the complete rejection of the "locationist account of emotions which, despite its numerous shortcomings is the still-entrenched viewpoint of most neuroscientists":

> The architectural features of the brain are such that they provide massive opportunity for cognitive-emotional interactions. These interactions are suggested to involve all brain territories. For example, extensive communication between the amygdala and visual cortex exists, and efferent amygdala projections reach nearly all levels of the visual cortex. Thus, visual processing takes place within a context that is defined by signals occurring in the amygdala (as well as the orbitofrontal cortex, pulvinar, and other regions), including those linked to affective significance. Therefore, vision is never pure vision, but is affective vision—even at the level of primary visual cortex. Cognitive emotional interactions also abound in the prefrontal cortex, which is thought to be involved in abstract computations that are farthest from the sensory periphery. More generally, given inter-region interactivity, and the fact that networks intermingle signals of diverse origin, although a characterization of brain function in terms of networks is needed, the networks themselves are best conceptualized as neither "cognitive" nor "emotional."[55]

For anthropologists and historians, the appeal of this new critical assault on the basic emotions view is obvious. The anthropological literature has turned up few communities that distinguish between thinking and feeling in the way that is common in Western-influenced contexts. If the two are one and the same as far as brain structure is concerned, this fact licenses one to regard this distinction as simply ethnocentric and helps explain the great variety of emotional vocabularies, emotional values, and concepts of the self that have been turned up by interpretive research.

Indeed, Lindquist et al. are perfectly aware that their findings make room for interpretive work as integral to emotional experience and make explicit reference to it with their concept of psychological constructionism, according to which "emotions emerge when people make meaning out of sensory input from the body and from the world using knowledge of prior experiences."[56] Neuroscientists who follow their lead may thus be focusing on how people make meaning and doing so in a way that carefully avoids the outmoded ethnocentric distinction between reason and passion. One could not ask more of the most well-informed cultural historian or ethnographer.

Conclusion

In responding to this acknowledgment of convergence implicit in Lindquist et al.'s theory, would humanists be guilty of leaping across epistemological incommensurabilities and of selecting (for no good reason internal to neuroscience) those neuroscience trends most suited to our purposes? Three considerations suggest otherwise.

1. We cannot afford to ignore the fact that human experience is part of the world or that there is a remarkable articulation of relationships between brain structures and events, on the one hand, and human experience and behavior, on the other. The linguistic theory that undergirded the linguistic turn provided, via structural linguistics, a quasi-physical real-world basis on which elaborate metaphysical understandings were later built by poststructuralist philosophers who relied on the jargon of linguistics to distinguish their metaphysical speculations from the tradition of metaphysics—which they, in traditional fashion, damned for its groundless and illusory drifting.[57] But structural linguistics was, after all, a product of social science research, a product that has since been shown to be profoundly wrong.[58]

Following the examples of Ruth Leys and Daniel M. Gross, I believe that we have no choice but to critically engage with neuroscientific research work.[59] Already, numerous authors are rushing to inform the general reading public about the brain structures supposedly underlying their anxieties, their love affairs, their reactions to the stock market.[60] *Neuro-* has become the indispensable prefix to a new wave in numerous disciplines (neuroeconomics, neurohumanities, etc.). This discussion is under way, and the only question is whether it will be carried on with a minimum of epistemological responsibility.

One widespread fallacy is especially important to avoid: the over-simplified treatment of causality that suggests that what happens in the brain causes what happens in our own interiors. Brain architecture itself supports a very different view, which is that causation flows in many directions at once. There are as many efferent pathways as there are afferent ones; in other words, as the remark by Pessoa quoted above suggests, every process is connected, at least indirectly, to every other process through elaborate two-way channels. Research in pain perception, speech recognition, and object recognition reveals a brain architecture suggesting that our most concrete sensory experiences of the world are shaped as much by what we believe and expect (much of this habituated belief and expectation operating unconsciously) as by what we encounter.[61]

Reflecting on such findings, many experimentalists have adopted an agnostic attitude toward causation, simply pointing out when conscious, behavioral, or verbally reported phenomena correlate with physically measurable state changes in the nervous system. These associations are not regarded as cause or effect. Salzman and Fusi, for example, carefully speak of brain structures as a "substrate" for personal events or as "mediating" such events.[62]

It is obvious, finally, that neuroscientists are engaged in interpretive work whenever they interact with experimental participants or devise tests for such participants. The validity of all their data relies on the accuracy of their interpretations of the intentions of participants. Were participants, as instructed, trying to respond accurately to word or object recognition tests? Were they distracted by being asked to count backward from a hundred while viewing frightening faces? It is crucial, in carrying out such interpretive work, to deploy a concept of the person that is as culturally neutral as one can get.

2. The basic emotions view appears to be entrenched for cultural reasons, not scientific ones. The reluctance of researchers to treat cognition and emotion as inseparable facets of every process derives in part from a folk theory of the self that still has great resonance in Western contexts. Trends in philosophy and in the humanities since World War II have rendered this folk theory untenable; it is time for experimentalists to get the word. And it is in terms of the concepts of the person deployed in interpreting participants and interacting with them that one can quite legitimately draw a line of incommensurability *within* neuroscience, between adherents of an older folk theory of the person and those who are able to shake free of this theory.

3. The movement within neuroscience that is critical of the basic emotions view is significant, not marginal. *Behavioral and Brain Sciences* is ranked as the most influential journal, by impact factor, in all fields of psychology, according to ISI Web of Knowledge.[63] The *Annual Review of Psychology*, which published Salzman and Fusi's review, is ranked as the second most influential journal. Opponents of the basic emotions view publish routinely in many other high-impact journals. Luiz Pessoa has published in *Proceedings of the National Academy of Sciences, Na-*

ture Reviews Neuroscience, and *NeuroImage.* Members of the Lindquist team of researchers have published in *Science*[64] and the *Annual Review of Psychology*[65] as well as in new journals such as *Emotion, Emotion Review,* and *Social Cognitive and Affective Neuroscience.* This last journal, founded in 2006, already ranks twelfth in impact factor among all psychological journals. If the basic emotions view is "still entrenched," as Pessoa notes, it is perhaps more a question of an older generation and perhaps also of the beliefs of neuroscience experts who do not themselves work on affect and are therefore not up to date on developments in affect research.[66]

Thus, humanists with the epistemological commitments and training associated with interpretive method and with ethnographic, cultural, and literary readings must recognize that their own research and critical reflection on their own methods align them closely with the points of view of that stream of research, or those streams of research, that has supported various versions of appraisal theory, emotion regulation theory, and nonmodular understandings of neural functioning. We run the risk of placing ourselves in performative contradiction if we step back and view the modern science of emotion strictly as historians.

NOTES

1. Nicole Eustace, Eugenia Lean, Julie Livingston, Jan Plamper, William M. Reddy, and Barbara H. Rosenwein, "*AHR* Conversation: The Historical Study of Emotions," *American Historical Review* 117.5 (December 2012): 1487–1531, 1502.
2. For this use of *incommensurability,* see, e.g., Antony S. R. Manstead and Agneta H. Fischer, "Beyond the Universality-Specificity Dichotomy," *Cognition and Emotion* 16 (2002): 1–9.
3. See, e.g., Jacques Derrida, *De la grammatologie* (Paris: Minuit, 1967); and Michel Foucault, *L'archéologie du savoir* (Paris: Gallimard, 1969).
4. To mention just two classic texts that affirmed the primacy of the cognitive or symbolic dimension of culture, see E. E. Evans-Pritchard, *The Nuer: A Description of the Modes of Livelihood and Political Institutions of a Nilotic People* (Oxford: Oxford University Press, 1940); and Clifford Geertz, "Person, Time, and Conduct in Bali," in *The Interpretation of Culture* (New York: Basic, 1973), 360–411.
5. On this work, there is a large corpus of scholarly discussion. For starting points, see Sarah Grace Heller, "Light as Glamor: The Luminescent Ideal of Beauty in the 'Roman de la Rose,'" *Speculum* 76 (2001): 934–59; and C. S. Lewis, *The Allegory of Love* (Oxford: Oxford University Press, 1936), 112–56.

6. James A. Brundage, *Law, Sex, and Christian Society in Medieval Europe* (Chicago: University of Chicago Press, 1987); Damien Boquet, *L'ordre de l'affect au Moyen Âge: Autour de l'anthropologie affective d'Aelred de Rievaulx* (Caen: Publications du CRAHM, 2005); Robert R. Edwards, *The Flight from Desire: Augustine and Ovid to Chaucer* (New York: Palgrave Macmillan, 2006).
7. John Jeffries Martin, *Myths of Renaissance Individualism* (Basingstoke: Palgrave Macmillan, 2004).
8. Bernard Baertschi, *Les rapports de l'âme et du corps: Descartes, Diderot, et Maine de Biran* (Paris: Vrin, 1992); Daniel M. Gross, *The Secret History of Emotion: From Aristotle's "Rhetoric" to Modern Brain Science* (Chicago: University of Chicago Press, 2006); Philip Stewart, *L'invention du sentiment: Roman et économie affective au XVIIIe siècle* (Oxford: Voltaire Foundation, 2010).
9. See M. R. Ayers, "Mechanism, Superaddition, and the Proof of God's Existence in Locke's *Essay*," *Philosophical Review* 90 (1981): 210–51; Todd Ryan, "Bayle's Critique of Lockean Superaddition," *Canadian Journal of Philosophy* 36 (2006): 511–34.
10. David Gordon White, *Kiss of the Yoginī: "Tantric Sex" in Its South Asian Context* (Chicago: University of Chicago Press, 2003).
11. Ibid., 76–77.
12. Ibid., 8.
13. Frédérique Apffel-Marglin, *Wives of the God-King: The Rituals of the Devadasis of Puri* (Delhi: Oxford University Press, 1985), 217; White, *Kiss of the Yoginī*, 83–84.
14. White, *Kiss of the Yoginī*, 100, 12.
15. June McDaniel, "Emotion in Bengali Religious Thought: Substance and Metaphor," in *Emotions in Asian Thought: A Dialogue in Comparative Philosophy*, ed. Joel Marks and Roger T. Ames (Albany: State University of New York Press, 1995), 39–63.
16. Daud Ali, *Courtly Culture and Political Life in Early Medieval India* (Cambridge: Cambridge University Press, 2004), 75.
17. "Et nos dicimus illam concupiscentiam semper malam esse, quia foeda est, et poena peccati; sed non semper peccatum est. Saepe enim delectatur vir sanctus secundum carnem in aliqua re, ut requiescendo post laborem, edendo post esuriem; nec tamen talis delectatio est peccatum, nisi sit immoderata." Peter Lombard, *Sententiarum libri quatuor*, in *Patrologiae cursus completus . . . series Latina* (221 vols.), ed. J.-P. Migne (Paris: J.-P. Migne, 1844–65), vol. 192, col. 921. See John W. Baldwin, *The Language of Sex: Five Voices from Northern France around 1200* (Chicago: University of Chicago Press, 1994), 120.
18. Unni Wikan, "Managing the Heart to Brighten Face and Soul: Emotions in Balinese Morality and Health Care," *American Ethnologist* 16 (1989): 294–312; idem, *Managing Turbulent Hearts: A Balinese Formula for Living* (Chicago: University of Chicago Press, 1990).

19. Andrew Beatty, "Emotions in the Field: What Are We Talking About?" *Journal of the Royal Anthropological Institute*, n.s., 11 (2005): 17–37, 28; Barbara Koziak, "Homeric *Thumos*: The Early History of Gender, Emotion, and Politics," *Journal of Politics* 61 (1999): 1068–91.
20. See Jürgen Habermas, *The Philosophical Discourse of Modernity: Twelve Lectures*, trans. Frederick G. Lawrence (Cambridge, MA: MIT Press, 1990), 238–65 (chap. 9, "The Critique of Reason as an Unmasking of the Human Sciences: Michel Foucault"), 266–93 (chap. 10, "Some Questions concerning the Theory of Power: Foucault Again"). For example, Habermas remarks that in his early work on madness Foucault posed without answering the question "how a history of the constellations of reason and madness can be written at all, if the labor of the historian must in turn move about within the horizon of reason." Habermas also points out: "The danger of anthropocentrism is banished only when, under the incorruptible gaze of genealogy, discourses emerge and pop like glittering bubbles from a swamp of anonymous processes of subjugation." Thus, the discourses of the sciences lose their privileged status, with other discourses forming power complexes that offer a domain of objects sui generis. However: "Of course, *Foucault only gains this basis by not thinking genealogically when it comes to his own genealogical historiography and by rendering unrecognizable the derivation of this transcendental-historicist concept of power.*" Ibid., 247, 268, 269 (emphasis added).
21. Hilary Putnam, "Richard Rorty on Reality and Justification," in *Rorty and His Critics*, ed. Robert B. Brandom (Oxford: Blackwell, 2000), 81–87.
22. See esp. Michel Foucault, *The Order of Things* (New York: Pantheon, 1971), 303–87.
23. John A. Bargh and Tanya L. Chartrand, "The Unbearable Automaticity of Being," *American Psychologist* 54 (1999): 462–79, 468.
24. For example, Edgar Kiser and Michael Hechter, "The Debate on Historical Sociology: Rational Choice Theory and Its Critics," *American Journal of Sociology* 104 (1998): 785–816.
25. See Maeve Cooke, *Re-Presenting the Good Society* (Cambridge, MA: MIT Press, 2006), esp. 47; Jürgen Habermas, "Communicative Rationality and the Theories of Meaning and Action" (1986), in *On the Pragmatics of Communication*, ed. Maeve Cooke (Cambridge, MA: MIT Press, 1998), 183–213, and "Richard Rorty's Pragmatic Turn," in Brandom, ed., *Rorty and His Critics*, 31–55, 43; Christian Jacobs-Vandegeer, "Insight into the Better Argument: Habermas and Lonergan," *Revista portuguesa de filosofia* 63 (2007): 1223–47; and Davide Panagia, "The Force of Political Argument," *Political Theory* 32 (2004): 825–48.
26. Linda Martin Alcoff, *Real Knowing: New Versions of the Coherence Theory* (Ithaca, NY: Cornell University Press, 1996), 120. See also Jensen, chapter 10 in this volume; Leys, chapter 2 in this volume; and Gere, chapter 5 in this volume.
27. Peter Dear, *Revolutionizing the Sciences: European Knowledge and Its Ambitions, 1500–1700* (Princeton, NJ: Princeton University Press, 2001), 2.

28. Daniel Wegner, "Who Is the Controller of Controlled Processes?" in *The New Unconscious*, ed. Ran R. Hassin, James S. Uleman, and John A. Bargh (Oxford: Oxford University Press, 2005), 19–36, 24–25.
29. Jürgen Habermas, "The Entwinement of Myth and Enlightenment: Rereading *Dialectic and Enlightenment*," *New German Critique* 26 (1982): 13–30.
30. R. B. Zajonc, "Feeling and Thinking: Preferences Need No Inferences," *American Psychologist* 35 (1980): 151–75, 151.
31. Roy Porter, *The Creation of the Modern World: The Untold Story of the British Enlightenment* (New York: Norton, 2000), 178.
32. See Ruth Leys, "How Did Fear Become a Scientific Object and What Kind of Object Is It?" *Representations* 110 (1020): 66–104, and chapter 2 in this volume. For a critical review of the Ekman-inspired research, see James A. Russell, "Is There Universal Recognition of Emotion from Facial Expression? A Review of the Cross-Cultural Studies," *Psychological Bulletin* 115 (1994): 102–41.
33. Richard S. Lazarus, "Thoughts on the Relations between Emotion and Cognition," *American Psychologist* 37 (1982): 1019–24.
34. J. J. Campos and C. R. Sternberg, "Perception, Appraisal, and Emotion: The Onset of Social Referencing," in *Infant Social Cognition*, ed. M. Lamb and L. Sherrod (Hillsdale, NJ: Erlbaum, 1981), 273–313. The use of the term *appraisal* for certain self-relevant or subjective cognitions was of long standing, but Lazarus was the first to introduce the term into this specific debate. See, e.g., R. G. Hopkinson, "The Multiple Criterion Technique of Subjective Appraisal," *Quarterly Journal of Experimental Psychology* 2 (1950): 124–31; and Magda B. Arnold, *Emotion and Personality*, vol. 1, *Psychological Aspects* (New York: Columbia University Press, 1960).
35. Andrew Ortony and Terence J. Turner, "What's Basic about Basic Emotions?" *Psychological Review* 97 (1990): 315–31.
36. Leys, "How Did Fear Become a Scientific Object?"; Russell, "Is There Universal Recognition?" See also Brian Parkinson and A. S. R. Manstead, "Appraisal as a Cause of Emotion," in *Review of Personality and Social Psychology*, vol. 13, *Emotion*, ed. Margaret S. Clark (Newbury Park, CA: Sage, 1992), 122–49.
37. James A. Russell, Jo-Anne Bachorowski, and José-Miguel Fernández-Dols, "Facial and Vocal Expressions of Emotions," *Annual Review of Psychology* 54 (2003): 29–49.
38. Kevin N. Ochsner and James J. Gross, "The Neural Architecture of Emotion Regulation," in *Handbook of Emotion Regulation*, ed. James J. Gross (New York: Guilford, 2007), 87–109, 87.
39. Ibid., 87.
40. Luiz Pessoa, "Emotion and Cognition and the Amygdala: From 'What Is It?' to 'What's to Be Done?'" *Neuropsychologia* 45 (2010): 3416–29, 3417–18; C. Daniel Salzman and Stefano Fusi, "Emotion, Cognition, and Mental

State Representation in Amygdala and Prefrontal Cortex," *Annual Review of Neuroscience* 33 (2010): 173–202, 178.

41. Ochsner and Gross, "Neural Architecture," 89.
42. Ibid., 95.
43. Ibid., 98, 99.
44. This proposed structure is compatible with the theory of emotives proposed in William M. Reddy, *The Navigation of Feeling: A Framework for the History of Emotions* (Cambridge: Cambridge University Press, 2001).
45. Heather L. Urry, Carien M. van Reekum, Tom Johnstone, and Richard J. Davidson, "Individual Differences in Some (but Not All) Medial Prefrontal Regions Reflect Cognitive Demand While Regulating Unpleasant Emotion," *NeuroImage* 47 (2009): 852–63.
46. Sharon M. Thomas and Timothy R. Jordan, "Contributions of Oral and Extraoral Facial Movement to Visual and Audiovisual Speech Perception," *Journal of Experimental Psychology: Human Perception and Performance* 30 (2004): 873–88; Eiling Yee and Julie C. Sedivy, "Eye Movements to Pictures Reveal Transient Semantic Activation during Spoken Word Recognition," *Journal of Experimental Psychology: Learning, Memory, and Cognition* 32 (2006): 1–14; Moshe Bar et al., "Top-Down Facilitation of Visual Recognition," *Proceedings of the National Academy of Sciences* 103 (2006): 449–54.
47. Marco Iacoboni and Mirella Dapretto, "The Mirror Neuron System and the Consequences of Its Dysfunction," *Nature Reviews Neuroscience* 7 (2006): 942–51.
48. David A. Havas, Arthur M. Glenberg, Karol A. Gutowski, Mark J. Lucarelli, and Richard J. Davidson, "Cosmetic Use of Botulinum Toxin-A Affects Processing of Emotional Language," *Psychological Science* 21 (2010): 895–900.
49. Sophie K. Scott, Carolyn McGettigan, and Frank Eisner, "A Little More Conversation, a Little Less Action—Candidate Roles for the Motor Cortex in Speech Perception," *Nature Reviews Neuroscience* 10 (2009): 295–302.
50. Salzman and Fusi, "Emotion, Cognition, and Mental State Representation," 175–76, 195.
51. Kristen A. Lindquist, Tor D. Wager, Hedy Kober, Eliza Bliss-Moreau, and Lisa Feldman Barrett, "The Brain Basis of Emotion: A Meta-Analytic Review," *Behavioral and Brain Science* 35 (2012): 121–43.
52. Ibid., online supplementary materials, p. 3, http://journals.cambridge.org/action/displaySuppMaterial?componentId=8601296.
53. *Behavioral and Brain Sciences* 35 (2012): 144–72. While this is the journal's standard format, the sweep of the Lindquist et al. review essay and the large number of commentators appear unusual.
54. Ross W. Buck, "Prime Elements of Subjectively Experienced Feelings and Desires: Imagining the Emotional Cocktail," *Behavioral and Brain Sciences* 35 (2012): 144.

55. Luiz Pessoa, "Beyond Brain Regions: Network Perspective of Cognition-Emotion Interactions," *Behavioral and Brain Sciences* 35 (2012): 158–59.
56. Lindquist, Wager, Kober, Bliss-Moreau, and Barrett, "The Brain Basis of Emotion," 123.
57. William M. Reddy, "Postmodernism and the Public Sphere: Implications for an Historical Ethnography," *Cultural Anthropology* 7 (1992): 135–68.
58. William M. Reddy, "Saying Something New: Practice Theory and Cognitive Neuroscience," *Arcadia* 44 (2009): 8–23.
59. Leys, "How Did Fear Become a Scientific Object?" Gross, *The Secret History of Emotion*.
60. Two recent French examples: Pierre Cassou-Noguès, *Lire le cerveau: Neuro/science/fiction* (Paris: Seuil, 2012); and Christophe André, *Les états d'âme: Un apprentissage de la sérénité* (Paris: Odile Jacob, 2009).
61. For further discussion, see William N. Reddy, "Review Essay: Daniel Lord Smail, *On Deep History and the Brain*," *History and Theory* 49 (2010): 412–25.
62. Salzman and Fusi, "Emotion, Cognition, and Mental State Representation."
63. Web of Knowledge (www.webofknowledge.com) journal impact factor for *Behavioral and Brain Sciences*: 25.056; article influence score: 10.76. (For comparison: *Proceedings of the National Academy of Science*: 9.681, 4.896; *Nature*: 36.280, 20.373; *Nature Reviews Neuroscience*: 30.445, 16.117; *NeuroImage*: 5.89, 2.162.
64. Eric Anderson, Erika H. Siegel, Eliza Bliss-Moreau, and Lisa Feldman Barrett, "The Visual Impact of Gossip," *Science* 332 (2011): 1446–48.
65. Lisa Feldman Barrett, Batja Mesquita, Kevin N. Ochsner, and James J. Gross, "The Experience of Emotion," *Annual Review of Psychology* 58 (2007): 373–403.
66. That neuroscientists often read narrowly and do not keep up on the whole field is noted in Gross and Preston, chapter 3 in this volume.

TWO

"Both of Us Disgusted in *My* Insula": Mirror-Neuron Theory and Emotional Empathy

RUTH LEYS

This chapter is undertaken as a contribution to the history of attempts to understand the nature of empathy. It is not surprising that today's researchers try to analyze empathy in neuroscientific terms. It is often said by such investigators that our knowledge of the neural basis of empathy is in its infancy, the suggestion being that it is only a matter of time before problems will be solved, as if the difficulties facing the research field are merely technical. But the implication of my analysis is that the issues confronting empathy theorists are as much theoretical or, say, philosophical as they are technical or scientific. Adam Smith's name is today routinely evoked in introductory remarks on the nature of empathy. But how many people realize that for Smith empathy (or sympathy) was not a natural phenomenon or an automatic process of resonance with the feelings of another? Rather, according to him sympathy was conditioned by an inherent theatricality that, by making persons into actors and spectators who distance themselves from each other and even from themselves, forestalls the possibility (the dream) of complete sympathetic merger or identification.[1] Freud expressed the same difficulty, indeed impossibility, in his own way when he made psychic ambivalence—the constitutive impossibility of separating eros and thanatos, love and hate, immersion and distance—central to his understanding of the sympathetic-identificatory phenomenon. According to Freud, rivalry with the other is as inherent in human nature as is love and indeed is inseparable from love: the taming of these emotions is the necessary but endless task of civilization.[2] For such thinkers, then, our knowledge of other minds cannot be explained by an appeal to a simple mechanism of mutual resonance or mutual attunement. Yet such a mechanism is precisely what recent neuroscientific theorists of empathy posit. As I hope to show, however, the problem of emotional empathy will never be solved if investigators persist in adopting the mirror-neuron theory and associated assumptions about the noncognitive, categorical nature of our emotions.

Introduction

In a report on the results of an experiment pertaining to a common neural basis for seeing and feeling the emotion of disgust, the claim is made that, just as we have mirror-neuron mechanisms for understanding other people's intentional actions, so we have mirror-neuron mechanisms for understanding or empathizing with other people's emotions. The paper was published in 2003 by a group of scientists that included first-author Bruno Wicker and coauthors Giacomo Rizzolatti and Vittorio Gallese, the last two being well-known as members of the research team that discovered mirror neurons in monkeys. I consider their paper a telling example of what can go wrong in emotion research today, and in the following discussion I shall try to say why.[3]

A mirror neuron is a neuron that fires both when an animal enacts a movement and when that animal merely observes the same action by another (especially a conspecific). In other words, mirror neurons appear to mirror the behavior of another animal by a kind of motor simulation or motoric resonance. Mirror neurons were first detected in the 1990s in experiments using electrodes directly implanted in the premotor cortex of the macaque monkey. Although they are often assumed to exist in humans and other species, the evidence is scant.[4] In humans, for example, the only published direct evidence of mirror-neuron activity exists in the form of single-neuron electrode recordings from the brains of epileptic patients, and even that evidence is equivocal.[5]

From the start, the function of mirror neurons has been the topic of much speculation and controversy. Many researchers have claimed that they provide a mechanism for an animal's ability to grasp the motor-intentional actions of others without the intervention of higher cognitive or sensory processes. Dysfunction in the mirror-neuron system is also thought to explain mind-reading failures associated with autism. The art historian David Freedberg and Vittorio Gallese have recently applied the idea of mirror neurons to the field of neuroaesthetics by claiming that our empathic responses to works of art as well as to everyday images depend on the activation of embodied, noncognitive mirror-neuron mechanisms.[6] They thus follow the trend in the neurosciences to expand the role of mirror neurons to include the capacity for emotional empathy.[7]

In their 2003 paper, Wicker et al. described the results of a functional magnetic resonance imaging (fMRI) study in which experimental subjects were asked to inhale odorants selected to produce strong feelings of disgust. The same subjects were also asked to observe video clips of other individuals exhibiting or showing the facial expression of disgust. Wicker et al. reported that the same sites in the anterior insula (and to a lesser extent in the anterior cingulate cortex) were activated both when the experimental subjects themselves experienced disgust and when they observed the filmed expressions of disgust on the faces of others.[8] They therefore concluded in reference to the mirror-neuron matching system: "Just as observing hand actions activates the observer's motor representation of that action, observing an emotion activates the neural representation of that emotion. This finding provides a unifying mechanism for understanding the behavior of others."[9] The Wicker et al. study has generated considerable interest in the neuroscientific community (a recent Google search indicates that the paper has now been cited in over nine hundred research articles). In subsequent publications, Rizzolatti, Gallese, Keysers, and others have cited Wicker et al.'s experiment on disgust as confirming the idea that there is a common neural basis for emotional empathy.[10] That experiment has also provided confirming evidence for Alvin I. Goldman's influential approach to the "problem of other minds." In his 2006 *Simulating Minds*, Goldman begins his review of the empirical evidence supportive of his Simulation Theory of mind reading by focusing on the "low-level" task of recognizing the emotional expressions of others because, as he observes with reference to the findings of Wicker et al. and others, the case for simulation here is "very substantial."[11]

The Wicker et al. Experiment in Detail

Let me begin by providing some details about the Wicker et al. experiment. First, the experimenters recruited from a Marseille theater school six actors (male and female) who agreed to be filmed while smelling either neutral, pleasant, or unpleasant odors. The actors were presented with a glass containing either pure water (for the neutral expression), water with an added pleasant odor (perfume designed to produce the pleased expression), or water with an added unpleasant odor (the content of "stinking balls" from a local toy store designed to induce the disgust expression). They were asked to display the relevant emotional reactions in a "natural but clear way." Each emotional reaction was filmed three times for each actor, and the "most natural example" was selected by one of the experimenters. These filmed enactments served as the visual "stimuli" for the experiment that followed.

The experiment itself was conducted with fourteen males, each of whom was asked to participate in two "visual runs" and two "olfactory runs" while undergoing fMRI. In the "visual runs" the participants passively viewed the film clips that had been made of the actors smelling the contents of the glass. They were not informed of the aim of the study and were not explicitly instructed to empathize with the actors. In the "olfactory runs" the participants themselves inhaled the same pleasant, disgusting, and neutral odors that had been smelled by the filmed actors.

Wicker et al.'s central finding was that the anterior insula was *not* activated during the participants' observation of happy expressions or during their experience of pleasant odors. But it *was* activated both during their observation of the actors' disgusted facial expressions and during the feeling of disgust evoked in the participants themselves when they smelled the foul odors. The investigators suggested that two different hypotheses might explain our ability to recognize and understand emotions in other people. According to the "cold hypothesis," we recognize the affects of others by using our perceptual and cognitive mechanisms without ourselves experiencing or sharing the same emotions and without activating the same causal mechanisms. But Wicker et al. claimed that their findings appeared to confirm the "hot hypothesis," according to which observing the emotions of others automatically generates the same emotion in ourselves because of a shared neural basis for seeing and feeling. In the case of disgust, the authors stated: "[T]his automaticity may explain why it is so hard to refrain from sharing a visceromotor response (e.g., vomiting) of others when observing it in them." They suggested that in evolutionary terms hot activation is likely to be the oldest form of emotion understanding, permitting a form of primitive empathy that may protect monkeys and young human infants from food poisoning even before sophisticated cognitive skills develop.[12]

Goldman considers Wicker et al.'s disgust experiment an original contribution to his Simulation Theory because it provides evidence for an "Unmediated Resonance (Mirroring)" model of simulation, according to which the perception of the target's facial expression "directly" triggers subthreshold activation of the same neural substrate of the emotion in question. He states that this model does not require the cognitive "pretend" or "off-line" states on which higher-level mind reading is theorized to depend, but only a minimum automatic matching between the pair of emotion events in the target and the observer. "The observer's emotional system 'resonates' with that of the target," he writes, "and this is the matching event on which the attribution is based."[13]

Background Assumptions

Wicker et al.'s experiment and the conclusions drawn from it presupposed a set of theoretical and methodological premises that are deeply entrenched in the field of emotion research today, and it is important to be clear about them from the outset. The main assumptions informing their work can be briefly summarized as follows:

1. There exists a small set of "basic emotions" defined as pan-cultural categories or "natural kinds."[14] These basic emotions are evolved, genetically hardwired, reflex-like responses of the organism. Disgust is one such basic emotion, as are fear, sadness, anger, joy, surprise, and perhaps contempt. The evolved status of the emotions implies some degree of emotional commonality between human and nonhuman animals, although the similarities and differences are rarely articulated.

2. Each basic emotion manifests itself in distinct physiological and behavioral patterns of response, especially in characteristic facial expressions. When not masked by cultural or conventional requirements of display or by deliberate deception, the face "expresses" the affects, which is to say that under the right conditions facial displays are authentic "readouts" of the discrete internal states that constitute the basic emotions.

3. The facial expressions associated with the basic emotions can be posed or portrayed by actors in a natural way so as to convey the authentic truth of the affects.

4. Each basic emotion is linked to specific neural substrates in the brain, an assumption that implies the embrace of some degree of modularity and information encapsulation in brain functions. Whereas (at least until very recently) the amygdala has been pinpointed as the neural seat of fear, insula activation has been especially implicated in the response to facial expressions of disgust, a finding that Wicker et al. claim not only to have confirmed but also to have extended, such that the insula is activated both during the experimental subject's observation of the facial expressions of disgust in others and during the subject's own experience of disgust.[15]

5. Emotional processes occur independently of "cognitive" or "intentional" states.[16] As Ekman has declared: "Emotional expressions are special . . . because they are involuntary, not intentional. . . . [E]motional expressions occur without choice. . . . [W]e trust them precisely because they are unintended."[17] According to this view, the basic emotions do not involve "propositional attitudes" or beliefs about the emotional objects in our world. Rather, they are rapid, phylogenetically old, automatic responses of the organism that have evolved for survival purposes and lack the cognitive characteristics of higher-order mental processes. The tendency in the recent literature on empathy to distinguish between "cognitive empathy," our ability to identify someone else's intentional actions, and "emotional empathy," our ability to sympathize with or match someone else's feelings, helps reinforce a noncognitive (or nonintentionalist) theory of the affects by suggesting that our affects occur independently of our cognitions. Wicker et al. give a definition of *disgust* that conforms to the noncognitive model by pigeonholing disgust as in essence a sensory or corporeal phenomenon—a point to which I will return.[18] They explicitly present the "hot hypothesis" as a noncognitive theory of emotional empathy.

6. Grasping another person's emotional state is also a noncognitive process. It is just a matter of responding automatically to the triggering effect of another person's facial displays. Goldman has criticized those whose view of empathy involves imputing purposive states to others. He argues that the kind of low-level faced-based emotion recognition that occurs in emotional empathy recruits a simulation mechanism that operates automatically and subpersonally, without the necessity of propositional contents, desires, or beliefs of any kind. The comparative simplicity of faced-based emotion recognition, he writes in this regard,

"consists in recognizing emotion types (e.g., fear, disgust, and anger) without identifying any propositional contents, presumably a simpler task than identifying desires or beliefs with specific contents."[19] On this view, reading someone's emotional expressions has survival value, and specialized mirroring mechanisms have evolved for this primitive kind of emotion detection.

Now, to anyone knowledgeable about the history of research on the emotions, the assumptions I have just summarized will be familiar, belonging as they do to an emotion theory or paradigm that has had tremendous success over the last thirty years in the United States and, to a large extent, in Europe as well. Specifically, the presuppositions of Wicker et al. can be traced most directly to the work of the American psychologist Silvan S. Tomkins and especially to that of his follower Paul Ekman, both of whom have proposed an evolutionary-classificatory approach to the affects.[20] Key features of their approach include the claims that there exists a small number of basic emotions, such as disgust, that can be defined in evolutionary terms as universal or pancultural, adaptive responses of the organism; that these emotions are discrete, innate, reflex-like "affect programs" located in subcortical parts of the brain; that the basic emotions manifest themselves in distinct patterns of physiological arousal and especially in characteristic facial expressions; that according to Ekman's "neurocultural" model for explaining commonalities and variations in human facial displays socialization and learning may determine the range of stimuli that can "trigger" the emotions and can moderate facial movements according to social norms or "display rules" but that under the right conditions the underlying emotions can nevertheless leak out; and that the more complex or "higher" emotions are made up of blends of the basic emotions. This view of the emotions has been given a variety of names; in this chapter I shall refer to it as the Basic Emotions View.[21]

A further claim associated with the Basic Emotions View, one that we have already seen in both Wicker et al.'s and Goldman's work, is that, although the emotions can and do combine with the cognitive systems in the brain, they are essentially separate processes. For Freud and the "appraisal theorists" such as Richard Lazarus, Robert Solomon, Martha Nussbaum, Phoebe Ellsworth, and others, emotions are embodied intentional states that are directed toward objects and depend on our beliefs and desires. But the Basic Emotion View denies this by interpreting the affects as nonintentional responses. It thus posits a constitutive disjunction between our emotions, on the one hand, and our knowledge of what causes and maintains them, on the other, because feeling and cognition are two separate systems. On this conceptualization, disgust concerns not the meaning of the objects or situations that disgust us but the inherent noxiousness or offensiveness of physical objects (such as animal and body wastes or contaminated foods) that are capable of automatically triggering an adaptive disgust response.

The Basic Emotions View has been extremely influential, especially because Ekman's strategy of using pictures of posed facial expressions as stimuli to test the responses of subjects in experimental situations is so easy to use and conforms so well to the requirements of the newer imaging methods of research. Hundreds of experiments have now been performed using as emotional stimuli a standard set of photographs of posed expressions that Ekman and Friesen first made available for research purposes as far back as 1976.[22] In order to give the appearance of greater ecological validity to their study, Wicker et al. used moving rather than still pictures of actors posing expressions, but this does not alter the fact that their assumptions and research methods fundamentally adhere to the norms of the Basic Emotions View.

But are those assumptions and research methods valid? There are serious reasons to doubt it. Not only have appraisal theorists questioned the validity of the Basic Emotions View by emphasizing the role of perceptual-cognitive evaluation of the situation in emotional processing.[23] For other reasons as well, it is doubtful whether the Basic Emotions View can withstand critical scrutiny. In recent years especially, investigators such as Alan J. Fridlund, James A. Russell, José-Miguel Fernández-Dols, and Lisa Feldman Barrett have published cogent criticisms of the Ekman paradigm.[24] The net result of those criticisms has been to directly challenge from within the emotion research field the empirical and theoretical validity of the Basic Emotions View. Nevertheless, for reasons I cannot examine here, that paradigm continues to dominate the field. Indeed, it currently represents the orthodox position.

Critique

Against this background, I now want to raise certain questions about Wicker et al.'s experiment and the uses to which it is being put to explain emotional empathy. I cannot do justice to the entire range of issues that interest me but will restrict myself to the following points:

1. My first question concerns the validity of Wicker et al.'s assumption that there exists a small set of basic emotions and that under the right conditions facial expressions can be viewed as involuntary read-

outs of internal emotional states. This is an assumption that Fridlund, Russell, Fernández-Dols, Feldman Barrett, and others have criticized. If we are to take their criticisms seriously, as in my view we must, then we need to reject the presuppositions underlying Wicker et al.'s analysis of emotional empathy. The idea that there exists a set of basic emotions that manifest themselves in characteristic patterns of physiological reactions and facial movements has been shown to be erroneous and the "readout" view of the affects mistaken.[25] Not that the reader would learn anything about those criticisms from Wicker et al.'s paper, which simply ignores them. The authors' failure to acknowledge the work of critics or to admit the existence of dissent exemplifies what I regard as a striking fact about the current situation of research on emotion, namely, that most scientists committed to the Basic Emotions View feel free to cite selectively and mention only the work of others that supports their views. The result is that objections are not allowed to disturb the investigators' basic premises or their experimental approach. Simply put, the network of presuppositions and methods associated with the Basic Emotions View is too attractive and the laboratory methods too convenient to be given up.

2. The experiment on disgust by Wicker et al. was based on a belief central to the Basic Emotions View, namely, that under the right conditions the face reliably and sincerely reveals the truth about the subject's inner feelings. Put slightly differently, the body does not lie. The facial displays performed by the actors as "stimuli" for the participants in the experiment were assumed to be authentic emotional expressions of this kind. It is because Ekman thinks that under the right conditions the face is bound to reveal the authentic truth of a person's feelings that since 9/11 he has been developing methods of surveillance designed to read the telltale involuntary signs he believes will identify terrorists. His goal is to reassure us that we do not have to be frightened by the tendency of human beings to dissimulate because trained observers can be counted on to reliably distinguish authentic facial expressions from false ones, the genuine from the feigned. His speculations have recently led to his involvement with the fanciful television series *Lie To Me* in which the lead character, a jet-setting Ekman surrogate named Lightman, oversees a large firm of beautiful men and women, reads faces to solve crimes, and routinely makes the police and the FBI look foolish.

But what if his assurance that the face reliably divulges the truth of our emotions is false? What if, as critics have argued, there is no simple one-to-one relationship between a person's facial behavior and his or her emotional state? What if facial displays cannot be considered simple readouts of underlying "basic emotions" because they are intentional communicative signals that aid in the negotiation of social encounters? As Fridlund has pointed out in this regard: "[A]ny reasonable account of signaling must recognize that signals do not evolve to provide information detrimental to the signaler. Displayers must not signal automatically, but only when it is beneficial to do so, that is, when such signaling serves its motives. Automatic readouts or spill-outs of drive states (i.e., 'facial expressions of emotion') would be extinguished early in phylogeny in the service of deception, economy, and privacy. Thus, an individual who momentarily shows a pursed lip on an otherwise impassive face is not showing 'leakage' of anger but conflicting intentions . . . for example, to show stolidity *and* to threaten."[26] In short, what if deception is widespread in nature and can be advantageous for the displayer?[27] Would this not imply that Wicker et al. were wrong to take for granted the meaning of the posed facial expressions used in the experiment because they ignored the potential for a mismatch between facial displays and their subjects' actual emotional experiences?

Wicker et al. were so confident that faces normally and automatically express the truth of the hypothesized basic emotions that it did not occur to them to ask the actors what they themselves were feeling when they sniffed the various odorants. Of course, it is possible that they really did feel the emotion of disgust they were exhibiting on their faces; the smell in question was selected because it was vile and was assumed to be intrinsically disgusting. But possibly they did not. To repeat, no one asked. Nor did the investigators make any effort to find out or discern whether the participants in the experiment felt disgust when they observed the actors posing facial expressions of disgust. This omission is all the more striking because, according to the "hot hypothesis," individuals recognize emotions in others by actually experiencing the same emotions themselves. But how do we know that this was the case in the absence of any effort to find out? In the Wicker et al. experiment, any attempt to discern what the participants were feeling was ruled out from the start. All they were asked to do was to passively witness the actors' facial displays or to smell the various odors themselves while submitting to brain imaging: emotion was equated with brain activation, with the result that the distinction between subjective experience and neuronal response was elided. In order to induce disgust in the participants, the investigators puffed the unpleasant and other odors into an anesthesia mask. Moreover, the subjects' mouths and eyes were closed throughout the olfactory runs, and during the experiment itself they did not speak or report on their feelings. In other words, it is as if the irrelevance of their subjective state was assumed from the outset.[28] I could apply to the experiment what Despret has said more generally about such methods in psychology: "The subject proves the scientist's point so well only because the latter has managed to keep him from speaking."[29]

3. My third question concerns the strategic role played in their experiment by Wicker et al.'s definition of disgust as simply and primordially a visceromotor reaction. For these authors, primitive or "core" disgust does not involve any cognitive-interpretive dimension entailing, as an intentionalist might argue, an embodied revulsion against appraised objects of various kinds, whether real or symbolic. Rather, they assumed that disgust is essentially a reflex response of the body to repulsive smells. They treated the more familiar or ideational forms of disgust as elaborated forms of the more fundamental olfactory and gustatory reflexes that serve to protect the organism against poisoning by preventing the ingestion or inhalation of harmful substances and smells. Disgust was therefore viewed as a food-related sensation involving a reflex revulsion at the incorporation of revolting or noxious foods. On this interpretation, derived from the work of Paul Rozin, disgust just *is* the sensation of a bad smell or bad taste (as Rozin points out, the word *dis-gust* simply means "bad taste"): in human development distaste may become linked cognitively, ideationally, and symbolically to an array of non-food-related items and objects, but at its core disgust is "a type of rejection primarily motivated by sensory factors."[30] The fact that the anterior sector of the insula is an olfactory and gustatory center that appears to control visceral sensations and related autonomic responses helps support this sensory-corporeal definition of disgust (the anterior insula region is known as the gustatory cortex).[31]

One can see the point of Wicker et al.'s definition. If disgust is just a bodily sensation with visceromotor manifestations, then the subjective-experiential dimension can be collapsed into the corporeal by studying brain activation directly, without any apparent conceptual loss. If both of us are disgusted in *my* insula because the activation of *your* insula when you experience an emotion is automatically duplicated by the activation of *mine* when I observe your disgust expression, then scientists do not have to worry about what I am feeling or what you are feeling because the neural mechanism we share will tell us everything they need to know.[32]

But is such a reflex definition of disgust valid? Fridlund, for one, does not think so. He concedes that the social disgust display resembles the protective gag reflex but thinks it more likely that the display is a convention or a "conversational icon," of the kind we see when a child sticks out her tongue in a display of defiance.[33] Nor does he believe that the disgust face should be considered an expression of a basic emotion. As he puts it, the gag reflex acts "not to 'express' sensory disgust but to abort it. Likewise, the social display signifies not 'you make me sick' so much as 'I want to do with you what I do with bad food (lest I get sick).' It thereby denotes an intention rather than an 'expression' of an emotion, and is therefore better named 'revulsion' or even 'rejection' than 'disgust.'"[34] In other words, Fridlund proposes that the disgust display should be regarded as an intentional movement subserving various social motives, which means not only that it is responsive to proximate elicitors but also that it is sensitive to those who are present, one's aim toward them, and the nature and context of the interaction.[35] He cites various experiments suggesting that facial responses to odors and tastes do not behave like simple reflexes but are influenced by the social situation in which they occur, including the presence of others.[36] Research on animal signaling has also suggested that many nonhuman facial and vocal displays likewise vary with the presence of interactants and with the relationship between the interactants and the displayer.[37]

Such findings are known collectively as *audience effects*, a characterization that has the virtue of drawing attention to the performative-transactional nature of facial and other displays.[38] It is precisely this performative-transaction dimension that Wicker et al. ignored. In their experiment, they treated both the actors and the experimental subjects or participants as if the latter were entirely alone in the room, which is to say as if they were completely liberated from the various cultural constraints that ordinarily guide people in any situation along a trajectory of social interaction with the expected and appropriate roles and expressions, with the result that they were free to exhibit their natural, innately determined expressions. In other words, these scientists forgot that the laboratory is a social space structured by conscious and unconscious or subconscious demands and expectations, including not only those of the experimental subject but those of the scientists involved as well. Fridlund has emphasized the "dramaturgical" dimension of such demands and expectations, suggesting that, when subjects are asked to pose or mimic facial displays to the point of being emotionally aroused themselves, the experimenter is actually a director and the subject-actor posing the expression is a Stanislavski actor who "slips into role." "It is the role or 'set' taken in the given social context that determines the emotion," Fridlund observes in this regard, "not the facial displays themselves."[39]

4. In light of such considerations, which emphasize the sociality of facial displays, Wicker et al.'s decision to define disgust as primordially a primitive reflex can be understood as a means of denying or suppressing the social-transactional character of the organism's emotional reactions. It is all the more interesting, then, to note that at one moment in their paper Wicker et al. themselves naively invoked Stanislasvki's acting theories in ways that unexpectedly redounded on themselves. The issue came up when they were discussing another experiment on emotional empathy, one that appeared the same year as their own and that covered somewhat similar ground. In the experiment in question, Laurie Carr and her associates asked the experimental participants—ordinary persons, not actors—to pose all six of the basic emotions, including disgust, in order to determine by fMRI whether the same neural substrate was activated both when the participants actually experienced emotions through posing or imitation in this way and when they observed the same emotions in others by observing a set of Ekman and Friesen's photographs of facial expressions on a computer screen. Carr et al. showed that both the imitation of emotions and their observation activated a largely similar network of brain regions, including the anterior insula, although activation was greater when the subjects imitated the expressions than when they merely passively observed them in others.[40]

In their paper Wicker et al. acknowledged the agreement between their own results and those of other researchers, including those of Carr et al. But they also drew attention to certain differences. They pointed out that no previous study of disgust, including that of Carr et al., had actually evoked the "sensation of disgust" in experimental subjects, as they themselves had done, in order to investigate whether the activated locations were common to both the experience of disgust and the perception of the same emotion in others. They stressed in this regard that merely imitating or posing an emotion, as Carr et al.'s experimental subjects were asked to do, does not require or guarantee that the poser subjectively feels the portrayed affect because "imitation usually does not require experiencing the imitated emotion." They thus declared that Carr et al. had demonstrated only that the insula was involved in imitation, *not* that it was directly involved in the "experience of emotions."[41] In other words, Wicker et al. claimed that, unlike the subjects in their own experiment, who by smelling a foul odor actually experienced the emotion or sensation of disgust, the participants in Carr et al.'s experiment might only have represented but not personally felt the emotions they were showing on their faces (as if Carr et al.'s subjects experienced only cold emotional responses).

Since Carr et al. found that the insula was nevertheless activated, their findings appeared to invalidate Wicker et al.'s claim that the insula is necessary for actual emotional experience. But Wicker et al. ingeniously proposed a solution to this apparent difficulty. They suggested that, like good method actors, some of Carr et al.'s participants must have been so swept up in their roles that they really must have felt the emotions they were portraying on their faces. "However," Wicker et al. observed in this connection, "in the light of our findings, it is possible that, during imitation, some of their participants felt the imitated emotion—as actors do when using the 'Stanislavsky' [*sic*] method of emotion induction."[42] I call the invocation by Wicker et al. of Stanislavski's theory of acting "naive" because they do not seem to have appreciated the problem of acting in their own case. The interesting question here is, Why in their own experiment did Wicker et al. use not ordinary people but precisely actors to perform expressions in front of the camera for the purposes of making portraits of disgusted, neutral, or pleased facial expressions to show to the participants in the experiment? If disgusting smells are disgusting to everyone and automatically induce the experience (or "sensation") of disgust, then ordinary volunteers could have served the investigators' purposes just as well. The fact that professional actors were used and that they were asked to display expressions in a "natural but clear way"[43] suggests that some degree of acting skill and "stage direction" was necessary to produce the required disgust display, or at any rate that Wicker et al. believed that to be the case—in other words, they believed, or proceeded as if they believed, that ordinary people are not very good at portraying such emotional expressions in the way scientists require. We might put it that, in their experiment, Wicker et al. functioned in part as directors of the facial displays, although it remains an open question whether the performers slipped into their role so deeply that, like good "method" actors, they really felt the emotion or "sensation" of disgust the investigators attributed to them—as I say, the actors were not asked. In any case, the notion of a "natural but clear" display begs every conceivable question, implying as it does that performers or actors are capable of producing on demand "natural" appearances (as opposed to what exactly?) and moreover that they can produce on demand emotional expressions that are "intense but natural" and not, let us say, overdone or exaggerated.[44] But the entire history of modern theories of dramaturgy testifies to the fact that nothing of the sort can be taken for granted.[45] All this may be summed up by saying that, in their appeal to the ideas of Stanislavski, Wicker et al. inadvertently drew attention to the contextual-social influences at work in the production of emotional expressions, influences that their Ekman-inspired reflex, corporeal, readout approach to the affects was meant to forestall.

New Findings

The disgust story does not end here. Perhaps aware that Wicker et al.'s experiment was defective in some respects, investigators returned to the fray with a follow-up experiment in 2007. In the new study, by Jabbi, Swart, and Keysers (the latter being one of the authors of the 2003 experiment), disgusting *tastes* rather than disgusting *smells* were the focus of inquiry, but the basic experimental setup remained the same. As before, actors were filmed while posing disgusted, pleased, and neutral expressions in a "naturally vivid manner," this time when sipping unpleasant (quinine), pleasant (sucrose), and neutral (artificial saliva) solutions from a cup, and the ten best clips for each emotional category were selected for use in the experiment. In the "Visual runs" the experimental participants (eighteen right-handed subjects, ten females and eight males) were asked to observe the movie clips of those posed expressions while they themselves underwent fMRI. In the "gustatory runs" the participants were asked to sip the same liquids as those the actors had tasted, again while undergoing brain scanning. (The solutions were delivered by an experimenter standing beside the fMRI scanner, using a tubing system consisting of a syringe connected to an infusion tube inserted into a pacifier.) Just as in the previous experiment, insula activation was reported in both the visual and the gustatory or "experiencing" condition. But this time the investigators added a new feature: they asked the participants to rate their own experiences both on tasting the solutions and on seeing the actors' emotional expressions when the latter posed their facial reactions to the same drinks. It is as if the researchers recognized that, without documenting the participants' actual subjective states, the "hot hypothesis" predicting that the participants would actually experience the same emotion as those whom they were observing had remained unproved. Put less critically, it is as if they wanted to document assumptions that the Wicker et al. experiment had apparently taken for granted but that skeptics could rightly question.[46]

It is worth remarking that the attempt to evaluate the participants' subjective responses raised some theoretical difficulties for Jabbi et al. The "hot hypothesis" claimed that observers experience emotions in an automatic, noncognitive way just by observing the facial expressions of others. That hypothesis cannot be supported without demonstrating that people really do experience disgust when they see disgust expressions in others—evidence of brain activation alone will not suffice. The dilemma Jabbi et al. faced was that the attempt to determine an observer's emotional experience required asking him or her to make conscious and explicit what, on the "hot hypothesis," had been theorized as an implicit, nonconscious, and subpersonal process. Evaluating a participant's subjective feelings therefore necessitated asking him or her to transform a hypothesized noncognitive experience or event into an actual cognitive one in order to articulate and report on it. In effect the hypothesis of emotional simulation could not be tested because the moment the researchers asked their subjects to report on their subjective experience those subjects were doing cognition and hence transforming what was understood to be a hot, noncognitive process into a cognitive one. In short, the "hot hypothesis" could not be confirmed without contradicting its basic, noncognitive premise.

Moreover, in designing their experiment Jabbi et al. appear to have been motivated by a further concern, namely, that, although the "hot hypothesis" could explain the observer's tendency to emotional "contagion" or emotional resonance, it could not in itself account for the empathic understanding of another, as the "hot hypothesis" had seemed to propose. As Jabbi et al. observed in this regard, infants contagiously cry when they witness the distress of other people but are presumably unable to distinguish their feelings from those of others. In contrast, more mature persons not only resonate contagiously to the emotions of others but also are able to interpret and attribute their subjective states to someone else while distinguishing their emotions from those of another, thereby acquiring genuine empathic understanding or conscious knowledge of the other. In short, Jabbi et al. now appeared to concede that mirroring or resonance or contagion of the kind proposed by the "hot hypothesis" might be a prerequisite for empathic understanding of another but is not sufficient for it, as the "hot hypothesis" had at first appeared to claim.[47]

Against this background of these issues and concerns, we can understand why Jabbi et al. made an effort to determine the subjective responses of their experimental subjects.[48] First, they rated the subjective reactions to the gustatory emotions of the actors in the movies by asking the participants how willing they would be to drink the beverages the actors had just tried (using a scale from −6 ["absolutely not willing"] to 6 ["very much willing"]). Second, they asked the participants to rate the solutions they themselves had to ingest during the experiment (on a scale ranging from "extremely disgusting" to "extremely delicious"). These scales were taken to be measures of the participants' evaluations of the beverages involved, in the third person ("He tastes") and the first person ("I taste") perspectives, thus allowing a direct comparison of these two perspectives. In other words, how willing the participants were to taste the drinks they witnessed the actors ingesting was taken to be a measure of the affective states the facial expressions induced in them. In addition, Jabbi et al. obtained the participants' self-reported empathy scores as measured by an Interpersonal Reactivity Index. They then correlated these scores with the insula activation that occurred during the same participants' witnessing the clips of the actors posing the pleased, disgusted, and neutral expressions.[49]

The main new result reported by Jabbi et al. was that for the first time it had been demonstrated that, during the observation of other people's "gustatory emotions" (i.e., the observation of other people's disgust expressions), the size of insula activation correlated with differences in self-reported interpersonal reactivity, or empathy. They took this finding to extend the previous demonstration by Wicker et al. that the insula was activated during the experience and observation of negative emotional states, such as disgust, and therefore to provide further support for the hypothesis that the anterior insula was involved in the transformation of emotional states into experienced ones. Their results also showed that the insula's involvement was not restricted to negative emotions but included the processing of positive emotions as well.

In light of the criticisms I have already offered in this chapter, many questions could be raised about this experiment and its purported findings, but I will limit myself to two.[50] First, Jabbi et al. appear to have made no attempt to ascertain whether the *participants* (observers) felt disgust or pleasure when they actually watched the actors' facial expressions, with the result that their effort to determine their experimental subjects' subjective experience seems to have fallen short.[51] Equally interesting from my point of view is the fact that Jabbi et al. made no attempt to discover what the *actors* experienced when they were asked to taste various liquids and produce the relevant facial movement or expression in order to be filmed. Why did they limit their inquiries in this regard? Perhaps they assumed that the liquids the actors were asked to sip inevitably aroused in them the relevant experience of disgust or pleasure and, according to the readout theory, therefore also produced the relevant facial expression.[52] Or maybe they took it for granted that facial mimicry of the kind involved in posing expressions automatically induces in actors the relevant internal emotional states (although the evidence on the topic of facial mimicry or facial feedback is mixed at best).[53] But they did not address this question at all.

Why does the omission matter? I think it matters because, by failing to determine the actors' personal or subjective experiences, Jabbi et al. left open the possibility that, just as in the earlier experiment by Wicker's research team, so in this experiment the actors might not have actually experienced disgust themselves but merely posed the facial expressions they were asked to represent on their faces. (Of course, as I have said, the quinine used to induce the actors' disgust expression was taken to be inherently disgusting, but this claim was not tested by asking the actors their subjective reactions, so it remains an open question whether such a response should have been taken for granted.) Actors do this all the time, and indeed Ekman's neocultural theory predicts insincerity or feigning in many social situations, of which the demand that actors pose an expression can serve as an example. But the effect of the omission is to suggest that, since, according to the "hot hypothesis" of emotional empathy, we automatically empathize with or resonate to the emotional expressions of others, we will do so whether or not the people we observe are really feeling what they show on their faces. The "hot hypothesis" therefore seems to imply that we are destined to spend our days resonating madly, nonselectively, immoderately, automatically to whatever facial signals someone else, anyone else, sends us, without our knowing whether those signals are telling us the truth about the latter's emotional state. If the mirror-neuron theory of simulation is true, we can be fooled—we *will* be fooled—about the emotional states of others all the time. Both of us disgusted in *my* insula? It might be more accurate to say that *I* will be disgusted in *my* insula as long as *you* display or perform an expression of disgust—regardless of whether you are sincere. But what kind of theory is *that*?

NOTES

My thanks to Michael Fried, Avery Gilbert, James A. Russell, and Rainer Reisenzein for their helpful comments on my chapter. A slightly longer version first appeared at nonsite.org.

1. David Marshall, *The Figure of Theater: Shaftesbury, Defoe, Adam Smith, and George Eliot* (New York: Columbia University Press, 1985), 167–92.
2. As Menninghaus has usefully reminded us, for Freud, the emotion of disgust is the direct opposite of a natural given since it is a result or symptom of the repression of archaic libidinal drives and hence of the passage into culture: as a transformation of eros, disgust is indissociable from pleasure or desire. Wicker et al.'s narrow definition of disgust as essentially a visceromotor reflex, like vomiting, contrasts with the long Western tradition of theorizing disgust that, since Kant, has been oriented toward questions of aesthetics and aesthetic judgment. In his discussion of this tradition, Menninghaus brings out the ways in which in earlier theorizing both attraction and revulsion—hence ambivalence or conflict—were seen to characterize the disgust experience, a dimension entirely lacking in contemporary scientific definitions of the kind adopted by Wicker et al. See W. Menninghaus, *Disgust: Theory and History of a Strong Sensation*, trans. Howard Eiland and Joel Golb (Albany: State University of New York Press, 2003); and Bruno Wicker et al., "Both of Us Disgusted in *My* Insula: The Common Neural Basis of Seeing and Feeling Disgust," *Neuron* 40 (2003): 655–64. For an account of the role of sympathetic identification or imitation in Freud's work on trauma and the affects, see Ruth Leys, *Trauma: A Genealogy* (Chicago: University of Chicago Press, 2000).
3. For the Wicker et al. study, see the previous note.
4. As Dinstein has observed, in the absence of direct evidence for mirror neurons in humans many researchers interpret any fMRI response by the relevant brain areas as being due to mirror-neuron activity. Ilan Dinstein, Cibu Thomas, Marlene Behrmann, and David J. Heeger, "A Mirror Up to Nature," *Current Biology* 18 (2008): R13–R18.
5. For an experiment claiming to demonstrate by direct recording the existence of mirror neurons in the human brain, see Roy Mukamel, Arne D. Ekstrom, Jonas Kaplan, Marco Iacobini, and Itzhak Fried, "Single-Neuron Responses in Humans during Execution and Observation of Actions," *Current Biology* 20 (2010): 750–56. For endorsements of such claims, see Christian Keysers and Valeria Gazzola, "Social Neuroscience: Mirror Neurons Recorded in Humans," *Current Biology* 20.8 (2010): R353–R354; and Giacomo Rizzolatti and Corrado Sinigaglia, "The Functional Role of the Parieto-Frontal Mirror Circuit: Interpretations and Misinterpretations," *Nature Reviews Neurosciences* 11.4 (2010): 264–74. For critiques, see esp. Greg Hickok, "Two New Ways the Mirror System Claim Is Losing Steam," *Talking Brains*, May 18, 2011, www.talkingbrains.org, "Eight Problems for the Mirror Neuron Theory of Action Understanding in Monkeys and Humans," *Journal of Cognitive Neuroscience* 21.7 (2008): 1229–43, and "What Mirror Neurons Are REALLY Doing," *Talking Brains*, September 17, 2009, http://talkingbrains.org; Greg Hickok and M. Hauser, "(Mis)understanding Mirror Neurons," *Current Biology* 20 (2010): 14; and Vittorio Gallese,

Morton Ann Gernsbacher, Cecilia Heyes, Gregory Hickok, and Mario Iacoboni, "Mirror Neuron Forum," *Perspectives in Psychological Science* 6.4 (2011): 369–407.

6. David Freedberg and Vittorio Gallese, "Motion, Emotion, and Empathy in Esthetic Experience," *Trends in Cognitive Sciences* 111.5 (2007): 197–203.
7. Vittorio Gallese, Stephanie Preston, and F. B. M. de Waal are among those who proposed early on that empathy depends on a perception-action model, according to which the perception of another's state automatically activates the observer's representations of that state, and that activation of those representations generates the autonomic and somatic responses associated with the emotion. See Vittorio Gallese, Luciano Fadiga, Leonardo Fogassi, and Giacomo Rizzolatti, "Action Recognition in the Premotor Cortex," *Brain* 119 (1996): 593–609; and Stephanie Preston and F. B. M. de Waal, "Empathy: Its Ultimate and Proximate Bases," *Behavioral and Brain Sciences* 25 (2002): 1–72.
8. The insula (or insular cortex) is a portion of the cerebral cortex folded deep within the brain; the cortical area overlying it toward the lateral surface of the brain is the "operculum" (meaning "lid"). The insula is divided into two parts, the larger anterior insula and the smaller posterior insula. The anterior insula appears to be involved in a variety of functions, including emotion regulation and physiological homeostasis. In the original formulation, the mirror-neuron system was considered a strictly neocortical system, so the assumption that mirror neurons exist subcortically in the anterior insula is a theory, one that relies only on indirect evidence of the kind ostensibly provided by Wicker et al.'s experiment. For an interesting debate on the topic of mirror neurons and emotional processing, see the 2008 exchange between Jan Panksepp, Ross Buck, and others at the Web site of the International Society for Research on Emotion (ISRE): isre.org.
9. Wicker et al., "Both of Us Disgusted," 655.
10. Vittorio Gallese, Christian Keysers, and Giacomo Rizzolatti, "A Unifying View of the Basis of Social Cognition," *Trends in Cognitive Sciences* 8 (2004): 396–403; Christian Keysers and Valeria Gazzola, "Toward a Unifying Theory of Social Cognition," *Progress in Brain Research* 156 (2006): 379–401; Giacomo Rizzolatti and Corrado Sinigaglia, *Mirrors in the Brain: How Our Minds Share Actions and Emotions*, trans. Francis Anderson (Oxford: Oxford University Press, 2006).
11. Alvin I. Goldman, *Simulating Minds: The Philosophy, Psychology, and Neuroscience of Mindreading* (Oxford: Oxford University Press, 2006), 113.
12. Wicker et al., "Both of Us Disgusted," 661.
13. Goldman, *Simulating Minds*, 137. Goldman argues that, on the minimum Unmediated Resonance (Mirroring) model of simulation, the subthreshold tokening of the same emotion experienced in the target serves as the matching or mirroring event on which the subsequent attribution

or imputation (projection) of the emotion to the other is based. It is, however, only the first stage of a two-stage routine comprising simulation and projection. In Wicker et al.'s experiment, as Goldman acknowledges, the participants were not asked to judge the actors' emotion displays, so the second, attribution stage of the simulation routine was not tested. However, Goldman thinks that lesion studies of the kind he reviews in his book do clearly suggest the existence of an association between damage to substrates for the experience of disgust and impaired interpersonal judgments of disgust. Ibid. (see also 40). For a critique of some of the lesion studies on which Goldman relies, see Ruth Leys, "How Did Fear Become a Scientific Object and What Kind of Object Is It?" *Representations* 10 (Spring 2010): 66–104.

14. Wicker et al., "Both of Us Disgusted," 658.
15. In fact, Wicker et al. showed that, in the gustatory runs, the amygdala was activated along with the anterior insula; they reported that, in the visual runs, only the disgust expression activated the insula and there was no activation of the amygdala in response to seeing disgust. For a critique of claims that the emotions are located in functionally specialized brain regions that are independent of cognition and a review of the relevant literature, see Luiz Pessoa, "On the Relationship between Emotion and Cognition," *Nature Reviews Neuroscience* 9 (February 2008): 148–58.
16. The term *cognitive* can mean different things to different theorists. Because it is often associated with the "cognitive revolution" in psychology and with computer models of the mind, the emotion theorist Paul Griffiths prefers to use the term *propositional attitude* theory to describe the position of appraisal theorists who stress the role of beliefs, appraisals, and meaning in emotion. See Paul E. Griffiths, "The Degeneration of the Cognitive Theory of the Emotions," *Philosophical Psychology* 2.3 (1989): 297–313, and *What Emotions Really Are: The Problem of Psychological Categories* (Chicago: University of Chicago Press, 1997). I use the term *intentionalist* to describe these theorists as a way of signaling the importance of (conscious or unconscious) intentionality in emotion, without regard to the role of human speech. Intentionalism (or cognitivism) in affect theory is often represented by its critics as offering a peculiarly disembodied view of the emotions, but it is perfectly compatible with the claim that the emotions involve an organism's embodied disposition to act in certain ways toward the objects in its lifeworld.
17. Paul Ekman, "Universality of Emotional Expression? A History of the Dispute," in *The Expression of the Emotions in Man and Animals* (3rd ed.), by Charles Darwin, ed. Paul Ekman (Oxford: Oxford University Press, 1998), 373.
18. It is an interesting question whether intentional states of the kind involved in the understanding of actions can in fact be explained by the firing of mirror neurons or whether a philosophical confusion is involved

here. For challenges to mirror-neuron theory along these lines, see Emma Borg, "If Mirror Neurons Are the Answer, What Is the Question?" *Journal of Consciousness Studies* 14.8 (2007): 5–19; Pierre Jacob, "What Do Mirror Neurons Contribute to Human Social Cognition?" *Mind and Language* 23.2 (2008): 190–223; and Shaun Gallagher, "Simulation Trouble," *Social Neuroscience* 2–3 (2007): 353–65.

19. Goldman, *Simulating Minds*, 10–11, 113 (quote).
20. Ultimately, the Basic Emotions View can be traced even further back, to the work of William James and others. But that story must be told on another occasion.
21. James A. Russell and José Miguel Fernández-Dols use the term the *Facial Expression Program* to describe the Tomkins-Ekman position. See James A. Russell and José Miguel Fernández-Dols, "What Does a Facial Expression Mean," in *The Psychology of Facial Expression*, ed. James A. Russell and José Miguel Fernández-Dols (Cambridge: Cambridge University Press, 1997), 3–30, 7. Griffiths uses *affect program theory*. See Griffiths, *What Emotions Really Are*, 77. Alan Fridlund uses *Emotions View*. See Alan J. Fridlund, *Human Facial Expression: An Evolutionary View* (San Diego, CA: Academic, 1994), 124.
22. Paul Ekman and Wallace V. Friesen, *Pictures of Facial Affect* (Palo Alto, CA: Consulting Psychologists Press, 1976). In an earlier study of neural responses to disgust, subjects were shown pictures of neutral, disgusted, frightened, and mildly happy facial expressions—pictures taken from Ekman and Friesen's standard set of posed expressions—while they underwent fMRI. The experiment confirmed the involvement of the anterior insula in the recognition of disgust displays. Mary L. Philips et al., "Neural Responses to Facial and Vocal Expressions of Fear and Disgust," *Proceedings of the Royal Society of London* B 265 (1998): 1809–17.
23. For a recent review of appraisal theories, see Klaus R. Scherer, Angela Schorr, and Tom Johnstone, eds., *Appraisal Processes in Emotion: Theory, Methods, Research* (Oxford: Oxford University Press, 2001).
24. Much, if not most, current work on the emotions adopts the tenets of the Basic Emotions View. In a large literature, see Paul Ekman, "An Argument for Basic Emotions," *Cognition and Emotion* 6 (1992): 169–200, and "Basic Emotions," in *Handbook of Cognition and Emotion*, ed. Tim Dalgleish and Michael Power (London: Wiley, 1999), 45–60; Paul Ekman and Daniel Cordaro, "What Is Meant by Calling Emotions Basic," *Emotion Review* 3 (2011): 364–70; Griffiths, *What Emotions Really Are*; and Andrea Scarantino and Paul Griffiths, "Don't Give Up on Basic Emotions," *Emotion Review* 3.4 (2011): 444–54. A recent book on disgust by the philosopher Daniel Kelly adopts the Ekman view of disgust as a basic emotion but fails to consider alternative approaches. See Daniel Kelly, *Yuck! The Nature and Moral Significance of Disgust* (Cambridge, MA: MIT Press, 2011). For criticisms of the Basic Emotions View, see esp. Fridlund, *Human Facial Expression*;

Russell and Fernández-Dols, eds., *The Psychology of Facial Expression*; Brian Parkinson, "Do Facial Movements Express Emotions or Communicate Motives?" *Personality and Social Psychology* 9.4 (2005): 278–311; Lisa F. Barrett, "Are Emotions Natural Kinds?" *Perspectives on Psychological Science* 1 (2006): 28–58, and "Solving the Emotion Paradox: Categorization and the Experience of Emotion," *Personality and Social Psychology Review* 10 (2006): 20–46; Lisa F. Barrett et al., "Of Mice and Men: Natural Kinds of Emotion in the Mammalian Brain? A Response to Panksepp and Izard," *Perspectives on Psychological Science* 2 (2007): 297–312; and Christine D. Wilson-Mendenhall, Lisa Feldman Barrett, W. Kyle Simmons, and Lawrence W. Barsalou, "Grounding Emotion in Situated Conceptualization," *Neuropyschologia* 49 (2011): 1105–27.

25. In addition to the reference given in n. 17 above, see Ruth Leys, *From Guilt to Shame: Auschwitz and After* (Princeton, NJ: Princeton University Press, 2007), "The Turn to Affect: A Critique," *Critical Inquiry* 37 (2011): 434–72, "Affect and Intention: A Reply to William E. Connolly," *Critical Inquiry* 37 (2011): 799–805, "Facts and Moods: A Reply to My Critics," *Critical Inquiry* 38 (Summer 2012): 882–91, and "How Did Fear Become a Scientific Object?"; and Marlene Goldman, "Navigating the Genealogies of Trauma, Guilt, and Affect: An Interview with Ruth Leys," in "Models of Mind and Consciousness," special issue, *University of Toronto Quarterly* 79.2 (2010): 137–49.
26. Fridlund, *Human Facial Expression*, 32.
27. Fridlund sees deception as omnipresent in nature, but he does not treat all signals as deceptive or manipulative because cooperation between signaler and receiver is also important. He thus agrees with animal communication experts who propose the existence in any signaling system of a dynamic equilibrium between cooperative and exploitative signals. Fridlund, *Human Facial Expression*, 137–39. The question of reliability and deception in animal communication, which has played an important role in the history of debates over the nature of the emotions, is a large topic that deserves separate discussion.
28. But the researchers could have asked the participants about the feelings elicited by the odors by letting them press buttons while they were inside the fMRI tube, or by asking them about their feelings before or after the fMRI session, or by presenting the odors a second time outside the imaging process, and so on. But none of this was done. My thanks to Rainer Reisenzein for these suggestions.
29. Vinciane Despret, *Our Emotional Makeup: Ethnopsychology and Selfhood* (New York: Other, 2004), 92. See also Gallese, Keysers, and Rizzolatti, "A Unifying View of the Basis of Social Cognition," 396, in which it is claimed on the basis of Wicker et al.'s findings that the mirror-neuron system gives us "direct experiential understanding" of the actions and emotions of others without the intervention of conceptual reasoning or

reflective mediation. But the participants' emotional experiences of disgust when observing the disgust faces of the actors were not measured by Wicker et al.

30. Paul Rozin and April E. Fallon, "A Perspective on Disgust," *Psychological Review* 94.1 (1987): 23–41; Paul Rozin, Jonathan Haidt, and Clark R. McCauley, "Disgust," in *Handbook of Emotions* (2nd ed.), ed. Michael Lewis and Jeannette M. Haviland-Jones (New York: Guilford, 2000), 637–53.
31. In support of their interpretation, Wicker et al. cite electrical stimulation experiments on the anterior section of the insula conducted during neurosurgery. The stimulations induced nausea and unpleasant sensations in the throat and mouth, suggesting a role for the anterior insula in transforming unpleasant sensory input into visceromotor reactions and the accompanying feeling of disgust. See Wicker et al., "Both of Us Disgusted," 658. For an interesting debate over the nature of disgust between the researchers Royzman and Kurzban, who defend Fridlund's strategic signaling position, and Chapman and Anderson, who defend a readout view of disgust, see Edward Royzman, Robert Leeman, and John Sabini, "'You Make Me Sick': Moral Dyspepsia as a Reaction to Third-Party Sibling Incest," *Motivation and Emotion* 32 (2008): 100–108; Edward Royzman and Robert Kurzban, "Minding the Metaphor: The Elusive Character of Moral Disgust," *Emotion Review* 3 (2011): 269–71, and "Facial Movements Are Not Goosebumps: A Response to Chapman and Anderson," *Emotion Review* 3 (2011): 274–75; Hanah A. Chapman, David A. Kim, Joshua A. Susskind, and Adam K. Anderson, "In Bad Taste: Evidence for the Oral Origins of Moral Disgust," *Science* 323 (2009): 1222–26; and Hanah A. Chapman and Adam K. Anderson, "Response to Royzman and Kurzban," *Emotion Review* 3 (2011): 272–73. Royzman and Kurzban's "Facial Movements Are Not Goosebumps" is especially useful for its brief discussion of how problematic the evidence is for the existence of characteristic disgust facial expressions in congenitally blind people, children, and other individuals.
32. It appears that for Rozin disgust is a more cognitively sophisticated emotion than Wicker et al. take it to be, even if food rejection is central to it. On this point, see William I. Miller, *The Anatomy of Disgust* (Cambridge, MA: Harvard University Press, 1997); and Edward Royzman and John Sabini, "Something It Takes to Be an Emotion: The Interesting Case of Surprise," *Journal for the Theory of Social Behavior* 31 (2001): 29–60.
33. Fridlund, *Human Facial Expression*, 120. Darwin described the disgust response in these terms: "The term 'disgust,' in its simplest sense, means something offensive to the taste. It is curious how readily this feeling is excited by anything unusual in the appearance, odour, or nature of our food. In Tierra del Fuego a native touched with his finger some cold preserved meat which I was eating at our bivouac, and plainly showed utter disgust at its softness; whilst I felt utter disgust at my food being touched by a naked savage, though his hands did not appear dirty." Darwin, *The*

Expression of the Emotions in Man and Animals, 255. As Ahmed has pointed out, despite Darwin's apparent emphasis on the self-evident nature of the disgust reaction, his own description points to the complexity of the emotion in its mediated entanglement with questions of familiarity vs. unfamiliarity, purity vs. impurity, proximity vs. distance, white man vs. native, and so on. Sara Ahmed, *The Cultural Politics of Emotion* (New York: Routledge, 2004), 82–84. Recently, Rizzolatti and Sinigaglia naively quote this passage from Darwin as if its meaning is self-evident because for them disgust is one of the basic emotions they regard as visceromotor reflex responses with characteristic facial movements, emotions to which we empathically respond in a mirror-like simulation process. See Rizzolatti and Sinigaglia, *Mirrors in the Brain*, 175–78.

34. Fridlund, *Human Facial Expression*, 121.
35. Wicker et al. regard the contagiousness of vomiting as further evidence of the automaticity of emotional empathy. See Wicker et al., "Both of Us Disgusted," 661. Fridlund, however, argues that only when odors are very strongly unpleasant or irritating to the nose and throat do expulsive facial reflexes occur (as when the trigeminal nerve is irritated by ammonia) and considers these brainstem-mediated, protective reflexes whose actions imply neither hedonics nor emotion. So for him the issue is whether, apart from such supranormal stimulation, patterned faces automatically accompany the hedonics of odor or taste. His conclusion is that they do not. He regards the contagiousness of retching not as a matter of the automaticity of simulation but as an aversive reaction caused not only by the sight of the face but also by the sound and posture of vomiting as well as by the sight and smell of the vomitus. Fridlund, *Human Facial Expression*, 108–22, 153–54.
36. For a review of the experimental literature up to 1994, not mentioned by Wicker et al., see Fridlund, *Human Facial Expression*, 155–57; Russell and Fernández-Dols, eds., *The Psychology of Facial Expression*; Peter Marler and Christopher Evans, "Animal Sounds and Human Faces: Do They Have Anything in Common?" in Russell and Fernández-Dols, eds., *The Psychology of Facial Expression*, 133–226; and James A. Russell, Jo-Anne Bachorowski, and José Miguel Fernández-Dols, "Facial and Vocal Expressions of Emotions," *Annual Review of Psychology* 54 (2003): 349–59. In a recent study Sandra Miener undertook a comparison between Ekman's and Fridlund's positions, focusing on the emotion of disgust, using as stimuli computer images of objects held to be inherently disgusting to the observer, and manipulating the conditions under which the experimental subject viewed these objects, such as being alone, or with friends, or with strangers. Sandra Miener, "Die Basisemotion Ekel: Untersuchungen zum Zusammenhang zwischen Gefühl und Ausdruck" (The basic emotion of disgust: Studies on the relation between feeling and expression) (Ph.D. diss., University of Bielefeld, Faculty of Psychology and Sports, 2007),

available at http://bieson.ub.uni-bielefeld.de/volltexte/2007/1128/index.html. Miener's findings were mixed, but a detailed discussion of her experiments lies outside the scope of my essay. I thank Markus Studtmann for drawing my attention to Miener's work.

37. Fridlund, *Human Facial Expression*, 142–52. In an interesting investigation the facial responses of female subjects were videotaped while they smelled six odors in each of three experimental conditions (spontaneous, posing to real odors, and posing to imagined odors). Videotaping was covert in the spontaneous condition (in other words, the subjects were unaware of being filmed) but overt in the posed conditions. Raters were then asked to identify the type of odor—good (cloves, roses), bad (urine, rancid sweat), and neutral (mineral oil)—from the poses. The findings demonstrated that subjects exhibited few facial responses to the odors when smelling them in private, despite dramatic differences in their hedonic ratings of the odors. In short, patterned faces did not automatically accompany the hedonics of odor, as the reflex theory of facial expression predicted, but were influenced by the social demand character of the setting. See Avery N. Gilbert, Alan J. Fridlund, and John Sabini, "Hedonic and Social Determinants of Facial Displays to Odors," *Chemical Studies* 12 (1987): 355–63. For a more recent discussion of experimental work on the facial expression of disgust, including a critical analysis of Rosenberg and Ekman's 1994 experimental work, see also José Miguel Fernández-Dols and Maria Angeles Ruiz-Belda, "Spontaneous Facial Behavior during Intense Emotional Episodes: Artistic Truth and Optical Truth," in Russell and Fernández-Dols, eds., *The Psychology of Facial Expression*, 255–74.
38. For a comparison between Ekman's Basic Emotions Theory and Fridlund's Behavioral Ecology Theory that finds little evidence to support Ekman's position, see Parkinson, "Do Facial Movements Express Emotions or Communicate Motives?" The issue of the sociality of expression is at the center of Fridlund's important critique of Ekman's canonical Japanese-American experiment, an experiment that is foundational for the Basic Emotions View. For an analysis of Ekman's effort to respond to Fridlund's criticisms, see Leys, *From Guilt to Shame*, 88–89. Likewise, Despret observes that leaving a subject alone in a room and thinking that he or she is not aware of being observed borders on the naive—and thinking that the subjects are naive. Despret, *Our Emotional Makeup*, 85.
39. Fridlund, *Human Facial Expression*, 179.
40. Laurie Carr, Marco Iacoboni, Marie-Charlotte Dubeau, John C. Mazziotta, and Gian Luigi Lenzi, "Neural Mechanisms of Empathy in Humans: A Relay from Neural Systems for Imitation to Limbic Areas," *Proceedings of the National Academy of Sciences of the United States of America* 100.9 (2003): 5497–5502. Carr et al. frame their results in terms of mirror-neuron theory by suggesting: "The type of empathic resonance induced by imitation does not require explicit representational content and may be a form of

'mirroring' that grounds empathy via an experiential mechanism." Ibid., 5502. The main difference between the findings of Wicker et al. and Carr et al. was that, according to the latter, the insula's importance was not restricted to its role in disgust but extended to the other basic emotions as well.

41. Wicker et al., "Both of Us Disgusted," 658–59.
42. Ibid., 659. It is possible that Carr et al.'s experimental subjects did in fact induce emotions in themselves in some method-acting way. Carr et al. simply report that their subjects were "asked to imitate and internally generate the target emotion on the computer screen, or to simply observe." Carr, Iacoboni, Dubeau, Mazziotta, and Lenzi, "Neural Mechanisms of Empathy in Humans," 5498. On the other hand, their subjects were not actors trained in method acting, so how good they were at generating internal emotions while imitating facial expressions on the computer screen is an open question. Moreover, we do not know what Carr et al.'s subjects actually felt when imitating the emotions because, as I say, they were not asked to report on their subjective experiences.
43. Wicker et al., "Both of Us Disgusted," 661.
44. In a study of emotional recognition in a patient with extensive brain damage from severe *Herpes simplex* encephalitis, including bilateral damage to the amygdala and insula, an effort was made to test the patient with "dynamic facial expressions" by having one of the experimenters pose the expressions held to characterize the basic emotions when seated across from the subject. The researchers report that, in each case, the experimenter produced an "intense but natural expression of an emotion." They also followed up on the patient's impaired recognition of disgust by "acting out" behaviors or acting "scenarios" normally associated with intense disgust, such as eating and then regurgitating and spitting out food, accompanied by retching sounds and facial expressions of disgust. See Ralph Adolphs, Daniel Tranel, and Antonio R. Damasio, "Dissociable Neural Systems for Recognizing Emotions," *Brain and Cognition* 52 (2003): 61–69, 63, 66. But, in the light of my criticisms, these methods seem naive.
45. Thus, for Denis Diderot, e.g., the threat of exaggeration or falseness or mannerism—in short, of theatricality—in acting is omnipresent. It may be that an inspired actor under exactly the right conditions can by some inner process of identification produce now and then the impression of authentic feeling, but the whole point of Diderot's well-known *Le paradoxe sur le comédien* is that this cannot be assured. Instead, he insists that what matters is simply how an actor's performance appears to the audience—with the further, crucial proviso that an authentic-seeming performance requires that the actor convey not the least suggestion that the audience has been taken into account. This explains his recommendation that the actor treat the audience as if it did not exist or as if the curtain separating the actor from the audience had never risen. In other words, the

actor must seek to create the illusion that he or she is entirely absorbed in the actions and situations taking place on the stage; only then will the performance have the look and ring of authenticity—but it is an illusion, not some ultimate truth. In other words, these are deeply complex issues that defy facile formulations. A further layer of complexity is implied by the fact that live performance is one thing and posing for a photograph or a film is something else again (two different somethings, as a matter of fact). Yet the scientists who have used posed expressions seem oblivious to this entire nuance.

46. Mbemba Jabbi, Marte Swart, and Christian Keysers, "Empathy for Positive and Negative Emotions in the Gustatory Cortex," *Neuroimage* 34.4 (2007): 1744–53. However, Jabbi et al. no longer imply or believe that insula activation is specific to negative emotions such as disgust or pain since they provide evidence of insula activation also in the case of positive feelings. More recently, Keysers and his colleagues have acknowledged the absence of a reliable mapping of particular emotions onto specific brain regions. Instead, they propose the existence of a "mosaic" of affective, motor, and somatosensory components involved in emotional processing while continuing to stress the role of motor simulation in triggering the simulation of associated feeling states. Jojanneke Bastiaansen, Marc Thioux, and Christian Keysers, "Evidence for Mirror Systems in Emotions," *Philosophical Transactions of the Royal Society* B 364 (2009): 2391–2404. For a general critique of the locationist approach to the brain basis of emotions, see Kristen A. Lindquist, Tor D. Wager, Hedy Kober, Eliza Bliss-Moreau, and Lisa Feldman Barrett, "The Brain Basis of Emotion: A Meta-Analytic Review," *Behavioral and Brain Sciences* 35.3 (2012): 121–43.
47. The problem Jabbi et al. were dealing with is one that has haunted Simulation Theory for a long time, as the philosopher Sean Gallagher has pointed out. Simulation depends on one's own first-person experience as the basis for what goes into the simulation: "We start with our own experience and project some tentative empathic conception of what must be going on in the other's mind. . . . The question is, when we project ourselves imaginatively into the understanding of the other, are we merely reiterating ourselves? Goldman describes simulation in the following way: 'In all these cases, observing what other people do or feel is therefore transformed into an inner representation of what we would do or feel in a similar situation—as if we would be in the skin of the person we observe.' . . . But [Gallagher goes on] how does knowing what we would do help us know what someone else would do? Indeed, many times we are in a situation where we see what someone is doing, and know that we would do it differently, or perhaps not do it at all." Gallagher calls this the *diversity problem* in order to stress the idea that most of the time we do not impute to others the exact same experience that we have undergone, nor do we automatically feel what others feel. Gallagher, "Phenomenology,

Neural Simulation, and the Enactive Approach to Intersubjectivity." He goes on in this and other publications to suggest that it is an error to call mirror-neuron systems *simulations* at all since such processes are neither pretend processes nor modes of instrumental actions, as the term *simulation* usually implies. In subpersonal mirror-neuron processes of the kind proposed by Wicker et al., there is no pretense since neurons either fire or do not fire: they do not pretend to fire. There is no as if involved at all.

48. Of course, like facial movements themselves, self-reports of emotion may also be closely attuned to the perceived interpersonal context and hence be sensitive to audience effects or experimental demand. For a discussion, see Parkinson, "Do Facial Movements Express Emotions or Communicate Motives?"
49. Jabbi, Swart, and Keysers, "Empathy for Positive and Negative Emotions in the Gustatory Cortex," 1747–48.
50. In a paper originally titled "Voodoo Correlations in Social Neuroscience" that has caused a lively controversy, Jabbi et al.'s 2007 experiment was included in a list of "non-independent" studies that were accused of exaggerating the correlations between emotional and other behaviors and measures of brain activity. See Edward Vul, Christine Harris, Piotr Winkielman, and Harold Pasher, "Puzzlingly High Correlations in fMRI Studies of Emotion, Personality, and Social Cognition," *Perspectives on Psychological Science* 4 (2009): 274–90, and "Reply to Comments on 'Puzzlingly High Correlations in fMRI Studies of Emotion, Personality, and Social Cognition,'" *Perspectives on Psychological Science* 4 (2009): 274–90. For replies to Vul et al., see Mbemba Jabbi, Christian Keysers, Tanya Singer, and Klass Enro Stephan, "Responses to 'Voodoo Correlations in Social Neuroscience' by Vul et al., Summary Information for the Press," http://cogns.northwestern.edu/cbmg/replyVul.pdf; and Matthew Lieberman, Elliot T. Berkman, and Tor D. Wager, "Correlations in Neuroscience Aren't Vodoo," *Perspectives on Psychological Science* 4.3 (2009): 299–307.
51. It is unclear to me from their description of their experiment whether Jabbi et al. determined whether the participants (observers) felt disgust when they were actually watching the actors' facial expressions.
52. But then would this not also be true for the participants? So why bother to ask them to rate their responses?
53. For a discussion of the origins of the facial feedback theory and a critique of the various attempts to prove its validity, a critique based in part on the claim that the experimental tests have been inextricably confounded with implicit suggestions to subjects about how they should act, see Fridlund, *Human Facial Expression*, 173–82. For a more recent discussion, see Jean Decety, "To What Extent Is the Experience of Empathy Mediated by Shared Circuits?" *Emotion Review* 2 (2010): 204–7.

THREE

Emotion Science and the Heart of a Two-Cultures Problem

DANIEL M. GROSS
STEPHANIE D. PRESTON

At least since the publication of C. P. Snow's Rede Lecture "The Two Cultures" in 1959, academics have worried about the divide between science and the humanities: the former striving to maintain clean, consistent results by controlling the environments in which phenomena are studied, the latter embracing natural complexities in the world, including culturally specific frames of reference that shape any particular study.[1] What has been the fate of human phenomena that—unlike particle physics or a poem—do not seem to fit comfortably into one culture or the other? Emotion, for instance, is a fundamental human phenomenon that should be available to laboratory study like anything else that transpires in our material lives; at the same time, it seems defined by complexities that disappear in the controlled environment of the laboratory. The *affective* or *emotional turn*[2] across the disciplines has been shaped by this problem at least since the rise of neuroimaging in the 1980s, with two typical responses.

The first response to this two-cultures divide either assumes or asserts incommensurability, whereby two fields go about their business in such different ways that they cannot be meaningfully compared to each other, despite the fact that each might proceed rigorously according to field-specific expectations, each might produce credible data, and each might use the same key terms. Take the example of neuroscience studies that examine empathy through the role of mirror neurons or automatic neural and physiological processes.[3] These seem incommensurable with empathy studies in professional histories, including the work of Carolyn J. Dean (see, e.g., the coda to this volume, as discussed below). Where this kind of neuroscience invokes a time frame that includes the end of the Pleistocene approximately ten thousand years ago and even the evolution of mammalian brains hundreds of millions of years ago, Dean ties the phenomenon of empathy to its 1873 appearance in the aesthetic/psychological vocabulary of German *Einfühlung* and tracks its much-discussed "exhaustion" after the Holocaust.[4] Incommensurability finds a semantic-theoretical corollary in the emotion studies of Scarantino and Griffiths, who propose distinctions that would at least mitigate errors of equivocation. They argue for a clear distinction between two projects: what they call the *folk emotion project*, which investigates how people actually categorize emotions, can be distinguished from the *scientific emotion project*, designated with subscripted versions of the relevant folk emotion categories (e.g., "$anger_B$"), which investigates how emotions should be categorized for scientific purposes.[5]

Deemed *consilience* by the biologist E. O. Wilson, a second influential response to the two-cultures divide has explicitly rejected incommensurability, arguing instead that we must be able to explain any human phenomenon in natural-scientific terms. Thus, according to Steven Pinker, for instance, a novelistic emotion like romantic love in Jane Austen's *Sense and Sensibility* should be read in terms of evolutionary psychology, including considerations of mating and sex differences, parent-offspring conflict, sibling rivalry, self-deception, taboo, coalitional psychology, and what, following Robert Trivers, Pinker calls the *moral emotions*, including sympathy, anger, gratitude, and guilt, which are supposed to modulate reciprocity in a manner compatible with reproductive success.[6]

Instead of proposing a third response to the two-cultures problem, we offer in this chapter a critique that points to its dissolution in the study of emotion. First, we explain in more detail how, since the early 1980s, emotion studies have been shaped by the two-cultures alternative. A key episode in this story is a relatively recent (2006) concern with the "reverse inference" problem, which renders causal arguments in neuroscience suspect and hence discourages the explicit process of inferring meaning into research studies and results, thereby sidelining a hallmark of the humanities and interpretive social sciences. Then, after introducing "situated cognition"—a recent approach to psychology and neuroscience that is compatible with humanities and social science research—we offer empathy as an important case study where laboratory, ethnographic, and historical work are integrated to treat socioemotional phenomena rigorously. The history of science and emotion post-1945, we finally argue, calibrates to broadly ecological concerns and, specifically, to situated models of cognition that have recently gained traction across the disciplines after a critique of the decontextualized brain.

Limits of Laboratory Science of Emotion

Where B. F. Skinner's psychology as an empirical behavioral science has been, at least since Noam Chomsky's famous 1959 review, subject to criticism for failing to study people in their natural contexts, the technical constraints of neuroscience have, of late, seemed to renew the abstraction problem in psychology.[7]

1. *Technological limitations*. Subsequent to technological advances in the neuroscience of the mind that began in the early 1980s, the key tools employed by neuroscientists to study the human brain have come to include functional magnetic resonance imaging (fMRI), positron emissions tomography (PET), transcranial magnetic stimulation (TMS), and external brain recordings (e.g., EEG, ERP, MEG). By and large, these techniques require subjects to sit or lie completely still during the recording, inside of or attached to machines that are cumbersome and sometimes also loud (especially fMRI, which is the most widely used). These logistical limitations make it very difficult to present subjects with naturalistic situations, scenarios, or decisions. For analytic purposes, typical stimuli must also consist of many, repeated trials of events that occupy no more than a few seconds each, delivered via visual slides or audio recordings, and that allow only simple finger movements as a relevant response. Some experimenters have devised creative ways to study more natural human behavior in these environments, for example, by having subjects experience the soft touch of a partner or stranger to an extended limb,[8] by showing or indicating the pain occurring to a loved one just outside the scanner,[9] or by presenting audio clips or images of one's own baby as opposed to an unfamiliar one.[10] But most researchers who study decisionmaking, emotion, and prosocial behavior (i.e., empathy and altruism) have tried to solve

the abstraction problem during brain scanning simply by using money as the dependent measure for value (see limitation 2 below).

No doubt these types of technical limitation on neural recording will diminish as devices allow for more ecological experimental protocols. For example, near infrared spectroscopy and wireless ERP electrodes would be less unwieldy and would permit more natural movement and interaction during the recording of neural signals. Currently, however, technologies that allow for greater degrees of movement or indirect recording also sacrifice the specificity of neural localization. For example, one can report with techniques that measure brain activity from the surface that activation was in the middle front part of the cortex rather than the lateral or posterior portion, but one cannot specify that it was in BA 44. (In 1909 Korbinian Broadmann published a numbered labeling scheme for the brain that subdivided larger regions into subregions on the basis of their cytoarchitectural properties, i.e., cell types; this nomenclature is still widely used today.)

2. *Money and the limitation of a utilitarian assumption.* Money is considered useful in this research because the decisions are considered incentivized (meaning that subjects are expected to give more truthful answers if their own cash rewards are at stake for doing so); moreover, the dependent measure is discrete and easy to manipulate and measure. For example, whether I value an object can be indexed by how much I am willing to pay for it; how much I care for you is indexed by how much of my money I am willing to give to you; and how much I dislike something is indexed by how much I will pay to avoid it. This approach is problematic, however, because people do not actually value money linearly[11] and not all individuals value it in the same way.[12] Even more problematic for cases like altruism and empathy, money is known to produce a mind-set that specifically limits one's prosocial motives.[13] Thus, the processes involved in making a prosocial decision in the money-incentivized laboratory may not generalize to real-world situations and, therefore, should at least be augmented with ecological tasks that are performed outside the brain scanner, where the monetary decisions are particularly useful (e.g., helping someone pick up dropped papers or responding to a baby's cry).[14]

3. *Professional journal impact factor.* A researcher's ability to manage inferences is constrained by a few factors, including recent economic pressure to reach broader audiences and augment a journal's impact factor, which does not regularly distinguish between the quality and the quantity of citations of a paper. In order to reach a broad audience

on publication or to garner media attention, researchers sometimes intentionally oversimplify while sacrificing professional knowledge about the complexity and indeterminacy of the system they study. For example, if you entitle your article something accurate but detailed like "The Temporal-Parietal Junction (TPJ) Appears to Be Involved during Theory of Mind (TOM) Processing, but Maybe Only Because the Task Requires Tracking Movement in Space over Time," you are unlikely to inspire professional journal or *New York Times* editors. Instead, you can entitle your article something more exciting and forceful—and also less accurate—such as "Brain Region Found for the Unique Human Ability to Read Minds." The exact same functional neuroimaging data can be marshaled either way, but the more appealing claim can practically obscure the more accurate claim.

4. *Mass media attention.* Increasingly, decisions about how much media attention a study will attract back-propagate to influence the decisions of academic editors, granting agencies, and the researchers themselves. No doubt, research grants are more often bestowed on experienced and capable researchers of basic and applied science, but the ability to quickly publish the results in a journal or to garner supplementary personal funding (e.g., from speaking events and book contracts) is more dependent on the ability to attract wide public attention, facilitated by simplicity and generalization that easily explain key problems of everyday experience such as how we feel and how we act toward others. These biases seem particularly pronounced in the human social and decision sciences, which pique the interest of lay audiences and may be responsible for increasing concerns about the validity and quality of research being published in these fields—evidenced by at least three recent data-fabrication scandals particularly implicating researchers who are widely cited for their exciting but oversimplified claims.[15]

5. *Localization.* Lay readers are provided oversimplified explanations of the brain that ostensibly map small regions or circuits to very specific, disciplinary tasks. With terms like the *social brain*,[16] *empathy circuit*,[17] and *self-related regions*,[18] neuroscientists admittedly simplify the inner workings of the brain and imply a straightforward mapping of tasks and functions to particular regions of the brain. However, such simplification is misleading because each brain region is associated with particular functions—such as tracking visual motion in space—and is presumably mobilized for any task that requires that function, including motor acts, person perception, and imagining the deceptions of another.[19] Even single cells as early as the visual cortex—a place where

neural processing is thought to be highly structured and dependent on the type of stimulus presented—show changes in the types of information that they process depending on the perceptual context.[20] Thus, while any one task can show activation of a specific region in the brain when compared to another task that is construed as the control, the precise level and pattern of activity will vary widely for similar processes depending on the nature of the task, the type of stimuli and decision process used, and the nature of the control task. Moreover, because that same neural region can be activated any time the related information process is invoked, it can be engaged by a wide variety of tasks, across diverse fields and phenomena. This complexity makes it all but irresponsible to ever infer that a cell or neural region is *for* a specific psychological task, which is virtually never the case.

Beyond the appeal of manageable laboratory procedures, what is the practical appeal of brain localization in this same historical period of, roughly, the last three decades? Here is one historicopolitical hypothesis. When, for instance, empathy, racial hatred, or a sense of self can be located in a particular brain region, then, realistically or not, prospects for intervention are in some ways easier to imagine, whether that means a therapeutic cure or, alternatively, resignation in the face of biology. At the same time, it can become harder to imagine social explanations and institutional responses that seem more difficult to implement and attributable to social welfare frameworks that lost much of their appeal in the Anglo-American 1980s.[21]

6. *The limits of professional expertise.* While any one neural region can be involved in a wide variety of disciplinary psychological tasks, researchers themselves typically read and follow only the research within their own field. This exacerbates the inferential problem of associating regions with tasks because researchers are often unaware that their pet brain region is also specifically implicated in relatively unrelated tasks studied in other fields, which in turn discourages more accurate and nuanced inferences about the function of the region. For example, if you follow only research in social psychology, you would perhaps never realize that the angular gyrus, which social psychologists associate with one's self or body concept, is in cognitive psychology also construed as the arithmetic region.[22] Indeed, this single parcel of cortex is known variously as the *arithmetic area*,[23] the *metaphor area*,[24] the *out-of-body-experience area*,[25] the *cross-modal area* for integrating visual and spatial information with the body,[26] or the *self-relevant area*, depending on the specialty and task of the researcher and the brain hemisphere being studied. Thereby cordoned off from relevant research in neigh-

boring disciplines, any particular research project lacks a perspective that could triangulate two functions and propose a more general function for a particular brain area, taking multiple types of tasks and multiple research fields into account.

7. *Reverse inference limitations.* This problem of associating a direct mapping of task to neural function has created a backlash from the more methodologically skilled cognitive neuroscientists against "reverse inference."[27] Reverse inference discourages the assumption that the cognitive process involved in a prior study that activated a particular brain area is also responsible for the activation of that region in a subsequent study. So, for example, the following inference is discouraged in the professional neuroscientific literature: since the dosolateral prefrontal cortex (DLPFC) is often associated with the ability to manipulate information and to control the direction of one's thoughts and efforts (i.e., "working memory" and "executive control," respectively), one can infer that a social psychology task that also activated the DLPFC requires working memory and executive control. No doubt, a ban on reverse inference of this sort is understandable in principle because weak inferences have been common in the neuroimaging literature, as demonstrated above and in the Poldrack article cited earlier in this paragraph (see n. 27). But, at the same time, such a ban enforced by editors and reviewers at prestigious journals forecloses any explicit consideration of the function of a brain area and why it was activated in the task at issue. Those who vocally oppose reverse inference, and the reviewers and editors who follow them, overlook the fact that interpretations based on extensive prior data would actually be *good* sources of information, as long as the inferences were sufficiently broad and judicious and took into account relevant data from neighboring fields and unrelated tasks. Broadly sampling from the world of data on the angular gyrus, for example, would require researchers to show more creativity and accuracy when inferring the role of this region in both arithmetic and self-reference phenomena, which could in turn tell us about the relationship between the embodied cognitive brain and psychological processes. Thus, by engaging in some responsible inference about the functions of a region, a researcher might actually be able to infer the underlying processes that various tasks require while generating causal arguments about how a higher-order phenomenon—namely, a "self" in this case—is produced, thereby implicating relevant humanistic and social scientific material.

So what makes the reverse inference problem stick? One reason is technocratic: vocal opponents are often from the same fraction of the profession with access to the tools and expertise required to build empirical inferences into the tasks. No one disagrees that it would be ideal to have built into the task conditions that empirically confirm the source of a task-region association. For example, in a typical theory of mind (TOM) task, the participant must infer the beliefs of another person—in this case Sally—on the basis of her unique second-person perspective in space and time. Thus, while the participant saw the experimenter move the cookies from the jar to the cupboard, Sally was not in the room at the time and will still assume that the cookies are in the jar. This inference requires the participant to track the movement and perception of Sally independently from the movement and perception of the cookies and is broadly referred to as *theory of mind*. However, TOM tasks do not simply require making inferences about Sally's mental state. They also logistically require mentally tracking her movements in time and space relative to the cookies. Thus, any brain activity associated with a TOM task could be due to purely mental inferential processes (what most people assume), or it could be due to the movement tracking that is required to make this inference but is not specific to TOM tasks. To avoid reverse inference, one could add to the scan a "functional localizer," in which the participants perform an additional block of trials that require movement tracking but not mental state attribution, such as tracking the movement of a ball behind an opaque wall.[28] One can then subtract the neural activity associated with the nonsocial ball-tracking task from the activity in the TOM task to determine the extent to which regions activated by the TOM task (e.g., the temporoparietal junction [TPJ] or angular gyrus) emanated from the perceptual-motor requirements of the task in a way that is not necessarily mental. Practically, however, the methods for generating such data are more difficult to implement, require specialized skills during analysis that many social scientists do not possess, and add time to the scan itself, which can reduce the power of the key contrast and costs significantly more money (money that the researchers opposed to reverse inference are likely to have but that others will not).

An additional problem with the tenet that all interpretations must be empirically backed by analyses is that sometimes the most interesting results from a study are those that were not anticipated and so were not built into the design. For example, a recent study in the Preston laboratory found activity in the insula associated with the decision to acquire the most valuable goods.[29] This was unexpected because the insula was hypothesized to be more active in the subjective and affective blocks when people chose or had to relinquish goods they really wanted. A review of the literature, however, confirmed that the insula is often activated during monetary cost-benefit decisions, perhaps either because it is part of the neural system for assessing value on the basis of internal and affective cues or because of affective responses to the surprisingly high- or low-value items that sometimes appeared.[30] These alternative possibilities could not be parsed in the study because the blocked design did not permit analysis of individual trials (e.g., separating valuable and cheap goods)—a design choice that is simple and statistically powerful but would not be utilized by the more cautious, sophisticated, and highly funded researchers who vocally eschew reverse or logical inference. Therefore, in this case, the authors could only speculate about the few likely reasons that the insula was active in the money condition. In fact, the inference could be made on the basis of extensive empirical evidence in various human studies that the insula is involved in monetary decision tasks that are not overtly emotional. The rules against making inferences about brain activity would require, however, that the authors say almost nothing about this finding if they report it at all, and a second study would have to be conducted to confirm the interesting result from the first study, doubling the cost of a methodology that is already exorbitantly expensive (at the time of writing, approximately $600 per subject, with an expected sample size of twenty to thirty participants per task, making it a minimum of $12,000 per experiment and usually over $100,000 owing to the ancillary costs of hiring a trained, full-time staff to run and analyze the imaging studies).

8. *Technocratic hierarchy.* This situation generates a sharp divide in which the few highly funded and technically proficient laboratories that manage multiple large federal grants and a large staff of personal secretaries, engineers, statisticians, and physicists control and generate the vast majority of published studies on the brain—these also happen to be the researchers who most often eschew the simpler techniques and logical inferences that nonetheless can produce usable information that moves the field forward if done responsibly.

Fear Situated

We have now outlined some important factors that have intensified the brain-abstraction problem along with the rise of neuroimaging over the last three decades, exacerbating a two-cultures problem in and around brain science. During this same historical moment, however, a powerful alternative has persisted in the philosophy of mind that references Wittgenstein and Vygotsky, phenomenological traditions after Martin Heidegger, and the perceptual ecology of the American psychologist J. J. Gibson. Now we will consider how this approach can meet the professional expectations both of neuroscience and of the humanities and interpretive social sciences. Case studies include first *fear* as it has been treated in the Phelps and Barsalou laboratories, then *empathy* as studied by Preston and her colleagues. In both cases, the emotional topics—fear and empathy—speak to pressing postwar social concerns: racism in the first instance, humanitarianism in the second. Hence our central historical claim: the *topics* of laboratory work on the emotional brain coincide meaningfully with particular sociohistorical concerns that show up at the margins of experiment design and interpretation.

In a widely cited 2000 article that appeared in the *Journal of Cognitive Neuroscience*, the New York University cognitive scientist Elizabeth A. Phelps and her colleagues used fMRI to examine the neural response of white American subjects who viewed images of African Americans and white Americans. Phelps et al. explain that their study in social cognition was motivated by the remarkable difference between, on the one hand, a steady decline in self-reported racial prejudice and, on the other hand, persistent "unconscious" racial prejudice directed at black people. One of two primary goals in this study was to examine the neural correlates of responses to racial groups, focusing on the amygdala because it appears to be involved in emotional learning, especially in relation to fear, memory, and evaluation. Using fMRI brain-imaging technology, Phelps et al. investigated correlations among scores on the Modern Racism Scale (commonly used to measure conscious, self-reported beliefs and attitudes toward black Americans), unconscious measures of startle response and bias, and fMRI-derived amygdala activity in white Americans responding to black and white male faces with neutral expressions.[31]

In experiment 1, the faces presented belonged to individuals who were unfamiliar to the subjects. In experiment 2, the faces belonged to "famous and positively regarded"[32] black and white individuals, including Arsenio Hall, Bill Cosby, Magic Johnson, Martin Luther King Jr., and Colin Powell, on the one hand, and Conan O'Brien, Tom Cruise, Larry Bird, John F. Kennedy, and Norman Schwarzkopf, on the other. According to Phelps et al., these two studies showed for the first time that members of black and white social groups can "evoke differential amygdala activity" and that "this activity is related to unconscious social evaluation."[33] In experiment 1, the strength of amygdala activation to black versus white faces was correlated with the two unconscious measures of race evaluation but not with the conscious expression of racial attitudes from the self-report questionnaire. In experiment 2, these patterns were not reproduced when the faces observed belonged to familiar and positively regarded black and white individuals.

Interestingly, Phelps et al. are cautious about making sociohistorical inferences connecting the indirect measures and the brain data while nonetheless placing the data within a larger context implicated by the provocative topic itself. They qualify: "These data cannot speak to the issues of causality." But they go on to posit: "Both amygdala activation as well as behavioral responses of race bias are reflections of social learning within a specific culture at a particular moment in the *history of relations between social groups*. . . . Unless one is socially isolated, it is not possible to avoid acquiring evaluations of social groups, just as it is not possible to avoid learning other types of general world knowledge."[34] So in one influential study we get a two-cultures impasse in dramatic form.

On the one hand, it is suggested that the history of relations between blacks and whites should count as a significant causal factor producing unconscious racial bias in a particular subject, as measured in amygdala response correlated with certain unconscious measures. On the other hand, Phelps et al. insist that their data *cannot* speak to issues of causality. Productively, this study comes right up to the brink of the two-cultures divide, then falters. What causes racialized fear? Phelps et al. gesture vaguely toward history but give us concretely only neuroscientific confirmation that such racialized fear exists (despite a subject's explicit protests to the contrary). The problem with this disconnect between vague history and concrete neuroscientific data is that the protocol for connecting the two—let us call it a *mediating field* of sociohistorical phenomena—is not sufficiently articulated to shape the laboratory work and its interpretation.

By drawing on critical race theory and rhetorical studies, we might say that the white, avowedly problack subject who startles more quickly and negatively associates more readily when presented with an unfamiliar black face belongs to a broader social phenomenon that must be analyzed at the historical level, which extends beyond any particular subject. The Phelps et al. study gives us a subject who demonstrates particular responses coded as racist, but, to understand or even to initially identify those responses, the phenomenon (racism) must have reference points beyond the brain physiology ultimately studied. Racism is not located primarily in the brain. It is located primarily in the racist institutions—like US race-based slavery and subsequent Jim Crow laws—that exceed any particular subject and can persist without any particular kind of incorporation or brain state. In this case, both sides of the equation are necessary for the study to matter. Without the physiology, no doubt, claims of declining racism might be overstated owing to the ideological prestige of antiracist discourse. And, without humanistic studies broadly conceived, such physiological evidence could never show up because we would not know what we were looking for, or at least we would not care. A more robust brain science of emotion—in this case racialized fear—would take seriously in design and interpretation the history of race relations invoked incidentally by Phelps.

Scholars in the humanities and interpretive social sciences might also be interested in how televisual and other sorts of mediation shows up in experiment 2, where faces belonging to familiar and positively regarded black and white individuals significantly mitigate the racism identified in experiment 1. But, instead of emphasizing like Phelps et al. the antiracist implications of this finding, which would suggest that familiarity mitigates racism, such scholars might underscore the socially mediated quality of each experimental stage. This would include category selection (*black* and *white*, which are sociohistorical categories, not natural kinds), image selection (black and white image attribution, which will vary significantly according to the cultural situation of the laboratory and its experiment and might look different in Brazil, e.g.), and also the very concept of familiarity, which in this case has nothing to do with face-to-face encounters, personal acquaintance, or family or communal relationships and everything to do with mediation as *media* locating and dis-locating images according to complex regimes of publicity and capital of the sort studied thoroughly by media and rhetoric scholars. Hence, one might question, for instance, the Phelps et al. presentation of stimuli that equates a list of black and white individuals whose faces are supposed to portray rough equivalents in degrees of fame, age, and achievement. Only a broader perspective can foreground the fact that, although Martin Luther King Jr. and John F. Kennedy might provide facial images roughly equivalent within strict parameters defined by the experiment, fame and achievement and even longevity are unevenly distributed across black and white populations, which means at least that a famous and familiar black face carried with it into the laboratory certain kinds of markedness, whereas the supposedly equivalent white face did not. To put this another way, Phelps et al. may be correlating *celebrity* with unconscious racial attitudes, not face-to-face familiarity. Their study does not adequately specify *famil-*

iarity, so no meaningful analytic distinction can be drawn between the familiarity of an Arsenio Hall face seen by a white subject only in the context of television and accompanying media and the familiarity, say, of Dred Scott to his slave master, John Sandford. In the context of US racialized slave culture, for instance, familiarity did not mitigate racism, which suggests that *familiarity* must be deployed with some more historical rigor if it is to be socially useful in the way suggested by Phelps et al. More broadly, the social cognition study of racialized fear must be situated with greater rhetorical specificity if it is to produce results that speak to causality. When it comes to emotion studies, it would appear, both sides of the two-cultures divide suffer the shortcomings of a shared phenomenon not treated as such.

Referencing Wittgenstein on natural language, Lawrence Barsalou and his colleagues also study fear as a concept, in the effort to critique and dissolve aspects of the two-cultures problem that face Phelps et al. and others. In a summative essay, "Grounding Emotion in Situated Conceptualization" (2011), Wilson-Mendenhall, Barsalou, et al. critique the "basic emotions" approach made famous by Paul Ekman, which assumes that "emotions reflect inborn instinct, and that the mere presence of relevant external conditions triggers evolved brain mechanisms in a stereotyped and obligatory way (e.g., a snake triggers the fear circuit)." Wilson-Mendenhall et al. instead assume that emotions are not fixed, discrete categories but rather concepts that refer to entire situations and thereby represent "settings, agents, objects, actions, events, interoceptions, and mentalizing." Thus, the *fear* of a runner who becomes lost on a wooden trail at dusk, for instance, differs from the *fear* of someone unprepared to give an important presentation at work. In this latter situated conceptualization, they explain, "a different set of concepts represents the situation, including *presentation, speaking, audience, supervisor*, and many others." In essence, their laboratory experiment explores this hypothesis that the basic emotion approach is wrong since, in fact, a situated conceptualization "produces the emotion." Or, to put this another way, the situation in which an emotion concept is experienced shapes how the emotion is instantiated in the brain. For example, the observation that social fear activated clusters in the left dorsolateral prefrontal cortex and inferior frontal gyrus suggests that social fear, as opposed to physical fear, requires more executive control to cope with threatening social evaluations.[35]

The situated emotion approach of Barsalou, Barrett,[36] Campos,[37] and others does challenge the two-cultures problem by explicitly incorporating language and audience considerations into experimental design and into interpretation. However, Barsalou's modular, functionalizing focus on situated conceptualization is forced to rely on elliptical figures ("and so forth," "and many others") to articulate a background that recedes indefinitely into what we would consider *non*conceptual affordances located in the environment, including the social environment, like Phelps et al.'s history of race relations. *Racialized fear,* for example, is not just a concept that refers to an entire situation and thereby represents "settings, agents, objects, actions, events, interoceptions, and mentalizing." It is, our next theorists would argue, actually built into a social situation that can affect people directly, without requiring a persistent or pervasive conceptual representation.

Instead of working exclusively in the laboratory, Paul Griffiths and Andrea Scarantino track "emotions in the wild" as they try to shift theoretical focus to neglected phenomena such as anger in a marital quarrel or embarrassment while delivering a song to an audience. Emotional content, they argue, has a "fundamentally pragmatic dimension, in the sense that environment is represented in terms of what it affords the emoter in the way of skillful engagement with it." In an effort to theorize cognition beyond the brain, they consider the "active contribution of the environment," such as confessionals in churches that enable certain kinds of emotional performance, which they call *synchronic scaffolding* after Vygotsky, and the broader Catholic culture that supports the development of the ability to engage in the emotions of confession, which they call *diachronic scaffolding.* They also consider material factors, including "emotional capital," such as the emotional resources associated with having a specific social status, gender, disability, etc.[38] But what would the science of emotion look like if we genuinely tried to integrate laboratory, ethnographic, and historical considerations? To address this question, we will conclude with an example from the Preston Ecological Neuroscience Lab (ENL), which specializes in studying how empathy is instantiated in the body and the brain. In doing so, we try to learn from, and advance, the emotion research of Phelps, Barsalou, Griffiths and Scarantino, and others toward the dissolution of a two-cultures divide.

Empathy Situated

What is empathy? Mirror-neuron research casts it as a noncognitive and transprimate phenomenon whereby observing an action (e.g., cup raising) fires a significantly similar set of neurons in the observer as when he or she performs that same action. But is a rigorous treatment of the phenomenon empathy not obligated to explain at least why certain actions elicit empathic responses while others do not? Otherwise one risks a circular definition where empathy is everything that involves mirror-neuron activity and all mirror-neuron activity is by definition empathic. To cite an obvious example, why does empathy appear in some suffering circumstances but not others, even when the objective conditions appear the same? At this point case studies become crucial.

A recent series of studies by Preston, Hofelich, and their colleagues involves videos of real hospital patients who are discussing their experience with serious or terminal illness.[39] While all these patients are hospitalized for a very serious medical condition, they vary widely in terms of the emotions they invoke during the interview, from distraught crying throughout, to laconic resistance to talk, to resilient sadness potentiated by affiliation. But it is not only the patients who differ in terms of their emotional profile; observer reactions also differ significantly. Some patient-observer combinations produce empathy and the desire to help; others do not. The variance across observers is particularly pronounced toward the most distraught patients, on one side of the spectrum, and the most reticent patients, on the other. Preston and her colleagues found that most observers in the study empathize with the former and not the latter. But observers who are currently experiencing their own intense negative affect owing to depression report less sympathy for the distraught patients—understood as empathy plus the desire to help—while those who report that they often "take another's perspective" in daily life report greater sympathy for the reticent patient who is less demonstrative but does appear to have deep emotions and need.

Why so? If empathy were a noncognitive, automatic, and context-insensitive phenomenon, would pain not trigger pain, sadness not trigger sadness, and in a linear fashion that could make both empathy reports and sympathetic behaviors more predictable over human subjects and situations? How does one explain, for instance, the reticence factor that correlates in a complicated fashion with apathy and empathy?

At this point we consider situational factors that include a high degree of historical specificity. In a coda to this same volume, for instance, Carolyn J. Dean carefully tracks how reticence became an important quality in the recent production of empathy, or compassion, in one crucial postwar emotional domain: Holocaust victimization. Her key point relevant to the Preston study is that—as a result of false victim controversies and the surrounding melodrama—reticence has become a key marker of authentic suffering and hence a minimum requirement for empathic or compassionate response in the case of survivor stories. However, Dean is careful to point out that, in this case and others, a situational formula is not forthcoming if we want to treat an emotional phenomenon accurately. Instead, a range of humanistic and social scientific tools must be brought to bear if we want to understand with some accuracy how empathy does or does not function in any particular case. "The preference for minimalist style in victim testimonies," Dean reminds us, "is but one component of a much broader discourse about the exemplary victim. Over and over, the presumptive cultural demands of false victims (here defined by the dramatic contrast between them and the putatively 'extraordinary reticence' shown by many survivors) trump any inquiry into how some victims are deemed more credible than others since we are presumed to know that 'real' survivors would by definition be reluctant to seek attention. Reticence thus distinguishes real and false victims" (p. 396).

So the larger point is not that the emotional reticence of Preston's hospital patients and the emotional reticence of Dean's survivors are equivalent. At least one significant difference is that, across the Preston and Hofelich studies, the distraught patients actually receive on average the most sympathy and help—though they can also elicit strong negative responses from minorities of the subjects. Presumably, their high and overt negative emotion is both necessary to know that they are in need and deemed permissible since they are currently experiencing the distress of illness or pending death. However, the resilient patients who exhibit little distress and much more positive emotion reliably receive the second most sympathy and the reticent the least—in contrast to a situation-insensitive supposition that reticence necessarily fosters sympathy.

More important, for us, is the general injunction to consider the relevant emotional genres and styles and the specific injunction to consider how a research question about empathy is located in a particular historical and cultural situation where we cannot take for granted the transparency of our key term. For instance, in the hospital study, it turns out that gender expectations play a key role in differentiating the extent to which people want to help the distraught as opposed to the resilient patients, even though all observers agree that the former need more help. This hospital study thus cannot be gender neutral without distorting the apathy-empathy phenomenon of interest in the first place. Meanwhile, in the reception of a survivor's tale, anxiety of transmission now becomes prominent as historians and their audiences worry about losing the subjects of study. In both cases, but in different ways, any neuroscience of empathy would have to take these historical considerations seriously at the level of both experimental design and interpretation. At the same time, a humanistic perspective on the shared key term *reticence* invites a line of inquiry that does not assume the link to be accidental but instead tracks the interference of suffering situations across particular case studies. Finally, with Dean, we can point toward humanist constructions of dignity that lay out broadly how victims—of systematic oppression or natural circumstances beyond one's control—are constituted, along with the agencies that are supposed to respond.[40] At the very least, Preston's study suggests how it is not necessarily impractical to consider history and situation when studying empathy scientifically, and the hope is that such considerations enhance the quality of the conclusions.

Summary and Conclusion

We began with a description of the two-cultures divide, in which academics of natural science, on one side, and the humanities and interpretive social sciences, on the other, appear to be at an impasse *explained* by incommensurability but *addressed* most obviously by consilience in the reductive tradition of E. O. Wilson. However, experiments—even within the context of neuroscience, which is focused on the mechanisms underlying behavior—can proceed more ecologically and thereby provide new insight into brain-body-world dynamics, especially as they are implicated in important social phenomena.

Significant barriers remain. Scientists must come up with experimental protocols that are ecological, situated, and statistically powerful, within the confines of existing technology. At the same time, humanists must avoid the all-or-none proposition that their topics—for example, racial hatred, empathy, or the self—must be completely available to experimental protocol or completely unavailable. Because of the nature of the experiment, which holds most things constant in order to be able to make causal claims about the role of the key variable of interest, one usually cannot study many, possibly competing and interacting variables at the same time. Thus, for example, to address our critique of Phelps et al. would take dozens of studies instead of just the one, which was already suitable for publication in the flagship journal for the cognitive neurosciences.[41]

We hope that this chapter, written by a rhetorician and a neuroscientist, can serve as an example of two-cultures collaboration at its early stages. In situated cognition we identify an approach that invites such collaboration. Given scientific concerns, it would be misleading to study cognition without considering work in neuroscience. At the same time, science cannot have the last word without incorporating some of the situational concerns of the humanities, because situations are environmental, including a social environment that can be quite specific like empathy exhaustion, and temporal, which sometimes means *historical* in the precise, professional sense of the word. Ultimately, the two-cultures divide dissolves when shared phenomena like human emotions are treated with the methodological diversity and cross-disciplinary conversation their complexity deserves.

NOTES

1. *Geisteswissenschaften* is a term of the middle nineteenth century established by Schiel, Helmholtz, and Droysen as a catch-all contrary to the natural sciences, or *Naturwissenschaften*. See, e.g., Hans Georg Gadamer, *Truth and Method*, 2nd rev. ed., trans. Joel Weinsheimer and Donald G. Marshall (New York: Crossroad, 1989), 3–42. Anglo-American and German distinctions overlap significantly, with a tripartite classification system—(1) humanities, (2) human sciences, and (3) natural sciences—complicating the transatlantic academic traditions. Interpretive social sciences would be grouped with humanities in a binary classification and social science fields anchored in quantitative methodologies with natural sciences. For discussion of the two- or three-cultures distinction, see also Jensen, chapter 10 in this volume.
2. Two examples among many: Patricia Clough and Jean Halley, eds., *The Affective Turn: Theorizing the Social* (Durham, NC: Duke University Press, 2007); and the proceedings of the conference "The Emotional Turn in the Social Sciences," University of California, Los Angeles, November 2011. For a discussion of the terminology *affect* and *emotion*, see the introduction to this volume.
3. A summary of mirror-neuron research can be found in Marco Iacoboni, *Mirroring People: The New Science of How We Connect with Others* (New York: Farrar, Straus & Giroux, 2008).
4. Carolyn J. Dean, "Empathy, Pornography, and Suffering," *Differences: A Journal of Feminist Cultural Studies* 14.1 (2003): 89–123. For a brief history of the term *empathy*, see ibid., 112–13n5.
5. Andrea Scarantino and Paul Griffiths, "Don't Give Up on Basic Emotions," *Emotion Review* 3.4 (2011): 444–54, 452. More compatible with our

project is the scientific psychologist Jerome Kagan's call for a moratorium on the use of single words such as *fear,* urging experts to use instead full sentences to write about the emotional process. See Jerome Kagan, *What Is Emotion? History, Measures, and Meanings* (New Haven, CT: Yale University Press, 2007), 216.

6. Steven Pinker, "Toward a Consilient Study of Literature," *Philosophy and Literature* 31.1 (2007): 162–78, 176. The problem with Pinker's approach, and others like it, is the vast distance between the evolutionary-psychological model and the work of literature itself. Indeed, Pinker explicitly distances the analysis from the local culture of its object: "One has to show—*independently of anything we know about the human behavior in question*—that X, by its intrinsic design, is capable of causing a reproduction-enhancing outcome in an environment like the one in which humans evolved. This analysis can't be a kind of psychology; it must be a kind of engineering—an attempt to lay down the design specs of a system that can accomplish a goal (specifically, a subgoal of reproduction) in a particular world (specifically, the ancestral environment)." Ibid., 170.
7. Critiques of the brain science of emotion include the following: Daniel M. Gross, *The Secret History of Emotion from Aristotle's "Rhetoric" to Modern Brain Science* (Chicago: University of Chicago Press, 2006); Kagan, *What Is Emotion?* Alva Noë, *Out of Our Heads: Why You Are Not Your Brain, and Other Lessons from the Biology of Consciousness* (New York: Hill & Wang, 2009); Paolo Legrenzi and C. A. Umiltà, *Neuromania: On the Limits of Brain Science* (Oxford: Oxford University Press, 2011); Raymond Tallis, *Aping Mankind: Neuromania, Darwinitis and the Misrepresentation of Humanity* (Durham: Acumen, 2011); and Ruth Leys, "The Turn to Affect: A Critique," *Critical Inquiry* 37 (Spring 2011): 434–72, and chapter 2 in this volume.
8. Hau Olausson et al., "Unmyelinated Tactile Afferents Signal Touch and Project to Insular Cortex," *Nature Neuroscience* 5 (2002): 900–904.
9. Tania Singer et al., "Empathy for Pain Involves the Affective but Not Sensory Components of Pain," *Science* 303 (2004): 1157–62.
10. James Swain, Pilyoung Kim, and S. Shaun Ho, "Neuroendocrinology of Parental Response to Baby Cry," *Journal of Neuroendocrinology* 23 (2011): 1036–41.
11. Daniel Kahneman and Amos Tversky, "Prospect Theory: An Analysis of Decision under Risk," *Econometrica: Journal of the Econometric Society* 47.2 (1979): 263–91.
12. Scott Rick, Cynthia Cryder, and George Loewenstein, "Tightwads and Spendthrifts," *Journal of Consumer Research* 34 (2008): 767–82.
13. Kathleen Vohs, Nicole Mead, and Miranda R. Goode, "The Psychological Consequences of Money," *Science* 314 (2006): 1154–56.
14. Stephanie Preston and F. B. M. de Waal, "Altruism," in *The Handbook of Social Neuroscience,* ed. Jean Decety and John Cacioppo (New York: Oxford University Press, 2011), 565–85.

15. Uri Simonsohn and Ed Yong, "The Data Detective: Uri Simonsohn Explains How He Uncovered Wrongdoing in Psychology Research," *Nature* 487.7405 (July 5, 2012): 18–19.
16. Chris D. Frith, "The Social Brain?" *Philosophical Transactions of the Royal Society B: Biological Sciences* 362 (2007): 671–78.
17. Simon Baron-Cohen, *Zero Degrees of Empathy: A New Theory of Human Cruelty* (London: Allen Lane, 2011).
18. Hannah Faye Chua, Israel Liberzon, Robert C. Welsh, and Victor J. Strecher, "Neural Correlates of Message Tailoring and Self-Relatedness in Smoking Cessation Programming," *Biological Psychiatry* 65 (2009): 165–68.
19. For a review of the neural theory of information processing, see Michael Anderson, "Neural Reuse: A Fundamental Organizational Principle of the Brain," *Behavioral and Brain Sciences* 33 (2010): 245–66. For a critique of information-processing terminology and the computational model of mind, see Noë, *Out of Our Heads*.
20. Florentin Wörgötter and Ulf Eysel, "Context, State and the Receptive Fields of Striatal Cortex Cells," *Trends in Neurosciences* 23 (2000): 497–503.
21. Lauren Berlant tracks a parallel process when compassionate conservatism relocated a particular zone of intimacy in the 1980s and 1990s: "Compassionate conservatism advocates a sense of dignity to be derived from labor itself—of a particular sort. No longer casting a living wage, public education, affordable housing, and universal access to economic resources as the foundation of the individual and collective good life in the United States, the current state ideology sanctifies the personal labor of reproducing life at work, at home, and in communities. That is, income-producing labor is deemed valuable chiefly in the context of its part in making smaller-scale, face-to-face publics." Lauren Berlant, "Introduction: Compassion (and Withholding)," in *Compassion: The Culture and Politics of an Emotion*, ed. Lauren Berlant (London: Routledge, 2004), 1–13, 3. Although locating an "empathy circuit" in the brain is not exactly the same thing as locating compassion primarily within the scope of personal expression, each of these explanations for pro- or antisocial behavior attributes the phenomenon to personhood fundamentally, thus mitigating against explanations that are fundamentally social: e.g., empathy requires the perception of similarity, which is socially determined and therefore can be socially transformed (see the discussion below).
22. Vinod Menon, "Dissociating Prefrontal and Parietal Cortex Activation during Arithmetic Processing," *NeuroImage* 12.4 (2000): 357–65.
23. S. Dehaene, E. Spelke, P. Pinel, R. Stanescu, and S. Tsivkin, "Sources of Mathematical Thinking: Behavioral and Brain-Imaging Evidence," *Science* 284 (1999): 970–74.
24. Vilayanur Ramachandran and Edward Hubbard, "The Phenomenology of Synaesthesia," *Journal of Consciousness Studies* 10.8 (2003): 49–57.

25. Olaf Blanke, Stephanie Ortigue, Theodor Landis, and Margitta Seeck, "Neuropsychology: Stimulating Illusory Own-Body Perceptions," *Nature* 419 (19 September 2002): 269–70.
26. For example, Eva Bonda, Michael Petrides, Stephen Frey, and Alan Evans, "Neural Correlates of Mental Transformations of the Body-in-Space," *Proceedings of the National Academy of Sciences of the United States of America* 92 (1995): 11180–84; and Gemma Calvert, Ruth Campbell, and Michael Brammer, "Evidence from Functional Magnetic Resonance Imaging of Crossmodal Binding in the Human Hetero-Modal Cortex," *Current Biology* 10 (2000): 649–57.
27. Russell Poldrack, "Can Cognitive Processes Be Inferred from Neuroimaging Data?" *Trends in Cognitive Sciences* 10.2 (2006): 59–63.
28. Simon Baron-Cohen, Alan Leslie, and Uta Frith, "Does the Autistic Child Have a 'Theory of Mind'?" *Cognition* 21 (1985): 37–46.
29. John M. Wang, Rachael D. Seidler, Julie L. Hall, and Stephanie D. Preston, "The Neural Bases of Acquisitiveness: Decisions to Acquire and Discard Everyday Goods Differ across Frames, Items, and Individuals," *Neuropsychologia* 50 (2012): 939–48.
30. Antonio Damasio, *The Feeling of What Happens: Body and Emotion in the Making of Consciousness* (New York: Houghton Mifflin Harcourt, 2000), 400.
31. Elizabeth Phelps et al., "Performance on Indirect Measures of Race Evaluation Predicts Amygdala Activation," *Journal of Cognitive Neuroscience* 12.5 (2000): 729–38. A portion of this experiment description and critique appeared in Daniel M. Gross, "Response: Toward a Rhetoric of Cognition," in *Neurorhetorics*, ed. Jordynn Jack (New York: Routledge, 2012), 53–62.
32. Phelps et al., "Performance on Indirect Measures of Race Evaluation," 730.
33. Ibid., 734.
34. Ibid. (emphasis added).
35. Christine D. Wilson-Mendenhall, Lisa Feldman Barrett, W. Kyle Simmons, Lawrence W. Barsalou, "Grounding Emotion in Situated Conceptualization," *Neuropyschologia* 49 (2011): 1105–27, 1105, 1107, 1108, 1110.
36. See, e.g., Lisa Feldman Barrett, Kristen A. Lindquist, and Maria Gendron, "Language as Context for Emotional Perception," *Trends in Cognitive Science* 11.8 (2007): 327–32.
37. See, e.g., James F. Source, Robert N. Emde, Joseph J. Campos, and Mary D. Klinnert, "Maternal Emotional Signaling: Its Effect on the Visual Cliff Behavior of One-Year-Olds," *Developmental Psychology* 21.1 (1985): 195–200.
38. Paul Griffiths and Andrea Scarantino, "Emotions in the Wild: The Situated Perspective on Emotion," in *The Cambridge Handbook of Situated Cognition*, ed. Philip Robbins and Murat Aydele (Cambridge: Cambridge University Press, 2009), 437–54, 438, 441, 443, 444.

39. Alicia J. Hofelich, "The Role of Unique Personal Representations in Understanding and Responding to the Emotions of Others, in Psychology" (Ph.D. diss., University of Michigan, 2012), 154. Stephanie Preston, Alice J. Hofelich, and R. Brent Stansfield, "The Ethology of Empathy: A Taxonomy of Real-World Targets of Need and Their Effect on Observers," *Frontiers in Human Neuroscience* 7 (488): 1–3. doi: 10.3389/fnhum. 2013.00488.
40. Dean continues: "The idea that we have embraced victims in this era of the witness (at least Jewish victims of the Holocaust) evades to what extent this new moral economy is a part of the ongoing historical refashioning of cultural attitudes to victims. These attitudes can be understood only in relational and affective terms in which suspicion, envy, attachment, and aversion are all in the process of being reformulated in reference to a new historical context in which the Holocaust of European Jewry has become problematically the paradigmatic catastrophe. It is crucial to recognize the power of a moral economy in which victims who demonstrated no agency under pressure of circumstances were least respected, least dignified, and perhaps even suspect. The narrative of the Holocaust is one of human willfulness against the odds, of human willfulness and its power to make or unmake the world, that is always implicated in any discussion about victims and victimization, including a profound cultural investment in the moral soundness of those who suffer. The introduction of the traumatized victim of catastrophe in his or her extreme disempowerment and traumatic, deferred, and thus often empirically indecipherable grief appears to have generated as much aversion and suspicion as sympathy" (coda, p.406).
41. One alternative: the stable-state, dynamic-systems approach that is already common in engineering and computer science. However, the inferences one can draw from such approaches are qualitatively different than those that can be drawn from the dominant approaches, a factor that will surely inhibit the spread of such techniques in psychology per se. See, e.g., Stephanie Preston and Alicia J. Hofelich, "The Many Faces of Empathy: Parsing Empathic Phenomena through a Proximate, Dynamic-Systems View of Representing the Other in the Self," *Emotion Review* 4.1 (2012): 24–33.

PART TWO

Medicine

FOUR

What Is an Excitement?

OTNIEL E. DROR

What is this excitement? Is it a distinct element in a state of mind, or can it be resolved into Cognition, Volition, or Feeling as Pleasure and Pain? FRANCES A. MASON (1888)

Excitement thus appears as the least specialized, the most generalized of all the emotions. GEORGE M. STRATTON (1928)

Though the experiments to be described are concerned largely with the physiological changes produced by the injection of adrenaline—which appear to be primarily the result of sympathetic excitation—the term physiological arousal is used in preference to the more specific "excitement of the sympathetic nervous system." STANLEY SCHACHTER AND JEROME E. SINGER (1962)

In this essay, I reinterpret post-1945 emotions in the sciences and the social sciences by presenting two challenges to a very large and dominant body of work. First, I challenge contemporary interpretations that position post-1945 emotions in terms of the (nineteenth-century) legacies of Darwin and James (and Freud). I argue that post-1945 emotions were (and are) significantly shaped and framed, not by Darwin, James, and Freud, but by the study of the physiology of emotions during the early twentieth century: by the study of adrenaline and "emotional excitement" inside physiological laboratories.

My second argument is that emotional excitement and adrenaline were laboratory creations. They were to a large extent products of late nineteenth- and early-twentieth century laboratory praxis. This second argument challenges the overwhelming dominance of intellectual approaches to post-1945 emotions. These latter approaches depict post-1945 emotions in terms of the intellectual legacies (and debates) of pre-1945 Darwinian and Jamesian emotions. My argument is that the *practices* of physiological laboratories construed basic assumptions and experiments of post-1945 investigators. My major aim is thus to recast post-1945 emotions by rewriting their genealogy from the perspective of pre-1945 physiological studies of emotions and in terms of the history of laboratory praxis.

I present these challenges by providing a succinct genealogy of twentieth-century *adrenaline excitement.* Adrenaline excitement was an emotional state, which was not an emotion, and an excited state, which was not "nervous" or "mental" excitement. Adrenaline emotional excitement came into its own during the early twentieth century and prefigured important dimensions of post–World War II "emotion," "activation," physiological "arousal," and "intensity." It embodied the shop-floor practices for studying emotions inside physiological laboratories. It was a laboratory creation.

I begin the chapter in the late nineteenth century. I briefly summarize British attempts to define and analyze *excitement.* These attempts and debates present the ambiguous, indefinite, and contested position of excitement in relation to feelings, cognition, and action.

After presenting the British attempts to explicitly theorize excitement, I shift to the laboratory and its practices for studying emotions from the late nineteenth century. I argue that one salient and new feature of the study of emotions inside the laboratory was the shift to the viscera. The shift constituted and operationalized emotion in terms of visceral actions, rather than in terms of facial expressions, bodily gestures, overt behaviors, or verbal locutions. This new mode of operationalizing emotions in physiology enacted and constituted the emergent twentieth-century embodied-visceral state of adrenaline excitement.

In the following section, I study several salient features of twentieth-century *adrenaline* excitement. After delineating some of the unique features of adrenaline excitement, I study the laboratory operations that construed it during the early twentieth century. Excitement embodied the unique arrangement of physiological studies of emotions, whose origins harked back to the late nineteenth century.

In the concluding section, I shift to post-1945 emotions in light of the new interpretation that I offer. I argue that early twentieth-century physiological excitement framed significant elements of post-World War II emotion. Major postwar enactments of emotion implicitly assumed the early twentieth-century construct of adrenaline excitement.

Since my major objective in this chapter is to provide a preliminary genealogy of postwar emotion from the perspective of the operationalization of emotion inside the laboratory, I do not broach the broader social and institutional contexts of emotions.

Excitement: Emotion

In attempting to define *excitement,* Alexander Bain wondered "whether it is anything more than the higher degrees of intensity of feeling itself."[1] Frances A. Mason asked similarly: "Is the excitement a cause of the volition? Or can we say that it is itself volition in its most rudimentary form—the raw material of what we recognise as volition, something implied in the Uneasiness of Locke and in the *Conatus* of Spinoza?"[2] According to the *Oxford English Dictionary,* the original and obsolete seventeenth-century meaning of *excitement* was "something that tends to excite (a feeling); a motive or incentive *to* action; an exhortation, encouragement." The *mental* meaning of *excitement* as a condition of being "mentally" excited appeared only during the mid-nineteenth century. During the latter half of the nineteenth century, *excitement, excite,* and *exciting* appeared in reference to emotions in some of the major texts of the nineteenth-century, including Alexander Bain's *The Emotions and the Will* (1859), Charles Darwin's *The Expression of the Emotions in Man and Animals* (1872), and William James's "What Is an Emotion?" (1884).

In *The Emotions and the Will,* Bain presented an in-depth analysis of excitement, distinguishing between two major categories of excitement. The first type—"mental excitement"—was cognitive. It was neither pleasure nor pain and linked the emotional and intellectual realms, as Bain explicitly argued.[3] This neutral excitement of "surprise" and "shock" harked back to the medieval concept of "wonderment," Descartes's "wonder," and the cognitive passions in general.[4]

The second type of excitement was the excitement of the chase. It was not analyzed in any great depth by Bain. It was exemplified by the sport of hunting *and* the botanist's chase after specimens and knowledge. As Bain observed in analyzing the "emotions of action," of "killing or catching some species of quadruped, bird, or fish": "All this applies eminently to the chase, guiltless of blood or suffering, maintained by the botanist."[5] It was in the context of this second type of excitement that Bain also referred to the excitement of battles, intellectual confrontations, and scientific disputations—all of which appeared on a continuum, despite the clear distinctions between them.[6]

Darwin and James also referred to *excite* in their writings on emotions, but they did not theorize or analyze excitement.[7] *Excite* appeared in their studies of the emotions in reference to an "excited" nervous system, "excited" sympathy, "excit[ing] a blush," "excit[ing]" an "interest," "an excited condition from joy," "exciting states of the mind," the "exciting nature of anger," "the excitement of strong feelings," and "exciting fear in an enemy" (Darwin) and "feelings" of "excitement," an "exciting train of ideas," an "exciting fact," "belligerent excitement," and "organic excitement" (James). Various late nineteenth- and early twentieth-century psychologists referred to *emotional excitement* in speaking of emotions, but they did not elaborate or analyze the different and multiple meanings and uses of *excitement*.[8]

Bain's first category of mental or "indifferent" excitement, which was neither pleasure nor pain, became the crux of a debate in the journal *Mind* during the 1880s. This debate focused on the meaning and nature of *excitement* and its relationships with feelings, emotions, and action. Four distinct questions emerged in the context of this debate. Was excitement a third category of feelings, neither pleasure nor pain? Was it "neutral"—or was it slightly pleasurable (or slightly painful)? Was the neutrality of excitement primary, or was it an outcome of an admixture of pleasure and pain? And what were the relationships between excitement and action vis-à-vis the dominant and prevailing theory of action, in which action was solely determined by pleasure and pain?[9]

The first three questions, which occupied the lion's share of the debate, pertained to the experiential-consciousness dimensions of excitement. These questions were not resolved by the interlocutors. They indicate the absence of consensus in respect to the very nature of excitement and its relation to feelings and action in the British context.

The question of whether "the neutral states [i.e., excitement] can operate in producing actions" was primarily analyzed by Bain.[10] He proposed an excitement theory of action that challenged the exclusivity of pleasure-pain in action by endowing neutral excitement with agency. This new form of agency, which was neither pleasure nor pain, was based on a "tendency to act out an idea" and the example of "imitation."[11] Bain distinguished between actions that belonged to the will, which were always products of pleasure and pain, and those that belonged to cognition (ideas) and were products of neutral excitement.[12]

Visceralization: Nineteenth-Century *Physiological* Emotion

In parallel with these intellectual formulations and debates with respect to excitement, laboratory-based physiologists transformed emotions into objects of laboratory knowledge. Though physiologists who studied emotions were centrally concerned with experience, experience was not the material with which they directly worked on a day-to-day basis. They worked with blood flows, contracting and dilating viscera, urine, sweat, saliva, blood—with the body and its visceral materiality. These visceral materialities were the substrates that they observed, touched, measured, recorded, and ultimately wrote about in studying the emotions.[13]

Angelo Mosso, the leading late nineteenth-century physiologist of emotions, introduced the basic modus operandi of the laboratory of emotion. In his laboratory, emotion was often known, defined, and identified in terms of visceral actions.[14] Mosso operationalized and enacted emotions inside the laboratory in terms that excluded any and all linguistic, overt behavioral, gestural, or facial expressions of emotions. His notion of emotion assumed and demanded an immobile, catatonic, and incommunicative overt body, which was juxtaposed to an animated and expressive viscera.[15] The animations of the viscera were recorded by scientific technologies that presented the internal actions of the viscera in terms of visible graphs—that is, emotion.

Numerous nineteenth-century experimenters adopted and internalized this emergent logic of operationalizing emotions. Technologies, laboratory protocols, and scientific narratives and reports constituted a visceral subject and visceral emotions by excluding, suppressing, and erasing the nonvisceral and presenting and exteriorizing only the visceral. This mode of studying emotions contrasted with numerous contemporary studies of emotions by, for example, Duchenne de Boulogne (1862), Charles Darwin (1872), and others in which the overtly gesticulating and animated body and its expressions and gestures of emotions were studied as the emotion.

The absence of the overt body as the signifier of emotions was even apparent when the body was visible. In these latter cases, the overt body was often present in terms of an isolated-detached petrified limb or finger. The inexpressive and detached appendage of the body was juxtaposed to a graph of the emotionally expressive internal-visceral activities that were taking place inside the inanimate appendage. The technologies for studying emotions and the instructions and stipulations of the protocol constituted this visceral ontology of emotion and the inexpressiveness, immobility, and incommunicability of the overt body.

From the shop-floor perspective of the physiological laboratory, visceralized emotions and subjects were necessary. They made good laboratory sense. Immobile subjects did not disturb the protocol or the delicate instruments during the emotion. The slightest overt movement or trembling during emotions introduced artifacts into the delicate measurements of the sensitive physiological apparatuses that gauged minuscule visceral actions beneath the skin. Ideally, subjects would also refrain from breathing since breathing skewed the capillary blood flow beneath the skin, which was the visceral movement and activity that constituted the emotion inside laboratories, as Alfred Binet and J. Courtier discovered.[16]

The visceral ontology of emotion also enabled the laboratory to study potentially disruptive, wild, or vehement emotions inside its orderly spaces. Unlike the enactments of emotions in terms of overt behaviors, actions, gestures, expressions, or locutions, visceral emotions did not disrupt the laboratory and its social order. Physiologists of emotions created and studied the most extreme and dangerous emotions without the disruptions and interferences of emotion. These dangerous and disruptive emotions were manifested in terms of controlled scientific representations—the harmless and well-behaved graph.

The suppression, immobilization, and control of the overt body during emotion was not conceived by physiologists in terms of an emergent *psychological* model of discontents. This latter psychological and psychologizing framing of emotions did not reflect the logic of physiology. The logic of physiology from the late nineteenth century and the early twentieth construed the viscera as the privileged site for authentic and trustworthy emotions. This physiological framing was highly critical of contemporary psychological modes of framing emotions in terms of the "superficial" indices of language and overt behaviors and gestures.

The operationalization of emotions inside the laboratory in terms of nonovert visceral actions went hand in hand with their theorization in terms of overt action. Emotions were thus *enacted* inside the laboratory in terms of vehement visceral actions and the inert overt body and *theorized* by physiologists in terms of models of overt action. These latter models of action explained the conjunction between visceral and overt actions in emotion. Visceral actions contributed to overt action, or overt actions were determined by nonovert motions of the viscera.

This late nineteenth-century enactment of emotions in terms of the viscera and the inert overt body inhered in physiological emotions for decades. It presents one important tributary of the genealogy of twentieth-century adrenaline emotional excitement. It was a laboratory creation.

Physiological Adrenaline Enactments

During the early twentieth century, physiologists progressively shifted to visceral indices of adrenaline-sympathetic activation in studying emotions. The extraction, purification, injection, transfer, measurement, visualization, and representation of adrenaline constituted the study of the physiology of emotions and emotional excitement.

Adrenaline emotional excitement was not an emotion. Unlike emotion, emotional excitement was never positioned in opposition to knowledge, seriousness, objectivity, or masculinity. It did not contradict, subvert, or skew rationality or the production of knowledge. Occasionally, it was experienced by practicing scientists and was presented as an integral—and positive—aspect of scientific investigation.

Excitement was a fuzzy and multifaceted state of the self. It could signify the excitement of a spectator at a football game (as measured by his rising level of adrenaline metabolites), the excitement aroused by consumerism (as manifest in a female consumer's psychogalvanic response to a consumer item *and* in a scientist's reaction to the arrival of a new technology), and the excitement of a student taking a college examination (as measured in terms of an elevated blood glucose level). It is thus not always clear what the physiologist meant when he referred to excitement and how he positioned excitement vis-à-vis emotion.

Adrenaline excitement was clearly, but also equivocally, distinguished from a myriad of seemingly similar other excitements. It was distinguished from the negative connotations of enthusiasm—it did not affect cognition, and it was not overt action.[17] It was far removed from the clinical-pathological nosology of "nervous" excitement—it was not in the nerves, being instead a visceral state of activation. And it was distinguished from sexual excitement—it did not behave like a *glandular* affect, as nineteenth-century writers referred to sexual desire, with its circulatory and predictable patterns of desire-satisfaction, and it did not have a "refractory period," like male-centered sex physiology. It was produced ad hoc and dissipated in different temporal curves. It was additive but did not seem to lead to satiation.

There was no physiognomy of emotional excitement. The individual's tissues, blood, urine, and viscera embodied excitement, but there were no gestures, behaviors, or locutions that could conclusively determine the state called *excitement.* In the absence of visceral embodiments, there was no excitement—that is, no possibility of defining excitement in terms of physiology and no possibility of experiencing this form of excitement.

Unlike emotion, adrenaline—and its excitement—was physically extracted from, and injected into, bodies. It also existed outside any body —inside the test tube. As Henry Rutgers Marshall suggested: "In future it . . . [may be] possible that directors of the drama would insist on their emotional artists being injected with adrenaline before a performance."[18]

Experimenters who worked with adrenaline literally transferred emotional excitement from one body to another and from humans to lower organisms. They studied human emotional excitement—but not human emotion—in *daphnia* (a small freshwater crustacean). In these latter experiments, investigators observed the visceral reactions of *daphnia* to blood samples that had been procured from human subjects. As George Gagnon explained in his early 1950s doctoral dissertation: "The blood of a manic depressive in a manic phase has the greatest effect on the heart rate [of a *daphnia*]. . . . [I]t would seem that the influencing factor is related to the emotional excitement rather than to the type of psychosis."[19]

The literal transferability of excitement across individuals contrasted with the nontransferability of feelings or emotions. In twentieth-century physiological literature, there was never a transfer of a feeling or an emotion between or across individuals. Excitement, however, easily traversed and closed the gap between embodied selves. Shared adrenaline levels created a community of shared visceral response. This shared visceral response, in turn, created a *shared*—intersubjective—embodied experience and a gamut of *different* emotions.

Transferring experiences and excitements across bodies and between individuals was not the physiologists' explicit intention. But they were "misinterpreted." These misinterpretations of physiological studies of emotions continued for several decades as physicians, psychologists, popular writers, and other commentators "misread" the work of physiologists. Walter B. Cannon was often—too often, as he expressed it—misinterpreted. "I have not shown that blood from an excited animal transferred to another animal will produce the effects of excitement," he wrote in the mid-1930s to the famous psychologist Joseph Jastrow, who had misinterpreted Cannon's experiments.[20]

Outside the laboratory, investigators collected urine, tissues, and blood from spectators at football games, soldiers in the throes of battle, and students taking examinations and observed that all produced measured quantities of adrenaline and all experienced what the physiologist defined in terms of emotional excitement. The emotions of religious fervor, of a movie, of purchasing a consumer item, of a football game, of a scientific protocol, and of a game of chess might differ substantially; their emotional excitements, materialized and embodied in terms of adrenaline, were indistinguishable.[21]

Conversely, the absence of adrenaline and of sympathetic activation did not affect the animal's *overt* behaviors and motivations. Experimenters produced "sympathectomized" animals with inactivated adrenaline-sympathetic systems and remarked on the absence of any modifications in overt behaviors—that is, conation, purpose, and motivation. As John F. Fulton was "remind[ed]" in corresponding with Walter B. Cannon: in attempting to study the reflexes of a sympathectomized cat, he was "almost eaten alive!"[22] The absence of adrenaline-sympathetic activation did not alter the animal's overt emotional behaviors and reactions.

This emergent operational enactment of adrenaline and excitement was epitomized in adrenaline injection experiments. Injecting adrenaline into human and animal bodies did not lead to any gesture, movement, action, or locution. It created the greatest amount of excitement inside the body while leaving the shell of the body, its gestures, and its behaviors under full control. Adrenaline did not influence thinking, and it did not induce disorderly conduct. Injecting adrenaline did not evoke dangerous, threatening, irrational, or uncontrollable gestures and behaviors. It did not disrupt or impinge on society. It did not disrupt the laboratory. It was the physiologist's way to study emotions, without the disturbing motions, and without generating *psychological* emotion inside the laboratory.

By injecting adrenaline, physiologists enacted and studied *physiological* emotions—that is, the *physiological* meanings and mechanisms of emotions—without the need to produce (psychological) emotions inside the laboratory. These latter were superfluous for studying emotion and often disrupted the production of knowledge.

The meaning of adrenaline as an intense excitement that was completely contained in the body—that is, was restricted to the viscera, lacked any direction, and was devoid of overt action—implicitly mimicked the basic template of protocols on emotions from the late nineteenth century: the production of extreme experiences without any overt—dangerous or uncontrolled—"motion" during the emotion. Emotional excitement and the meaning of injecting adrenaline were products of the laboratory's *operational* logic of studying embodied emotions.

Aldous Huxley's *Brave New World* reflected this new logic. As the protagonist explained:

> "Men and women must have their adrenals stimulated from time to time."
> "What?" questioned the Savage, uncomprehending.
> "It's one of the conditions of perfect health. That's why we've made the V.P.S. treatments compulsory."
> "V.P.S.?"
> "Violent Passion Surrogate. Regularly once a month. We flood the whole system with adrenin [adrenaline]. It's the complete physiological equivalent of fear and rage. All the tonic effects of murdering Desdemona and being murdered by Othello, without any of the inconveniences."
> "But I like the inconveniences."
> "We don't," said the Controller. "We prefer to do things comfortably."[23]

This operational construct was integrated post hoc into different teleological-functional narratives that gave meaning to *excitement*. Excitement was the "preparedness" of the body for action. It was "energy." It "exercised" the viscera and was beneficial for the internal hygiene of the viscera. And it was an anticipatory homeostatic mechanism.

Excitement was also a motive for action. The experience of "adrenalinemia"—adrenaline inundation of the blood—and of sympathetic activation could function as the *cause* for seeking excitement. As Walter B. Cannon, the leading physiologist of emotions during the early twentieth century, explained to a correspondent in 1942: "I have stressed the thrill of excitement with its energizing effect as something which men go out to seek in dangerous experiences in order that they may enjoy it."[24]

The experience of excitement explained why "men" seek excitement. This physiological framing of adrenaline excitement as a motive for action, for its *own* sake, prefigured important postwar conceptions of seeking arousal-excitement-stress. The physiological elaboration of excitement seeking was, however, marginal during the first half of the twentieth century. It would come into its own during the postwar period.

Postwar *Psychological* Adrenaline Enactments

Injecting adrenaline into human (or rat) bodies in order to test the James-Lange theory of emotions was an unusual practice for early twentieth-century physiologists—despite the disproportionate notoriety of a very few early twentieth-century adrenaline injection experiments in our intellectual histories of emotions (in and outside the discipline of history). Early twentieth-century physiologists of emotions injected adrenaline into numerous human and animal bodies in order to study a myriad of different aspects of the *physiology* of emotion, but not in order to create or study psychological emotions.[25]

During the immediate postwar period, adrenaline injection experiments shifted from physiology to psychology—from the study of the embodied-visceral physiology of emotion to the study of the qualitative experiences of emotions and their overt behaviors. This shift in intellectual context reframed the history of adrenaline injection experiments in terms of the legacy of the James-Lange theory of emotions. It erased and obfuscated the rich physiological-operational history of adrenaline and its injection.[26]

This disappeared history, nonetheless, partially determined the meanings of postwar adrenaline injection experiments and the discovered relationships between adrenaline, physiological arousal, excitement, and emotion. The prewar physiological history and enactments of adrenaline constituted postwar (psychological) emotion, despite the explicit framing of postwar emotion in terms of the long intellectual legacy of the James-Lange theory of emotions in adrenaline injection experiments.

Stanley Schachter and Jerome Singer's renowned study "Cognitive, Social, and Physiological Determinants of Emotional State" (1962) exemplifies the primacy of prewar enactments of adrenaline and excitement in constituting postwar discoveries. Schachter and Singer's adrenaline injection experiments were instrumental in the development of postwar cognitive-appraisal theories of emotions. These and other studies during the early 1960s demonstrated that injecting adrenaline into human (or rat) bodies did not produce a real emotion.[27] As Schachter and Singer demonstrated, physiological arousal or "sympathetic system" excitement—which they *operationalized* and enacted by injecting adrenaline—did not produce a real emotion.[28] A real emotion required physiological arousal or sympathetic excitement—that is, injecting adrenaline inside the laboratory—and a cognitive-interpretive component.[29]

Schachter and Singer's construal of adrenaline in terms of physiological arousal and/or the excitement of the sympathetic nervous system *and* their discovery that adrenaline (arousal/excitement) was not emotion stemmed from a long history in which adrenaline and excitement had been construed by physiologists as non-(psychological) emotion. The embodied-visceral state of adrenaline excitement was precisely that state, which was indistinguishable across a myriad of different emotions, was transferable across bodies and species, and lacked any overt direction, action, or agency. It was construed as a non*psychological* emotion by prewar physiologists.[30]

This nonemotion construct of adrenaline excitement was one important source for postwar "activation," "intensity," "energy mobilization," "excitement," and *physiological* "arousal."[31] Postwar excitement, physiological arousal, and intensity were often operationalized in terms of adrenaline or construed on the template of adrenaline excitement. Donald O. Hebb's formulations of "risk taking" and the "taste of excitement," Elizabeth Duffy's "activation," "energy mobilization," and "arousal," and Schachter and Singer's two-dimensional theory were partially construed on the template of adrenaline and excitement.[32]

Epilogue

The early shop-floor enactments of emotions in physiological laboratories mark the *true* beginnings of *noncognitivist* approaches to (and theories of) emotions in the modern history of the emotions. Physiologists enacted a nonevaluative physiological emotion inside the laboratory. This type of emotion prefigured William James's famous theory of (nonevaluative) emotion. Indeed, in presenting his theory of emotions, James explicitly drew on these early laboratory-physiological studies and enactments.

The early enactments of emotions also prefigured and embodied a second crucial and related aspect of James's theory of emotions. In his famous theory, James positioned the brain as a semipassive recipient of visceral reverberations that, on reaching the brain, created the emotion as experienced. The new physiologists of emotions studied emotions by gauging these very same visceral reverberations. The new machines gauged these visceral reverberations and presented them in terms of graphs of the subject's emotions. James was well aware of these instruments and their graphic output in proposing his theory.

There was a clear conceptual-structural parallelism between these machines and James's brain: both were passive recipients of visceral input. This input created the experienced emotion in the brain and the transcription of the emotion in the machine. This parallelism explains why some authors argued that the new machines could represent "an approaching emotional disturbance" *before* it was experienced in the brain.[33] The machine sometimes picked up the visceral perturbations and transcribed them into the emotion before the brain had time to experience the reverberation as an emotion.

Twentieth-century adrenaline excitement was the end product of these enactments and of many developments that are beyond the scope of my analysis in this chapter. These include the graphic representations that decontextualized emotion, the stimulus-response framework of laboratory experiments, the logic that eliminated teleological (anticipatory) responses during the protocol, the attempt to isolate/identify a pure-essentialist aspect of emotions that was independent of any particular context, the progressive intensification in studied emotions during the twentieth century, and major transformations and reorientations with regard to "impulse," "feelings," and "emotionology" in twentieth-century Western culture.[34]

Adrenaline excitement emerged as a way of speaking about experience itself, abstracted from any particular context and/or a moral or teleological narrative. It was an inner experience that did not lead to any particular action and had no particular expression, behavior, or articulation. It was an inner activation, energy, intensity, and physiological arousal, but it did not lead to action, and it was not a motive for any particular act. As Stanley Cobb put it in 1950: "Excitement is the least differentiated. There is no specific impulse to a certain kind of action. There is great alertness, action may take any direction."[35]

The early twentieth-century construct of adrenaline excitement was also an important source for postwar developments beyond psychology and physiology—in the social sciences. These latter theories drew on either the adrenaline injection experiments of postwar investigators—like Schachter and Singer—or the early twentieth-century physiology of adrenaline excitement.

Michelle Z. Rosaldo's pioneering studies in the anthropology of emotions and Norbert Elias and Eric Dunning's studies of the figurations of modern sports drew on the legacy of adrenaline excitement.[36] While Rosaldo borrowed directly from Schachter and Singer in arguing that, as Reddy put it, emotion "included a biological component (arousal) . . . coupled with an all-important interpretive component that was learned and therefore cultural" (this crucial cultural component of emotions became the new focus of an emergent anthropology of emotions and of social constructionist approaches to emotions in anthropology),[37] Elias and Dunning drew directly on the early twentieth-century physiology of adrenaline excitement in studying the excitements and the figurations of modern sports. Elias and Dunning explicitly drew on this latter physiology of adrenaline and excitement, rather than on a Freudian-inspired suppression model of civilization, in explaining the seemingly paradoxical (controlled) *de-controlling* of "emotional controls" in modern "civilized" sports.[38]

The study of the physiological science of emotions suggests that, in parallel with the development of a modern psychological model of the self, and in parallel with the Freudian, Jamesian, and Darwinian paradigms, a variety of physiologically inclined investigators and clinicians established a physiological model and logic of emotions. This physiological model was important for a myriad of contemporary developments, negotiations, and controversies that our histories have often attributed to the intellectual legacies of Darwin, James, or Freud rather than to the physiological laboratory. These latter postwar developments pertain to postwar discoveries in stress research, to contemporary debates with respect to the nature of emotions (from appraisal to affect and intensity models of emotions), and to contemporary brain mechanisms of pleasure or reward.

NOTES

My title obviously invokes William James's famous "What Is an Emotion?" *Mind* 9 (1884): 188–205. My chapter epigraphs are taken from, respectively, James Sully and Frances A. Mason, "On 'Feelings as Indifference,'" *Mind* 13.50 (April 1888): 248–55, 253; George M. Stratton, "Excitement as an Undifferentiated Emotion," in *Feelings and Emotions: The Wittenberg Symposium*, ed. Martin L. Reymert (Worcester, MA: Clark University Press, 1928), 215–21, 218; and Stanley Schachter and Jerome E. Singer, "Cognitive, Social, and Physiological Determinants of Emotional State," *Psychological Review* 69.5 (September 1962): 379–99, 380n2.

1. Alexander Bain, "Discussion: On Feeling as Indifference," *Mind* 14.53 (January 1889): 97–106, 100.

2. Sully and Mason, "On 'Feelings as Indifference,'" 253.
3. Alexander Bain, *The Emotions and the Will*, ed. Daniel N. Robinson (London: John W. Parker, 1859), 610–20.
4. For the history of wonder and the cognitive passions, see Caroline Walker Bynum, "Wonder," *American Historical Review* 102 (1997): 1–26; Philip Fisher, *Wonder, the Rainbow, and the Aesthetics of Rare Experiences* (Cambridge, MA: Harvard University Press, 1998); and Lorraine Daston and Katherine Park, *Wonders and the Order of Nature, 1150–1750* (New York: Zone, 1998).
5. Bain, *The Emotions and the Will*, 184–98 (chap. 10, "Emotions of Action-Pursuit"), 189.
6. For the excitement of "original research," which is "by no means a gentle excitement," see ibid., 193.
7. Charles Darwin, *The Expression of the Emotions in Man and Animals* (New York and London: D. Appleton, 1872); and James, "What Is an Emotion?" I note that James referred to *excitement* in other texts.
8. In the German context, Wilhelm Wundt introduced excitement as one of the hedonic qualities in his three-dimensional theory. That theory presented three dyadic pairs of oppositional hedonic qualities: pleasant vs. unpleasant, excitement vs. quiescence/calm/depression, and strain/tension vs. relief. Wundt's theory was taken up by laboratory-based psychologists, who attempted to measure the excitement dimension-quality of feelings in terms of physiological indices. Wundt had already observed that excitement (and depression) was visible in characteristic pulse and respiration curves. For an excellent example of multiple references to *emotional excitement* (in speaking of emotions, such as fear, or intellectual work), see James Rowland Angell and Helen Bradford Thompson, "A Study of the Relations between Certain Organic Processes and Consciousness," *Psychological Review* 6 (1899): 32–69. On Wundt's early work on emotions, see Claudia Wassmann, "Physiological Optics, Cognition and Emotion: A Novel Look at the Early Work of Wilhelm Wundt," *Journal of the History of Medicine and Allied Sciences* 64 (2009): 213–49.
9. A. Bain, "On Feeling as Indifference," *Mind* 12.48 (October 1887): 576–79; W. E. Johnson, "On Feeling as Indifference," *Mind* 13.49 (January 1888): 80–83; Sully and Mason, "On 'Feelings as Indifference'"; and Bain, "On Feeling as Indifference" (1889).
10. Bain, "On Feeling as Indifference" (1889), 104.
11. Ibid., 104–10.
12. Bain was well aware that his succinct and brief presentation of an alternative theory of action, which challenged John Stuart Mill, demanded further elaboration.
13. Otniel E. Dror, "Visceral Pleasures and Pains," in *Knowledge and Pain*, ed. Esther Cohen, Leona Toker, Manuela Consonni, and Otniel E. Dror (Amsterdam: Rodopi, 2012), 147–67.

14. I note and emphasize that Mosso's visceral physiology of emotions *preceded* William James's theory of emotions. In fact, James partially drew on Mosso's physiology of emotions in proposing his famous theory.
15. For this particular experiment, see Angelo Mosso, *Fear*, 5th ed., trans. E. Lough and Friedrich Kiesow (London: Longmans, Green, 1896), 79–80.
16. Alfred Binet and J. Courtier, "Circulation capillaire dans ses rapports avec la respiration et les phénomènes psychiques," *L'année psychologique* 2 (1895): 87–167.
17. On the two-edged sword of enthusiasm in twentieth-century medicine, see David Cantor, "Cortisone and the Politics of Drama, 1949–55," in *Medical Innovations in Historical Perspective*, ed. John V. Pickstone (Basingstoke: Macmillan/Centre for the History of Science, Technology and Medicine, 1992), 165–84.
18. Marshall quoted in Walter B. Cannon, "Some Disorders Supposed to Have an Emotional Origin," *New York Medical Journal*, October 28, 1916, 870–73, 871.
19. George Gagnon, "The Emotions and Some of Their Effects on the Blood" (Ph.D. diss., Catholic University of America, 1952), 19. Many experiments on adrenaline included such transfers between species.
20. Cannon to Joseph Jastrow, October 13, 1936, folder 1754, box 125, Walter Bradford Cannon Papers (H MS c40), Rare Books and Special Collections, Harvard Medical Library, Francis A. Countway Library of Medicine, Boston.
21. Physiologists thus studied "emotions" (like "fear" or "anger"), measured a variety of visceral indices of adrenaline (like heart rate, glucose metabolism, or the coagulation of blood), and invoked the "emotional excitement" (of "fear" or "anger").
22. John F. Fulton to Walter B. Cannon, February 9, 1929, folder 399, box 28, ser. I, manuscript group 1236, John Farquhar Fulton Papers, Yale University. In their renowned 1962 paper, Schachter and Singer were well aware of the normal reactions of sympathectomized cats and the challenge that these animals could present for their arguments. See Schachter and Singer, "Cognitive, Social, and Physiological Determinants of Emotional State." Schachter presented a response to this challenge. See Stanley Schachter, "The Interaction of Cognitive and Physiological Determinants of Emotional State," in *Advances in Experimental Social Psychology*, ed. Leonard Berkowitz (New York: Academic, 1964), 49–81.
23. Aldous Huxley, *Brave New World* (1932; New York: Harper Collins, 1998), 239–40.
24. W. B. Cannon to Carlos Kling, April 8, 1942, folder 1761, box 125, Cannon Papers.
25. One can easily compare the few—but repeatedly cited—experiments prior to the Second World War in which adrenaline was injected in order to test the James-Lange theory of emotions with the voluminous number of

experiments in which adrenaline was injected, extracted, transferred, and measured in order to study the *physiology* of emotions prior to the Second World War.

26. I note that various early twentieth-century experimenters (psychologists) continued to pursue the Jamesian question in terms of visceral measurements, but not in terms of the injection of adrenaline.
27. Stanley Schachter and Ladd Wheeler, "Epinephrine, Chlorpromazine, and Amusement," *Journal of Abnormal and Social Psychology* 65.2 (1962): 121–28 (physiological arousal is enacted in terms of adrenaline, and the blockage of arousal is enacted in terms of its antagonist chlorpromazine, which blocks the autonomic sympathetic system); Jerome E. Singer, "Sympathetic Activation, Drugs, and Fear," *Journal of Comparative and Physiological Psychology* 56.3 (1963): 612–15; Bibb Latané and Stanley Schachter, "Adrenaline and Avoidance Learning," *Journal of Comparative and Physiological Psychology* 55.3 (1962): 369–72. For a critique, see Albert F. Ax, "The Physiological Differentiation between Fear and Anger in Humans," *Psychosomatic Medicine* 25 (1953): 433–42.
28. Schachter and Singer, "Cognitive, Social, and Physiological Determinants of Emotional State"; Schachter, "The Interaction of Cognitive and Physiological Determinants of Emotional State."
29. Though Schachter and Singer suggest that other types of physiological arousal can function like adrenaline, they, and many other investigators, *operationalized* arousal by injecting adrenaline or by producing adrenalized states through physical exercise. In their renowned article, Schachter and Singer provide only one example of a nonadrenaline type of arousal, but one that is suggestive of their broader claims. See Schachter and Singer, "Cognitive, Social, and Physiological Determinants of Emotional State."
30. Schachter and Singer also frame their study in the context of (and as challenging) Lindsley's "activation theory" of emotion. See Donald B. Lindsley, "Emotion," in *Handbook of Experimental Psychology*, ed. S. S. Stevens (New York: Wiley, 1951), 473–516. Sometimes, the same individual who operationalized adrenaline in terms of emotional excitement—that is, in terms of a nonemotion—injected adrenaline in order to see whether it produced a real emotion in the context of the intellectual legacy of the James-Lange theory.
31. I emphasize *physiological* arousal in following the contemporary distinctions between different types of arousals, e.g., between physiological-embodied arousal and between EEG/reticular activating system arousal. On these distinctions, see also Robert B. Malmo, "Activation: A Neuropsychological Dimension," *Psychological Review* 66.6 (1959): 367–86.
32. See D. O. Hebb, "Drives and the C.N.S. (Conceptual Nervous System)," *Psychological Review* 62.4 (July 1955): 243–54; Elizabeth Duffy, "The Concept of Energy Mobilization," *Psychological Review* 58.1 (January 1951): 30–40,

and "The Psychological Significance of the Concept of 'Arousal' or 'Activation,'" *Psychological Review* 64.5 (September 1957): 265–75. Duffy explicitly refers to Cannon/adrenaline in speaking of mobilization. George M. Stratton's analysis of excitement during the late 1920s partially drew on adrenaline. See Stratton, "Excitement as an Undifferentiated Emotion." For another example of a direct link between arousal and adrenaline, see Morton D. Bogdonoff and E. Harvey Estes, "Energy Dynamics and Acute States of Arousal in Man," *Psychosomatic Medicine* 61.1 (1961): 23–32. For more recent literature, see Elizabeth S. Mezzacappa, "Epinephrine, Arousal, and Emotion: A New Look at Two-Factor Theory," *Cognition and Emotion* 13.2 (1999): 181–99; and Gary S. Krahenbuhl, "Adrenaline, Arousal, and Sport," *Journal of Sports Medicine* 3.3 (1975): 117–21. One could also suggest that aspects of Donald Lindsley's activation theory of emotion also drew on early twentieth-century adrenaline excitement. See Lindsley, "Emotion," esp. from 504.

33. Fred W. Eastman, "The Physics of the Emotions," *Harper's Magazine* 128 (January 1914): 297–303, 301.
34. Otniel E. Dror, "A Reflection on Feelings and the History of Science," *Isis* 100 (December 2009): 848–51. For the broader cultural contexts of adrenaline, see John Higham, "The Reorientation of American Culture in the 1890's," in *The Origins of Modern Consciousness*, ed. John Weiss (Detroit: Wayne State University Press, 1965), 25–48; Peter N. Stearns, *American Cool: Constructing a Twentieth-Century Emotional Style* (New York: New York University Press, 1994); T. J. Jackson Lears, *No Place of Grace: Antimodernism and the Transformation of American Culture, 1880–1920* (Chicago: University of Chicago Press, 1994); Norbert Elias and Eric Dunning, *Quest for Excitement: Sport and Leisure in the Civilizing Process* (Oxford: Blackwell, 1986); Bill Brown, *The Material Unconscious: American Amusement, Stephen Crane, and the Economics of Play* (Cambridge, MA: Harvard University Press, 1996); and Regenia Gagnier, *The Insatiability of Human Wants: Economics and Aesthetics in Market Society* (Chicago: University of Chicago Press, 2000).
35. Stanley Cobb, *Emotions and Clinical Medicine* (New York: Norton, 1950), 105.
36. I emphasize that my objective in this chapter diverges from recent criticisms of affect and emotion theorists for misappropriating the neurophysiology of emotions. My aim is not to critique these misappropriations but to examine the genealogy of the neurophysiology itself. For some recent criticisms of affect and emotion theorists in the humanities, see C. Papoulias and F. Callard, "Biology's Gift: Interrogating the Turn to Affect," *Body and Society* 16 (2010): 29–56; and Ruth Leys, "The Turn to Affect: A Critique," *Critical Inquiry* 37.3 (Spring 2011): 434–72.
37. William M. Reddy, *The Navigation of Feeling: A Framework for the History of Emotions* (Cambridge: Cambridge University Press, 2001), 35–36.
38. Elias and Dunning, *Quest for Excitement.*

FIVE

The Science of Pain and Pleasure in the Shadow of the Holocaust

CATHY GERE

Introduction

February 23, 1973, marked the beginning of the end of a long and productive research tradition in the science of the emotions. On that afternoon, in a Senate hearing on the ethics of human experimentation, Robert Heath of Tulane University showed some films of his human subjects undergoing electrical stimulation of their neural pain and pleasure pathways. This hedonic circuitry was, he believed, the key to emotion and behavior: "Our brain physiology studies in animals and patients have enabled us to localize pathways for emotion and feeling. With our techniques, we have demonstrated brain sites and pathways which are involved with pleasurable emotional states and those involved with painful emotional states—such as rage and fear, which are basic to violence and aggression." Heath explained how he had implanted electrodes deep in the brain tissue of asylum inmates and left them there for months and sometimes years for the purposes of experimenting with mood and behavior control. He showed three short films. The first depicted a patient in whom he "turned on the adversive brain circuitry to induce violent impulses." In the second, "pleasure sites of the brain are stimulated to relieve physical pain." And, in the third, "pleasure sites are stimulated to remove the emotional pain of episodic rage and paranoia." He then claimed that the technique could be used to treat behavior disorders of all kinds, including neurosis, psychosis, and addiction. (He did not mention his ongoing work to cure homosexuality by means of the same electrical stimulation of pain and pleasure circuits.) The film festival was followed by testimony from the behaviorist B. F. Skinner, who asserted that direct electrical stimulation of the brain was not necessary to achieve the same results and that ordinary pain and pleasure could be manipulated noninvasively for the purposes of behavioral and social control.[1]

Exposure to early 1970s public opinion proved fatal for electrical stimulation research. Heath hung on at Tulane, protected by his cloistered position within a wealthy private university, but he was targeted by student protests, and by the time of his retirement seven years later he found himself professionally isolated. Toward the end of his life he had to self-publish a manuscript defining his legacy, an ignominious conclusion to a research trajectory that had once been talked about in terms of a Nobel Prize.[2] His fall from grace was as inevitable as it was inexorable. After 1973, the whole structure of medical practice and clinical research was reformed around the principle of informed consent. This new order shifted the burden of medical decisionmaking from the doctor to the patient and from the researcher to the research subject. The sovereign individual rather than the greater good became the unit of moral reasoning in medical research. Vulnerable citizens such as the institutionalized patients who were Heath's subjects became the focus of special regulatory protections.[3] Now, forty years later, the films that Heath proudly played for Congress are considered so controversial that Tulane University responded to my request to view them by locking down the whole collection for seventy years on the grounds of patient privacy.[4]

Pace Tulane, I think there might be good reason to jam a foot in the door that was so briefly opened on that February afternoon. In directly experimenting with the primordial emotions of pain and pleasure, Heath inhabited a particularly concentrated and consistent version of the utilitarian ethos that motivated all American medical research in the period. His experimental research ethics, his psychological framework, his methods, and his results were all based in the utilitarian calculus. When American Cold War medical utilitarianism came under sustained and effective attack between 1963 and 1973, resulting in the mandating of informed consent across all types of medical practice,

Heath's research program was doomed to obsolescence. This chapter, in keeping with one of the dominant themes of the volume, will link these events to the ongoing assessment and reassessment of the lessons of the Holocaust. Specifically, I suggest that the trial of Adolf Eichmann in 1961 and Stanley Milgram's "obedience-to-authority" experiment that was inspired by it represent a turning point in the history of neuroscientific research on human emotions.

The Nuremberg Doctors' Trial and the Doctrine of Informed Consent

The reforms that closed down Heath's research were unquestionably a legacy of the German defeat of 1945. The proposal that informed consent should operate as a universal and absolute regulatory mandate for human subjects research originated at the Nazi Doctors' Trial.[5] It was the slow incorporation of informed consent into American medical research—explicitly and repeatedly cited as a lesson of the Holocaust—that brought Heath's research on pain and pleasure to its ignominious end. Our story therefore begins on December 9, 1946, when an American tribunal initiated proceedings against twenty-three German physicians and administrators who were charged with "plans and enterprises involving medical experiments without the subjects' consent . . . in the course of which experiments the defendants committed murders, brutalities, cruelties, tortures, atrocities, and other inhuman acts."[6]

Some eight months later, the judgment was handed down. Seven of the twenty-three accused were hanged; a further nine served long prison sentences; seven were acquitted. The judgment concluded with enumeration of a ten-point code of experimentation ethics now known as the Nuremberg Code, the first and primary provision of which reads:

> The voluntary consent of the human subject is absolutely essential. This means that the person involved should have the legal capacity to give consent; should be so situated as to be able to exercise free power of choice, without the intervention of any element of force, fraud, deceit, duress, over-reaching, or other ulterior form of constraint or coercion; and should have sufficient knowledge and comprehension of the elements of the subject matter involved as to enable him to make an understanding and enlightened decision. The latter element requires that before the acceptance of an affirmative decision by the experimental subject there should be made known to him the nature, duration, and purpose of the experiment; the

method and means by which it is to be conducted; all inconveniences and hazards reasonable to be expected; and the effects upon his health or person which may possibly come from his participation in the experiment.[7]

The doctrine of informed consent, exactly as laid out above, has since proved to be an extraordinarily robust and intuitive moral principle, one around which medical research the world over is gradually being restructured.

This origin story is a truism, but it can be misleading. Informed consent was not immediately integrated into research ethics, even in Germany. It took more than three decades to become a regulatory reality in Germany and the United States; other countries are slowly catching up. The Nuremberg Code itself was at once a statement of the obvious and a performative intervention of the utmost delicacy. One of the most ticklish matters facing the American prosecutors was how to counter the defendants' claim that the experiments they had conducted were ethically no different from research conducted by the Allies during the war. Both sides undertook high-altitude experiments, low-temperature experiments, and infectious disease research as part of the war effort. Both sides conducted research that was dangerous, deceptive, and coercive and justified doing so on the grounds of protecting the lives of their respective troops. The differences, the defense argued, were those of degree rather than kind. How, then, to bootstrap into existence a case against the Nazi doctors without catching American research practices in its coils?

The prosecutors consulted the American Medical Association (AMA), which voted to send in Dr. Andrew C. Ivy, a leading medical researcher in physiology and pharmacology, to make the preliminary report. Ivy had conducted human subjects research during the war with conscientious objectors and was an energetic defender of vivisection. Above all, he was perceived as someone who could be trusted to protect the interests of the American medical profession. Ivy's was a sensitive task: he had to formulate a code of ethics, assert that it was universally applicable, and make it seem as though the US wartime Committee on Medical Research had been adhering to it all along. In July 1946 he went to Germany to meet with the Nuremberg prosecution team. On his return he made a presentation to the AMA and submitted a report in which he laid out the rules of human experimentation, stating that these standards had been "well established by custom, social usage and the ethics of medical conduct."[8]

On December 11, 1946, two days *after* the opening statements in the

Nazi Doctors' Trial, the AMA adopted a skeleton version of Ivy's rules: "The voluntary consent of the person on whom the experiment is to be performed must be obtained; the danger of each experiment must be previously investigated by animal experimentation, and the experiment must be performed under proper medical protection and management." (The AMA did have a code of ethics, but it was a hundred-year-old manual of professional etiquette devoid of relevant content.) In June 1947, under cross-examination by the defense about this timing, Ivy asserted that, even though these standards had only recently been *formally* adopted by the AMA, they nonetheless represented "a matter of common practice" in American medicine. The Doctors Trial ended in August 1947 with a judgment that included the Nuremberg Code.[9]

Despite the fact that American judges had handed down the Nuremberg Code and the AMA had formally adopted its most important provisions, American researchers mostly continued to ignore it, extending World War II standards of research ethics deep into the Cold War. Ivy's assertion that informed consent was "a matter of common practice" among American medical investigators was downright false. In fact, the period between 1945 and 1972 was marked by tens of thousands of hazardous and invasive procedures conducted on uninformed and unconsenting human subjects and funded by the National Institutes of Health, the CIA, the Department of Public Health (DPH), and other American government agencies.[10] After scores of formal investigations, we do not yet know the full scope of such research; evidence is still surfacing four decades later.[11]

Various explanations have been offered for this flagrant example of victors' justice. The historian David Rothman has surmised that the American medical profession regarded the Nuremberg judgment as a "code for barbarians" that did not apply to their beneficent efforts to understand and conquer human disease.[12] The health-law scholar George Annas has suggested that Cold War imperatives such as techno-scientific competition with the Soviet Union may have played a part in trumping ethical niceties.[13] At least some of the scale of medical research after 1945 is also attributable to a more primordial appetite for its fruits. The mass production of penicillin, which was initiated as part of the war effort in 1943, created a sense of limitless therapeutic optimism. In his 1945 polemic *Science, the Endless Frontier,* Vannevar Bush opened his argument for research funding with the story of penicillin and called for a peacetime "war on disease."[14] In the mid-1950s, trials of the polio vaccine successfully enlisted 1.8 million children across the nation, a measure of their parents' trust in the beneficence of American medical research. The trials were successful, and the vaccination program was launched, eradicating one of the most dreaded diseases of the era. The war on disease looked set for a series of historic victories.

Starting in the early 1960s, however, a series of medical scandals broke in the media, belying Andrew Ivy's assurances at the Nuremberg Doctors' Trial that Allied researchers followed the rules of informed consent. It began to seem as if the heinous activities engaged in by the Nazis for the sake of racial hygiene were still being enacted by the Americans in the name of the greater good. The ideology may have been different, but the research ethics looked perilously similar. Eventually, the widening net of condemnation caught up Robert Heath and his fellow human behavior researchers. These scientists' materialist and functionalist approach to pain and pleasure was perfectly aligned with the utilitarian calculus that justified US Cold War medical research methods. Their science of motivation, no less than their consequentialist ethics, was underpinned by a hedonist psychology of appetites and aversions, rewards and punishments, satisfactions and irritations. In the context of the movement for informed consent, Heath and his colleagues' utilitarian psychology was therefore condemned as a totalitarian science of behavior control. The story of how pain-pleasure research was brought down by the belated American implementation of the Nuremberg Code brings into focus how the long shadow of events in Nazi Germany was cast over the science of affect in postwar America.

Comparisons with Nuremberg

The activist critique of pain-pleasure research came late in the movement for the reform of research ethics, after a decade of increasingly strident and devastating comparisons between American medicine and its Nazi counterpart. American medical research began to be called to question in the context of the intensifying public discussion of the crimes of the Third Reich during the trial of Adolf Eichmann. In May 1960, Israeli secret service operatives arrested the Nazi war criminal, and, in 1961, his trial in Jerusalem became a worldwide media sensation and a turning point in the historical understanding of the Holocaust. Not only was it on the front pages of the newspapers every day for its entire eight-month duration, an unprecedented media saturation, but the Israeli judges also placed the victims' testimony at the center of the proceedings, bringing a whole new urgency and vividness to the public understanding of events in Germany.[15] This increased sensitivity to the moral implications of Nazi atrocities quickly found its way into the conversation about the ethics of medical research. In 1963, a cancer researcher by the name of Chester Southam made the mistake of recruiting a group of Jewish doctors to help him with his studies, prompting, for the first time since the Doctors' Trial, a direct comparison between American utilitarian medicine and fascist human experimentation.

Pursuing a theory about the immunology of cancer, Southam was injecting live cancer cells into human subjects to see whether and how the cells were rejected. To test his theory, he calculated that he needed three different groups of subjects: cancer patients, healthy subjects, and sufferers from noncancerous terminal illnesses. The first group he found at his home institution, the Sloan Kettering Institute for Cancer Research. Southam told these patients that he was injecting them with "cells," not mentioning cancer. The second group—the healthy volunteers—he recruited at the Ohio State Penitentiary. These men were given the benefit of full informed consent and hailed in the local press as heroes. In embarking on the third phase of the research in 1963, Southam approached the administrators at the Jewish Chronic Diseases Hospital in Brooklyn to request access to a group of terminal noncancer patients. The director of medicine agreed, but, when he instructed his staff to give the injections without telling the patients that they were cancer cells, three young Jewish doctors refused. When the experiment went ahead, the doctors resigned.[16]

At a meeting of the hospital's board of directors, amid heated argument about what exactly the patients had been told about the injections, the chairman "called attention to certain of the Nuremberg Trials in which Nazi doctors were found guilty and some hanged . . . for using human beings for experimental purposes without their informed consent." After the meeting, a member of the board, a lawyer named William Hyman, sued the hospital for access to the records relating to the study. In his petition, Hyman stated that Southam was injecting cancer cells into noncancerous patients "for the purpose of determining whether cancer can be induced," a subtle but devastating misrepresentation of the research objectives.[17] Another affidavit in the lawsuit referred to "acts that belong more properly in Dachau."[18]

In 1965, the attorney general of New York State conducted a hearing to determine whether to suspend Southam's license. These proceedings, at the midpoint of the 1960s, show a hardening of the legal and ethical rejection of utilitarian research. The attorney general characterized Southam's research as "in no way therapeutic . . . an experiment relating to cancer research which had as its ultimate intention the benefit of humanity." In the context of the hearings, the "ultimate intention to benefit humanity" was framed as a ruthlessly utilitarian objective that made it "incumbent upon the respondents to have seen to it that ALL information connected with the experiment was given."[19] During the proceedings, Southam was asked about a statement that he had made to *Science* magazine, saying that, although he was sure that the research was without danger to the subjects, he did not inject himself because, "let's face it, there are relatively few skilled cancer researchers, and it seemed stupid to take even the little risk."[20] Under cross-examination, he quibbled about the exact wording but agreed that "the philosophy is correct." This "philosophy"—that risk to subjects should be weighed in the balance against the potential contribution to society—was perfectly consistent with a rigorously utilitarian approach to research. By the time of the licensing hearings, however, Southam's opinion of his own disproportionate social value ran up against a legal system radicalized by the civil rights movement. As the attorney general couched the case against Southam: "Every human being has an inalienable right to determine what shall be done with his own body . . . a right to know what was being planned . . . a right to be fearful and frightened and thus say NO."[21]

At the same time as Southam's licensing hearing, the crusading anesthesiologist Henry Beecher summoned members of the press to hear a paper that he was giving at a two-day symposium on clinical research. According to the *New York Times*, Beecher's paper "Ethics and the Explosion of Human Experimentation" cited twenty "breaches of ethical conduct" by American clinical researchers who had used "hospitalized patients, military personnel, convicts, medical students, and laboratory personnel for experiments in which the subjects are not asked for their permission."[22] Examples included the twenty-three charity patients who died when the treatment for typhoid fever was experimentally withheld from them, the fifty healthy inmates of a children's center who were given repeated doses of a drug that caused abnormal liver function, and the eighteen children undergoing heart surgery who received unnecessary experimental skin grafts. For reasons of professional solidarity and discretion, Beecher did not identify the scientists in question, but he hinted that other such cases would be easy to find. By the time the paper was published in the *New England Journal of Medicine* the following year, the number of cases Beecher presented had risen to twenty-two.

In his exposé, Beecher characterized the problem as "experimentation on a patient not for his benefit but for that, at least in theory, of patients in general."[23] Inasmuch as the researchers involved in these scandals advanced any ethical justification for their experiments, it took the form of a utilitarian calculus. Time and again, deceptive and coercive medical research on terminal, marginal, poor, and institutionalized human subjects was justified on the grounds of the benefit that might accrue to future generations. Under the terms of this calculus, anyone with little enough to lose was fair game in the race for scientific glory. It was this type of reasoning that the reform movement dedicated itself to overthrowing.

After the publication of Beecher's paper, various committees and commissions were charged with investigating the scandals that broke with disconcerting regularity in the mass media in the following decade. In 1970, the word *bioethics* was coined, and departments of law and philosophy at select universities began to offer courses in medical issues. Medical schools slowly started to incorporate ethics into their curricula. Whole new forms of medical governance came into existence, including the institutional review board and the hospital ethics committee. All these responses to the crisis of moral legitimacy in American medicine converged on one deceptively simple recommendation: the same principle of informed consent that topped the Nuremberg Code.

In July 1972, the most explosive outrage of all burst onto the pages of the newspapers. A venereal disease researcher at the DPH informed a journalist friend of his about a DPH study that by that point had been going on for forty years. After a screening process disguised as a medical outreach initiative in Macon County, Alabama, near Tuskegee, 399 disenfranchised African American sharecroppers suffering from syphilis had been selected for a study of the untreated disease. What was arguably justifiable in 1932 on the grounds of the dubious efficacy and severe side effects of the available treatments became unconscionable after penicillin went into mass production in the 1940s. Treatment was nonetheless still withheld for another thirty years, during which time some of the men died, some of their wives contracted the disease, and some of their children were infected at birth.[24]

In the newspaper coverage, a researcher from the Centers for Disease Control described the study as "almost like genocide."[25] On July 30, the *New York Times* science writer noted: "The Tuskegee Study . . . was begun the year Hitler came to power. It was Hitler's atrocious 'experiments' done in the name of medical science which led after World

War II to the promulgation of the Nuremberg Code."[26] In its targeted victimization of a specific ethnic group, the Tuskegee study set off a conflagration of public outrage, and in the aftermath many people condemned American medical research as no better than its wartime German counterpart. From the resignation of the doctors at the Jewish Chronic Diseases Hospital to the racism undergirding Tuskegee, utilitarian American research repeatedly invited comparison with medicine under National Socialism. It was this conflation between utilitarianism and fascism that would catch Robert Heath unawares and consign his research to the dustbin of history.

Obedience to Authority

To the background story of Heath's downfall I wish to add one more element. One of the many people whose understanding of the Holocaust was crystallized by the Eichmann trial was a young researcher at Yale by the name of Stanley Milgram, whose obedience-to-authority experiment recast the problem of totalitarianism in an American idiom and thereby taught the generation that came of age in the 1960s to see American psychology as latently fascist.[27] The child of Eastern European Jews who had come to the United States in the first decades of the twentieth century, Milgram produced a Ph.D. dissertation probing the differences in "social conformity" between Norwegians and French people. These experiments were cast as a series of investigations into "national character" but were also a manifestation of the free-ranging anxieties of the 1950s about organization men, authoritarian personalities, and the relationship of the individual to the crowd. Sometime in the spring of 1960, inspired by Eichmann's arrest, Milgram extended his conformity research into a new experimental design, wanting to see whether "groups could pressure a person into . . . behaving aggressively towards another person, say by administering increasingly severe shocks to him."[28] In October and November of that year, he sent preliminary letters of inquiry to three government agencies about the prospects of grant support for his research into obedience. His students built a fake "shock box," and he ran some preliminary studies using Yale undergraduates as subjects.

Subjects were told that they were involved in a study of punishment and learning, that they would be paired with another participant, and that lots would be drawn to determine who played the role of "teacher" and who played the role of "learner." What they did not know was that the other participant was a confederate of the experimenter's and that both lots indicated *teacher*, ensuring that they would be cast in this role. After the two men were treated to a brief theoretical lecture about the role of punishment in learning and memory, the learner was led away, and the subject was seated in front of the shock box, an authentic-looking piece of equipment with a series of thirty switches labeled from 15 to 450 volts, grouped into batches of four switches with labels running from SLIGHT SHOCK to SEVERE SHOCK, the last two simply and ominously labeled XXX. They were then instructed to administer a simple word-association test and to punish the learner's wrong answers with shocks of increasing severity. The subjects were unable to see the learner, but his cries of anguish (prerecorded for consistency) were clearly audible. Famously, about a third of participants continued with the experiment up to the last set of switches.

Milgram claimed: "[My experiment] is highly reminiscent of the issue that arose in connection with Hannah Arendt's book *Eichmann in Jerusalem*. Arendt contended that the prosecution's effort to depict Eichmann as a sadistic monster was fundamentally wrong, that he came closer to being an uninspired bureaucrat who simply sat at his desk and did his job. After witnessing hundreds of ordinary people submit to the authority in our own experiments, I must conclude that Arendt's conception of the *banality of evil* comes closer to the truth than one might dare imagine."[29] He maintained, with some justification, that he had designed an experiment revealing the universal psychological traits linking ordinary Connecticut folk to the defendant in Jerusalem. For the purposes of this chapter, what is interesting about his experimental protocol is its relation to Cold War American neuroscience.[30] The cultural authority of the scientist in the experiment was derived from his deployment of the psychology of reward and punishment. The subject was told: "Psychologists have developed several theories to explain how people learn various types of material. . . . One theory is that people learn things correctly whenever they get punished for making a mistake."[31] The banality of American evil, in Milgram's performance, was not the fascist preoccupation with racial hierarchies and the purity of the genetic pool but its ideological opposite, the dedicated environmental determinism of American pleasure-pain research. I want to suggest that Milgram's famous experiment laid the groundwork for the destruction of Heath's reputation by priming activists to see his research as latently or explicitly totalitarian.

The Kennedy Hearings

It was in the aftermath of the revelation of the Tuskegee study that Robert Heath was caught in the widening net of the movement for informed consent. Over the course of six days in February and March 1973, scores of witnesses were summoned to the Dirksen Senate Office Building near Capitol Hill to testify as to the ethics of human subjects research. Men who had participated in the Tuskegee study gave evidence about the deceptions and inducements that drew them into it. Women on welfare who had been forced onto the birth control drug Depo Provera described their experiences. The rabble-rousing British journalist Jessica Mitford gave evidence about experimentation on prisoners. The most frequently recurring theme was that of the sharp social inequities that marked the selection of human subjects for medical research. As one journalist noted during his testimony: "The earliest, riskiest and often shoddiest tests are conducted on the most helpless members of society: the poor, the retarded, the institutionalized."[32]

On February 23, the subject under discussion was research on the human brain, especially the recent development of techniques for behavior control. At 10 A.M., Edward Kennedy called the meeting to order and delivered his opening remarks: "The nature and functioning of the human mind has fascinated scientists for centuries. In recent years, they have begun to understand that this is the basis of behavior and have developed tools and techniques to modify and control it." He then summarized the controversy about these new tools and techniques: "There are those who say that the new behavioral research will enable us to realize our full potential as a nation and as a people. There are others who believe that the new technology is a threat to our most cherished freedoms." He closed his speech with the hope that "today's hearing will air both sides of the controversy and help us as a society to come to understand and master this new technology so as not to become the victims of it."[33]

In the hours that followed, some of the most famous neurologists in the country were summoned to give testimony about their research on behavior modification techniques—psychosurgery, electrical stimulation of the brain, and stimulus response conditioning. Summaries of research findings were presented. Amazing therapeutic results were recounted. Ethical analyses were conducted, ending in resounding justifications. Kennedy was repeatedly reassured that the patients always benefited from the treatment and that their consent was always sought, albeit informally.

In the afternoon, Robert Heath took the stand. He testified that a total of sixty-five human subjects had undergone the implantation of electrodes or cannulas deep in their brains in his laboratory at Tulane. The subjects were chosen on the grounds of "failure of all existing treatments and a prognosis, without this intervention, of long-term, virtually complete disability, usually requiring permanent institutionalization." The electrodes, he explained, "remain securely in the brain sites for many months," allowing him and his team "to build a meaningful bridge between mental activity and physical activity of the brain, the organ of behavior."[34] Therapeutic application of his technique consisted of artificial stimulation of the pleasure-pain pathways.

It had long been the practice of the Tulane laboratory to film the stimulation experiments, and, as part of his testimony, Heath screened three of these productions. As the first film began to roll, he explained to Kennedy what they were watching:

> This is the start of the stimulation. The patient begins to cry out with rage. He tells the doctor who is nearest to him, a neurosurgeon, that he is going to kill him; that he wants to murder him. As soon as the stimulus goes off, this emotional state disappears. We ask him why he wanted to kill the doctor, and he answers that he had no awareness of wanting to do this; that he likes him. What we were able to do was to induce this violent reaction that may occur spontaneously in association with some types of behavioral pathology.

It must have been a potent display. After watching the films, Kennedy continued with his questioning. "You have shown and testified about how you can replicate pain and pleasure by the implantation of these electrodes in different parts of the brain," he said. Heath assented. Kennedy went on: "This is behavioral control. As I understand it, you are trying to use this technique to treat people." Again, Heath agreed. Kennedy then asked: "Would it not be adaptable to treat other people as well, normal people?" To this Heath replied: "I think it would be, but I think normal adaptive people are already being treated. I am sure Dr. Skinner is going to talk on that. Our learning experiences, our attitudes are modified every day."[35]

Shortly afterward, B. F. Skinner took to the floor and elaborated on the continuity between Heath's control of behavior through direct stimulation of the brain and his own less invasive methods: "The control of human behavior through drugs, psychosurgery, or electrical stimulation naturally attracts attention, but far more powerful methods have been in existence as long as the human species itself and have been in existence throughout recorded history. . . . Behavior is selected and strengthened by its consequences—by what the layman calls rewards and punishments, and this fact has long been exploited for purposes of control."[36] Skinner then argued that punitive control had hitherto been the source of our insights into human conditioning but that the time had come to switch over to positive reinforcement, which was both more humane and more effective.

Arrayed against this remarkable concentration of neuroscientific expertise was one loudly dissenting voice, that of the antipsychiatry activist Peter Breggin. In 1970, Breggin had set up an organization called the Project to Examine Psychiatric Technology. In this capacity he embarked on a crusade against behavior modification techniques, especially psychosurgery and electrical stimulation of the brain. A barrage of articles in both the popular press and medical journals resulted in some significant political victories, and, by 1973, he had an impressive track record behind him of impeding the flow of funds for neuroscientific research. Against those who believed that the lobotomy operation had been shelved with the advent of antipsychotic drugs, Breggin argued that the late 1960s had seen a tremendous resurgence of psychosurgery and that hundreds, if not thousands, of patients every year, many of them children, were undergoing brain surgery and electrical stimulation of the brain as experimental treatments for sometimes fairly mild psychiatric disorders.

Before getting into the specifics of contemporary research, Breggin delivered a sermon on the ethical, spiritual, and political evils of behavior modification. His target, he announced, was the "mechanistic, anti-individual, anti-spiritual view," which "gives justification to the mutilation of the brain and the mind, in the interests of controlling the individual." For Breggin, the root of the problem lay in the "totalitarian" outlook of psychologists such as Skinner, whose work, he asserted, "ridicules the basic American values: Love of the individual, love of liberty, personal responsibility, and the spiritual nature of men." His language suggested that Milgram's performance of authoritarian behaviorist psychology was not just an ingenious experimental prank but an urgent contemporary reality: "These men, I believe, are doing nothing more than giving us a new form of totalitarianism 'in medical and ethical language.' The reliance on professional ethics and medical control over these issues leaves physicians in charge of the situation.

It creates for themselves an elitist power over human mind and spirit. If America ever falls to totalitarianism, the dictator will be a behavioral scientist and the secret police will be armed with lobotomy and psychosurgery."[37]

Breggin's misgivings about "totalitarian" neuroscience were widely shared at the time, but what is clearer in hindsight is the ideological distinctiveness of American Cold War brain science. What Heath's and Skinner's testimony made visible was the extent to which they operated not only according to the same utilitarian ethics as their colleagues in the medical sciences but also within a totalizing utilitarian conception of the human condition. Out of their commitment to neo-Darwinian materialism, these men conceived of pain and pleasure as the primary means by which all vertebrates, including humans, successfully negotiated their environments. This hedonist psychology of appetites and aversions was confirmed either by direct investigation of the affective circuits of the nervous system, as in Heath's electrical stimulation experiments, or by ordinary reward and punishment, à la Skinner. Underwriting all this activity was a utilitarian ethics in which the rightness or wrongness of a particular action was to be judged by a cost-benefit analysis of its consequences. The neat fit between his ethics, psychology, and methodology made Heath a particularly potent symbol of all that was wrong with Cold War medical research in the age of regulatory reform.

Conclusion

The consistency and completeness of Heath's and Skinner's psychological utilitarianism mark the apogee of a particular vision of the relationship between human emotions and scientific ethics. The informed-consent revolution in human subjects research broke apart that vision by insisting on a model of human rights and obligations at odds with utilitarianism. Under the philosophy of informed consent, each human person is a rights-bearing, rational, sacred individual whose autonomy must be respected before even the most beneficent physical intervention can begin. This is the Kantian antipodes of the Benthamite psychology that underpins behaviorism.

The informed-consent mandate originally emerged from a radical critique in which liberal denunciations of totalitarianism were turned against the liberal establishment itself. Over the course of the next two decades, as consent was integrated into the bureaucratic regimes of everyday medical practice, the critique lost its crusading edge. By the end of the 1980s, with neoliberalism in the ascendant, the circle had been neatly squared, and hedonist psychology was reconciled with individual rights under the sign of the ideal consumer: a pleasure-seeking individual exercising her freedom of choice through access to product information. When it came to reforming the science of the most primal emotions of pain and pleasure, the shadow of the Holocaust served to both inspire and obscure. Although invidious comparisons with Nazi medicine were necessary to shake Cold War American medical research out of its lethal complacency, the activists' conflation of totalitarianism and utilitarianism may have left the way open for a declawed utilitarianism to thrive unchallenged after the Cold War was won. Perhaps the time has come to revisit these debates. As global consumer culture threatens to tip us over into irreversible ecological crisis and the war on drugs undermines the legitimacy and stability of producer nations from Peru to Afghanistan, the politics of pleasure have surely never been more urgent.

NOTES

1. "Quality of Health Care—Human Experimentation," ed. Committee on Labor and Public Welfare, Hearings Before the Subcommittee on Health of the Committee on Labor and Public Welfare (Washington, DC: US Government Printing Office, 1973), 364, 366–67, 369–73.
2. Robert Heath, *Exploring the Mind-Brain Relationship* (Baton Rouge: Moran Printing, 1996).
3. Onora O'Neil, *Autonomy and Trust in Bioethics* (Cambridge: Cambridge University Press, 2002); Susan Singleton, *Data Protection: The New Law* (Bristol: Jordan, 1998); George P. Smith, *The New Biology: Law, Ethics and Biotechnology* (New York: Plenum, 1989); Cosimo Marco Mazzoni, ed., *A Legal Framework for Bioethics* (The Hague: Kluwer Law International, 1998).
4. Personal communications from Michael Bernstein, provost of Tulane University, September 2009–May 2010.
5. George Annas and Michael Grodin, eds., *The Nazi Doctors and the Nuremberg Code: Human Rights in Human Experimentation* (New York: Oxford University Press, 1992).
6. Library of Congress Federal Research Division, "Trials of War Criminals Before the Nuernberg Military Tribunals under Control Council Law No. 10, October 1946–April 1949," http://www.loc.gov/rr/frd/Military_Law/NTs_war-criminals.html.
7. Annas and Grodin, eds., *The Nazi Doctors*, 2.

8. Department of Energy, "Final Report of the Advisory Committee on Human Radiation Experiments," chap. 2, "The American Expert, the American Medical Association, and the Nuremberg Medical Trial," http://www.hss.energy.gov/healthsafety/ohre/roadmap/achre/chap2_2.html.
9. Ibid., 5–6.
10. The exact numbers are not known, but the government investigation into Cold War radiation experiments, which was undertaken in the 1990s, uncovered evidence of over four thousand separate studies. See ibid., "Executive Summary—Key Findings."
11. For example, in October 2010, Barack Obama formally apologized to the Guatemalan government for an American Public Health Department study conducted in 1946–48 in which more than thirteen hundred people were deliberately infected with syphilis and other venereal diseases in order to test penicillin's efficacy. Donald G. McNeil, "U.S. Apologizes for Syphilis Tests in Guatemala," *New York Times*, October 1, 2010.
12. David Rothman, *Strangers at the Bedside: A History of How Law and Bioethics Transformed Medical Decision Making* (New York: Basic, 1991).
13. George Annas, "Legacy of the Nuremberg Doctors' Trial," *Minnesota Journal of Law, Science and Technology* 10.1 (2009): 32.
14. Vannevar Bush, *Science, the Endless Frontier* (Washington, DC: US Government Printing Office, 1945).
15. Deborah E. Lipstadt, *The Eichmann Trial* (New York: Schocken, 2011), 192–93.
16. Elinor Langer, "Human Experimentation: Cancer Studies at Sloan-Kettering Stir Public Debate on Medical Ethics," *Science* 143.3606 (1964): 551–53.
17. Jay Katz, Alexander Morgan Capron, and Eleanor Swift Glass, *Experimentation with Human Beings: The Authority of the Investigator, Subject, Professions, and State in the Human Experimentation Process* (New York: Sage, 1972), 12, 11. While the injections did produce cancerous "nodules," Southam was confident that the cells would be rejected; he just wanted to see whether the noncancer terminal patients rejected them quickly, like his healthy subjects from the penitentiary, or slowly, like his cancer patients at Sloan-Kettering.
18. Ibid., 16.
19. Ibid., 47.
20. Langer, "Human Experimentation," 551.
21. Katz, Capron, and Glass, *Experimentation with Human Beings*, 49, 47.
22. John A. Osmundsen, "Physician Scores Tests on Humans: He Asserts Experiments Are Done without Consent," *New York Times*, March 24, 1965.
23. Henry Beecher, "Ethics and Clinical Research," *New England Journal of Medicine* 274 (1966): 1354–60, 1355.
24. Harriet Washington, *Medical Apartheid* (New York: Doubleday, 2006).

25. James T. Wooten, "Survivor of '32 Syphilis Study Recalls a Diagnosis," *New York Times*, July 27, 1972.
26. Jane E. Brody, "All in the Name of Science," *New York Times*, July 30, 1972.
27. Stanley Milgram, *Obedience to Authority: An Experimental View* (New York: Harper & Row, 1974).
28. Thomas Blass, *The Man Who Shocked the World: The Life and Legacy of Stanley Milgram* (New York: Basic, 2004), 62.
29. Ibid., 269.
30. Ian Nicholson, "'Shocking' Masculinity: Stanley Milgram, 'Obedience to Authority' and the 'Crisis of Manhood' in Cold War America," *Isis* 102.2 (2011): 238–68. This is the most historically contextualized treatment of Milgram's experiment, but it does not explore the critical implications of the behaviorist setup.
31. Blass, *Man Who Shocked the World*, 77.
32. "Quality of Health Care—Human Experimentation," 803.
33. Ibid., 337.
34. Ibid., 364.
35. Ibid., 367, 368.
36. Ibid., 369.
37. Ibid., 358.

SIX

Oncomotions: Experience and Debates in West Germany and the United States after 1945

BETTINA HITZER

In 1958, Pater Bolech, a Catholic priest working in an Austrian cancer ward, wrote an article to be published in a journal for oncologists describing what he perceived as peculiar with cancer patients. The essence of the article is summed up in the following words: "Attending to cancer patients, one is always amazed at their calm and quiet manner. One is astonished at their courage, their patience and the inner strength in enduring great pain. . . . Cancer patients are very good and compliant patients."[1] Fifty years later, in 2008, the German theater and art film director Christoph Schlingensief wrote in his cancer autobiography about his feelings some days after being diagnosed with lung cancer: "I am feeling aggressive, but actually I am dead. This evening, I feel I could walk the streets with a club and smash everything to bits. I am so angry, so very offended and hurt by this thing. At the age of 47. It's really incredibly insulting!"[2]

These are two very different, even contradictory emotional reactions to suffering from cancer. Of course, one could try to explain these striking disparities by pointing to the different social and cultural backgrounds of the two men or by pondering on their diverging perspectives, the first in a very circumscribed outsider position, the second affected in his sheer existence, moreover possibly in a state of shock directly after receiving his cancer diagnosis. One could also claim that being diagnosed with lung cancer with a particularly bad prognosis is different from being diagnosed with other types of cancer. While all this might be true, I will argue that there is a more general, sometimes very subtle story underlying and explaining these differences in "oncomotions" separated by half a century of shifting emotional underpinnings to cancer.

This holds true not only with regard to the experience of suffering from cancer or caring for the soul of cancer patients. Emotions or the way in which emotions have been tackled and conceptualized also influenced public health and research strategies as well as the doctor-patient relationship, therapeutic choices, and, eventually, public debates surrounding the "right" way to handle and speak of cancer and cancer-related experiences. I shall demonstrate that the feelings that can be identified in these different fields are by no means side effects or merely minor aspects. Quite the contrary, I will show that they are essential to understanding and connecting changes and continuities within the "unnatural history" of cancer.[3] But are these emotions and the manner in which they were dealt with specific to the history of cancer? I doubt it, and, therefore, I shall discuss whether and how they were also linked to more general trends in the history of emotions.

In my view, three major emotional regimes in the realm of oncomotions can be discerned in the years following 1945. As these shifting emotional regimes are very complex phenomena that deserve to be treated at length, I will here focus on a single regime, namely, that directly after World War II, and the first major shift, which occurred during the 1960s.[4]

I chose to investigate the entangled histories of West Germany and the United States as a particularly interesting and insightful example of what I termed *oncomotions*. The United States has had a very strong, outspoken, and influential "cancer culture" since ca. 1900 and especially since the 1940s, when the American Cancer Society was modernized. The American way of explaining, researching, and speaking of cancer has been ever since a kind of controversially discussed but, nevertheless, important model for most European societies. At least from the 1920s, there had been a close link to Germany after some German (or German-speaking) cancer researchers emigrated to the United States. After 1945, this relationship (with what was then West Germany) intensified even though research traditions and the more general approach to cancer in society had been very different in some respects in the two countries. In the postwar era, through transnational exchange these traditions partially merged and partially remained unchanged. At this point, it looks like these connections between West Germany and the United States were the strongest in Europe, stronger than those between neighboring European countries, even though further research is needed to determine the precise nature, direction, and volume of traffic between different countries and national scientific communities as regards cancer.

Who's Afraid of Fear? Postwar Emotions and the Cancer Experience

The period from roughly 1945 to the early 1960s was characterized by a cancer-related emotional regime that centered mainly on fear, insofar as fear became a greatly contested terrain. This was not completely new. Fear had been an important issue in dealing with cancer even earlier. From the very onset of early detection campaigns around 1900, cancer activists both in the United States and in Germany (as well as in other European countries) discussed whether it would be a useful tool in order to get the "Do not delay!" message through.[5] Nevertheless, fear was at that time more closely linked to moral conviction and blame because "excessive" or "unnecessary" cancer fear was considered the cause for delaying early detection. Since these campaigns, at least in the United States, focused very much on breast and cervical cancer, it was mostly women who were thought to be more vulnerable to cancer fears.[6]

After 1945, however, a conceptualization of cancer fear as potentially harmful to the body or even as pathological in itself became more accentuated and widespread. The then popular notion of "cancerophobia" was dominated by the paradoxical assumption that, on the one hand, instilling fear had to be prevented at all costs but, on the other hand, medical professionals and health officials were convinced that fear could not be avoided the minute the word *cancer* was uttered or even just insinuated. This paradox gave rise to a whole range of restraints and aporias that determined the handling of cancer by doctors, psychologists, priests and ministers, and cancer patients and their relatives. But it also defined the way in which one could speak or write of cancer in public and also influenced the design of public health campaigns focusing on early detection, which was a major issue in those days.

Those campaigns did not tire of claiming that cancer was curable if detected early in order to counteract popular beliefs and anxieties about the inevitable fatality of cancer. "Inventing a curable disease" was thus the major aim of almost all early detection campaigns, whether in the United States or elsewhere.[7] But, when asked in private, oncologists were much more pessimistic about the potential of cancer to be cured. Karl Heinrich Bauer, one of the founders of the Federal German Cancer Research Center in Heidelberg, admitted in a nonpublic memorandum in the early 1960s that more than 80 percent of cancer patients would ultimately succumb to the disease, while his colleague Heinrich Martius, the president of the German Society for Fighting Cancer, furiously argued against publishing these statistical data.[8] This secret knowledge should be taken into account when considering the way medical doctors acted toward cancer patients since it was not only the patients—as often argued—who received a cancer diagnosis as a kind of death sentence but most physicians as well.[9]

Equating cancer with death placed physicians in a burdensome situation when facing a patient's cancer diagnosis. However, as I would argue, it was a slightly different assumption that forced doctors to usually conceal a cancer diagnosis in the 1950s: they were convinced that suffering from cancer was a synonym for hopelessness, an emotion that no one could bear and, furthermore, contrasted sharply with a physician's self-perception of being in control of the situation. Carly Seyfarth, who wrote a widely used code of conduct for German physicians, published in several editions between 1935 and 1946, was unequivocal as to this point. He declared: "We should *conceal* the seriousness of the patient's condition and the prognosis that there is no hope for a cure. . . . *Under no circumstances should we reveal to a patient that their suffering is hopeless and that they are going to die.*"[10] Therefore, the physician should never ever utter the word *cancer* when talking to a patient. Instead, he or she should use the term *epitheliom* (a tumor that could be benign).[11]

Seyfarth's admonition was taken seriously. "Benign deception" was apparently common practice in the 1950s not only in West Germany but also in the United States.[12] Two American studies of 1953 and 1961 revealed that the overwhelming majority of US physicians did not tell their patients the truth about their cancer diagnosis.[13] Similarly to their German colleagues, American physicians relied on euphemisms like *lesion, mass,* or *growth*—in most cases reassuring their patients that this growth was benign even if they knew this was not the case. When asked why they would deceive their patients, physicians emphasized that they wanted to "sustain and bolster the patient's hope." Hopelessness scared physicians not only because of their self-perceptions; drawing on older, usually vague and unsubstantiated concepts of mind-body medicine, they argued that hopelessness would at best interfere with the healing process and at worst lead patients to commit suicide—even though very few could cite firsthand experience.[14] However, Donald Oken, who led the 1961 study and would later become the editor in chief of *Psychosomatic Medicine,* wondered in interpreting this data "whether the expectation of death insurmountably deprives the patient of hope."[15] This reasoning pointed to a new emotional understanding of cancer, death, fear, and hope that was well under way in the early 1960s.

Before further inquiring into this shifting emotional regime, it is worth asking whether this cancer taboo was something that would characterize the 1950s more than former times. Is it not just another place in the long-standing "silent world of doctor and patient" that Jay Katz tellingly described in his 1984 monograph?[16] With regard to the US context, Aronowitz has stressed the more modern character of a cancer taboo with surgeons starting to give evading diagnoses only in the early twentieth century.[17] As for Germany, this "silent world" suffered a leak during the 1930s and early 1940s, and it was mainly jurists who were responsible for this. Hence, several high court (*Reichsgericht*) decisions of the 1930s stipulated that, as a general rule, physicians had to disclose even fatal diagnoses, with only few acceptable exceptions.[18] The federal court (Bundesgerichtshof) continued with this line of argumentation in the 1950s, proclaiming that fear did not have to be avoided at all costs because it would not necessarily overcome and destroy a person's will to live.[19] However, federal judges were more reluctant when assessing priorities. Thus, judges from different jurisdiction levels conceded in 1958, 1959, and 1962 that, to a certain extent, truthful information about diagnosis and prognosis could be withheld if it endangered a patient's emotional stability, adding that patients should not be told the "naked truth" about their cancer diagnosis if it could be helped.[20]

Against this backdrop, one is amazed at the picture of cancer patients drawn by hospital staff in the 1950s. Usually, there is hardly any trace of fear or hopelessness to be found—quite the opposite. Charlotte

Neumann, for instance, a physician taking care of (female) cancer patients in a Munich sanatorium, described her patients as follows: "The female patients tended to play down even serious symptoms. They rarely complained. They were usually good-natured, obliging, and they quite often comforted other patients in a heartwarming manner. Only rarely did they act in an aggressive way, and even if they once complained with due cause, they ultimately asked me to forget about it and not to take any offense."[21] This description echoes the usual features found when examining how cancer patients had been characterized during those years. However, physicians, psychologists, and psychiatrists did not agree on how to evaluate and interpret these commonly noted psychological traits. Two contradicting narratives can be found indicating the existence of two more generally conflicting emotional regimes in the late 1950s and early 1960s. While one regime perceived the balanced and optimistic attitude of cancer patients as something basically positive, the other identified this attitude as a fundamentally harmful repression of one's inner feelings.

The first narrative dominated to a large extent the therapeutic encounter with cancer patients during the early postwar years, particularly in West Germany. It linked up with, and provided some justification for, the common practice of withholding the truth. Even though most medical practitioners assumed that patients would ultimately suspect the true diagnosis, they argued that knowing would be more disturbing than suspecting.[22] Not knowing would allow patients to more or less deliberately block out the diagnosis in order to achieve an otherwise unattainable peace of mind facing an extremely overwhelming disease. Some even claimed that the physiological changes brought on by the disease itself or else the "phlegmatic" temperament characteristic of individuals predisposed to cancer would enable them to do so.[23] The benign deception practiced by physicians, nurses, and others was thus suited to patients with a disposition to self-deception and seen as positive since it would allow patients to deal with fear and anxieties in a rational manner.

This form of dealing with emotions thought to be negative and disturbing was definitely not limited to hospital wards. Whether examining the role of fear in educating children, the rhetoric of 1950s peace movements, or civil defense brochures, one will encounter a similar paradox of fear.[24] On the one hand, fear was in a certain way omnipresent in the postwar era as it was widely accepted that society, politics, and family alike were dominated by a genuine and deeply felt fear.

But since—on the other hand—fear was branded as negative, irrational, pathological, or even harmful in a very concrete sense, "fear of fear" was a kind of guiding principle in dealing with potentially angst-inducing situations.[25] Thus, governments as well as political movements tried hard to avoid fear, either by distancing themselves from fear and emotions in general through an overly rationalistic rhetoric or by overemphasizing the merits of hope, avoiding all subjects that could be perceived as disturbing.[26] In addition to hope, both security and the notion of being in control were of crucial importance for the postwar years to counteract "irrational" feelings like fear.[27] Thus, there was no legitimate way to express fear or anxiety openly in the early postwar years. This silencing of personal fear was part of a more general culture of emotional restraint and anti-intensity that was fundamental in European societies during the 1950s. It was a kind of emotional lesson learned by Europeans who had experienced the mobilization of emotions during fascism and thus embraced partly an emotional style that had been dominant in the United States since the first half of the twentieth century, although for different reasons.[28]

In the US context, a different strand of pathologizing fear was perhaps more influential. For it was only in the late 1940s and early 1950s that psychoanalytic thinking became widely recognized, eventually contributing to the problematizing and semipathologizing of emotional states formerly regarded as normal and now classified in the new psychiatric category of disorder. Anxiety disorders were and are among the most prominent and ever more frequently encountered disorders.

However, a reevaluation of fear and other cancer-related emotions can still be traced back to the 1950s. After all, a competing and somewhat marginalized emotional regime existed even then, a regime that eventually became dominant, opposing all forms of concealment and emotional repression when facing cancer. Two factors contributed to that gradual shift. First, the emerging concept of informed consent changed the therapeutic encounter drastically. While informed consent was mainly about the patient's autonomy, it was nonetheless closely linked to both emotions within the doctor-patient relationship and the patient's emotional strength. A reassessment of the interrelation between fear, hope, and death was therefore an important part of this shift. The rise of psychosomatic medicine was a second major element in reconceptualizing the former emotional regime, calling into question the way in which emotions should be dealt with for medical, psychological, and, ultimately, political reasons.

Telling Straight Up and Acting Out: From Hidden Fears to Public Anger

The very idea of informed consent was first coined in the United States during the second half of the 1950s.[29] Obviously, obtaining the patient's consent for therapeutic interventions had been an issue in medical and legal debates before. The Nuremberg Doctors' Trial in 1946–47, conducted under American jurisdiction, was an important step toward informed consent because it led the Americans to formulate a code of experimentation ethics, the Nuremberg Code, that established the idea of the subject's voluntary consent.[30]

However, it was not until the late 1950s that this idea was implemented in therapeutic interventions, establishing in legal terms that the disclosure of certain kinds of information was a mandatory prerequisite to offering consent. One of the two cases from which the concept of informed consent is supposed to have originated was a cancer case, *Natanson v. Kline* (1960). Irma Natanson, a breast cancer patient, sued her physician, John R. Kline, for causing her to suffer severe burns from cobalt radiation therapy. Even though Dr. Kline was not found guilty of malpractice, he was found guilty of not having informed his patient about the possible risks and hazards associated with cobalt radiation therapy.[31] Hence, this sentence was important not only because it translated the ideas of autonomy and self-determination into legal practice; it also points to the fact that informed consent was connected to the implementation of new and extremely potent therapeutic agents such as cobalt radiation of which physicians did not have absolute control, knowing little about appropriate application doses and, therefore, experimenting in the wild, as one could say.

Obtaining informed consent did not necessarily entail disclosing the diagnosis. Numerous legal sentences as well as medical codes of ethics pondered the possibility (and legitimacy) of withholding a dreary diagnosis such as cancer while informing the patient about all risks associated with different therapeutic choices.[32] Thus, the stance toward informed consent in the early 1960s did not produce a definite decision regarding whether physicians should tell their patients the whole, naked truth.[33] However, benign deception became more difficult both to practice and to legitimize. How could one convince patients to undergo a grueling therapy without telling them they would otherwise almost certainly die? How could one recommend a particular therapy for curing a benign "growth" when early detection campaigns promoted the same therapy as a specific cure for cancer? How to concede self-determination to patients while at the same time denying them the right to know about their condition and possible future? Moreover, the issue of informed consent represented a fundamental shift in defining trust within the doctor-patient relationship: whereas trust was formerly based more or less exclusively on the physician's expertise, it now additionally required open and truthful communication and was, therefore, jeopardized when concealment was discovered or merely suspected.[34]

However, a different line of argument might have been more decisive for this shift, and this is related to a psychoanalytic reframing of emotions within the therapeutic process. One part of that argument concerned the doctor-patient relationship. Emphasis was now placed on the assumption that the physician's emotions were unwittingly mirrored by the inner feelings of the patient. Thus, the hidden fears of doctors would reinforce the patient's unspoken anxieties—an assumption that called for the physician's "emotional reeducation." This challenge was eventually tackled by the Hungarian-British psychoanalyst Michael Balint in his famous monograph *The Doctor, His Patient and the Illness* (1957), which described the doctor's personality as an important therapeutic asset.[35] The other part of the reframing of therapy addressed the patient's handling of his or her emotions. Uncertainty about one's "necessary end" was no longer regarded as beneficial, at least from a psychiatric or psychological point of view.[36] Quite the contrary, talking frankly about one's anxieties was thought to be a relief and was even supposed to assist the therapeutic process.[37]

Meanwhile, the idea that a central task of physicians treating the terminally ill was to sustain hope was not given up. Therefore, a reconfiguration of hope was desperately needed to redress the oxymoron of having to yield hope in cases where death was imminent. A Texas surgeon and a psychiatrist argued in 1966: "To tell a patient that his condition is hopeless is both cruel and technically incorrect. Incurability is a state of the body, whereas hopelessness is a state of mind, a giving-up—a situation that must be avoided at all cost. A patient can tolerate knowing he is incurable; he cannot tolerate hopelessness." They went on to explain how physicians could persist in raising hope despite ever-worsening prospects—to the point that the promise "I shall see you the first thing in the morning" would incorporate the very last flicker of hope.[38] This new understanding of sustaining hope without necessarily reverting to religion and transcendental hope turned out to be influential in the long run.[39]

Hence, medical practice surveys in the late 1970s revealed that a clear shift occurred regarding doctors telling the truth earlier and more frequently not only in the United States, where psychoanalytic thinking was much more widespread at that time, but also noticeably in West Germany.[40]

This shift was related to an additional development, namely, the emerging claims of psychosomatic medicine concerning "emotions as a cause of cancer."[41] In the past, only a small part of psychosomatic research was focused on cancer, whereas now psychological and psychosomatic cancer studies multiplied in postwar America.[42] Older assumptions had already identified the crucial importance of life events and the failure to deal with them at an emotional level, which would result in long-lasting grief and depression and eventually cause cancer.[43] The 1950s American studies, however, centered on personality patterns originating in early childhood experiences and used a psychoanalytic framework that was well received in US academic medicine. Based mainly on psychological tests performed with female cancer patients, a particularly influential 1951 study by the New York psychiatrists Milton Tarlau and Irwin Smalheiser concluded that breast cancer patients had little contact with their inner life, repressing disturbing emotions like anxiety and rejecting their femininity while at the same time maintaining a nice outer appearance.[44] Although these findings were criticized by subsequent studies for their methodological shortcomings, they informed popular notions of the "cancer personality," meaning a person who would suppress his or her own feelings and avoid conflict, driven by hidden fears and depression.[45]

In West Germany, a different strand of psychosomatic medicine was influential during the late 1940s and early 1950s. It was based on Viktor von Weizsäcker's concept of anthropological medicine, that is, considering the sick individual as an indissoluble biographical unity. In this perspective, everyone shapes his or her "personal" illness and not only suffers from a disease but is also in need of it.[46] By intently listening to patients' narratives about themselves and their illness without interpreting them in psychoanalytic terms, Weizsäcker's "biographical method" aimed at helping patients accept their suffering and, thus, heal the illness, which was only an external symptom of a suffering that had previously not been accepted.[47] Wilhelm Kütemeyer, one of Weizsäcker's disciples, made use of this method by focusing particularly on cancer. In his perspective, a cancer patient was a deeply ambivalent figure, in a state of complete alienation, submissive to the extreme, full of hatred and destructive forces that were dissociated from consciousness—the somatic reverse of psychosis. This notion of a cancer patient also had an implicit political dimension, suggesting broader struggles to come to terms with the National Socialist past, a fact that becomes obvious when reading Kütemeyer's case studies as well as his writings about the "European disease."[48]

However, Weizsäcker's and Kütemeyer's more philosophical understanding of illness and cancer was only partially adopted by later West German psychosomatic research and clinical practice. The first German psychosomatic institute opened its doors in 1950, later becoming the well-known and very influential Heidelberg Psychosomatic Clinic.[49] Viktor von Weizsäcker, at that time holding a professorship in internal medicine in Heidelberg, had taken the initiative together with his younger colleague Alexander Mitscherlich, a highly controversial figure, having served as official observer at the Nuremberg Doctors' Trial in 1946–47. Drawing on considerable funds given by the Rockefeller Foundation, they succeeded in conquering the antagonism of many of the Heidelberg University staff, especially in the psychiatry department. Only shortly afterward, Mitscherlich took the lead in German psychosomatic medicine, while Weizsäcker eventually fell more and more into oblivion, suffering from Parkinson's, which necessitated his retirement in 1952. He died in 1957.

Although Weizsäcker deeply venerated Freud and his psychoanalytic thinking, he—as well as Mitscherlich—never really integrated Freud's ideas into his medical theory. Yet, in 1951, Mitscherlich traveled to the United States and encountered Franz Alexander and other leading figures of American psychosomatic medicine, all of whom were deeply influenced by psychoanalysis. During his stay, he became aware of the close connection between American academic medicine and psychoanalysis, which he considered as a kind of leading example for future developments in Germany. On his return, he converted to psychoanalysis in a way. During the 1950s and early 1960s, he succeeded in "reimporting" psychoanalytic thinking into German psychosomatic medicine.[50] Frequent visitors to Germany from the United States such as the Danish-American psychologist Claus Bahne Bahnson served as important mediators between the two countries.[51] Nevertheless, traces of more anthropological or holistic visions of psychosomatic medicine remained present, as did a more political understanding that was crucial to German psychoanalysis and psychosomatic medicine.

Nonetheless, the psychosomatic explanation of cancer was picked up by German student activism in the late 1960s, in some ways drawing on Kütemeyer's concept of cancer. Hence, the disease came to be regarded as a symptom of the hidden morbidity and suppressed feelings of both individuals and society at large. The suppressed feelings and distorted emotionality had to be acted out, resulting in openly displayed personal and political anger toward parents and the establishment. A young Swiss man diagnosed with cancer of the throat at the age of thirty echoed this conviction in his personal diary, published shortly after his death in 1977. The book, entitled *Mars*, became a bible of the 1980s movements, opening with the famous sentence: "I am young, rich, and cultivated; I am unhappy, neurotic, and alone."[52] Its author, whose real name was Fritz Angst (Fritz Fear), chose for the publication the alias Fritz Zorn (Fritz Anger). With this choice he made it crystal clear that his aim was not to display fear but to transform hidden fears into public anger. Feeling and expressing anger equaled tapping into one's authentic inner feelings and—ultimately—succeeding in life. In this perspective, cancer became the chance to retrieve one's emotionality through self-scrutinizing—regardless of whether one won or lost the battle with the disease.[53] In this vein, cancer was now widely discussed in the German media, which placed a particular emphasis on it both as a symbol and an embodiment of the faults of society. Frank Biess has argued that this boom testified to a more general shift in the history of fear, anxiety, and subjectivity, namely, the internalization of anxiety formerly regarded and felt as a reaction to external dangers like the war or the Allied forces.[54] While this is certainly true, I would argue that this very same process contributed to blurring the line between the inner self and the outer world as well as between body and mind as society and environment were in a certain way integrated into the body. Therefore, one could argue that, while anxieties became internalized, the inner self itself was externalized in a paradoxical manner.

Mars was an enormous public success in the 1970s and early 1980s—not only in West Germany but also in all Western and Northern European countries. In contrast, the first American translation was published only in 1982, and it was never received with similar enthusiasm.[55] Although the notion that cancer is a psychosomatic disease was and still is very influential in popular self-help books for cancer patients such as Lawrence LeShan's *You Can Fight for Life* (1977) or Carl Simonton and Stephanie Matthews Simonton's *Getting Well Again* (1978), political activists in the 1970s did not make use of this argument.[56] Quite the contrary, some like Susan Sontag criticized all psychosomatic and other metaphoric explanations of cancer because they put the blame for the disease on the cancer patients themselves.[57]

However, there was a different link between cancer, anger, and the political arena in the United States, one that could be found in West Germany, too. Cancer treatment, especially the practice of radical mastectomy, widespread in the United States, triggered a protest movement against established medicine. This movement branded the establishment as paternalistic and devoid of emotions, as it figured in Rose Kushner's *Breast Cancer Report* (1975), Betty Rollin's *First, You Cry* (1976), and Audre Lord's *The Cancer Journals* (1980).[58] The scope of this movement was narrow at the beginning, gaining in importance only after the mid-1980s. Patient empowerment, along with feminist and environmental activism, then merged with the efforts of the environmentalist Rachel Carson, who died of cancer in 1964 having become a kind of integrating figure for all these movements.[59] Anger was used here as a kind of charge against the male medical and political establishment.[60] A similar line of argument can be found in the German actress Hildegard Knef's report of her breast cancer experience, published in 1975 as *The Verdict*, an enraged and expressive account that was quite well received in the United States.[61]

Epilogue: Positive Emotions and the Return of Fear

The 1980s and 1990s were marked by the onset of a new cancer-related emotional regime, this time ruled by the demand to get over feelings like fear, anger, and depression, which were again framed as negative and harmful in the long run. In a way, fear returned to the fore. In the emotional regime of the 1950s and early 1960s, fear was discussed extensively but in terms of being avoided; patients were seldom encouraged to talk it out. It was thought best to bolster hope and avoid fear by concealing the true diagnosis as the word *cancer* was tantamount to incurability and thus hopelessness. In the 1980s, however, *fear* acquired a different meaning and was dealt with in a different manner. In contrast to the late 1960s and the 1970s, the focus was no longer on retrieving hidden or suppressed emotions through psychoanalytically informed introspection in order to transform the formerly unacknowledged, sickening fear into a more extrovert and active anger and rage. Rather, the new emotional regime of the 1980s and 1990s focused on the emotional coping of patients living with, or dying of, cancer and other terminal diseases. Sustaining quality of life until the very end in terms of both pain and emotional management became a priority. The emerging hospice movement as well as the broadening field of psycho-oncology contributed to that shift.[62] While psychologists, psychiatrists, and physicians working within these contexts stressed the necessity of talking things out, they nonetheless claimed that patients ultimately must conquer or at least modulate their negative feelings into more positive emotions like hope or acceptance.[63] Medical findings about the impact of emotions on the immune system further enhanced this reasoning.[64]

This emotional regime is still valid. The strict criteria defining the so-called negative emotions become evident through the example of the American Psychiatric Association, which released the first draft of the fifth edition of its *Diagnostic and Statistical Manual of Mental Disorders* in 2012. According to the *DSM*, intense grief lasting more than two weeks after bereavement qualifies as major depression since "normal" grief is construed as *adaptive* response to a loss and therefore has to be transformed into more positive emotions—or at least less intense grief—in due time. However, the ensuing public outcry and heated media debate reveal that this way of dealing with negative emotions is riddled with controversy. A critical undercurrent is also visible with regard to oncomotions. On the one hand, psychiatrists, such as the pioneer in psycho-oncology Jimmie Holland, underline the imperative to respect one's own personal way of coping without following normative ideas about how to cope successfully.[65] Others criticize the dogma of positive thinking and point to the significance of fear and grief as forms of legitimate and authentic feelings in their own right.[66] In this sense, the public display of fear is regarded as a kind of protest against an overwhelming, politically naive, and false culture of coping and positive thinking—even when there is every reason to fear or mourn.

NOTES

1. Pater Bolech, "Die seelsorgerliche Betreuung der Krebskranken als besondere Aufgabe," *Der Krebsarzt* 13.8 (1958): 353–60, 355–56.
2. Christoph Schlingensief, *So schön wie hier kanns im Himmel gar nicht sein! Tagebuch einer Krebserkrankung* (Munich: btb, 2010), 71–72.
3. Robert Aronowitz chose this term as the title of a book studying the clinical experience of breast cancer patients in nineteenth- and twentieth-century North America. By this choice, he highlighted his assumption that the cancer experience was much more influenced by society, culture, and location than by biology itself. Robert A. Aronowitz, *The Unnatural History of Cancer: Breast Cancer and American Society* (Cambridge: Cambridge University Press, 2007), 7.
4. I am using *emotional regime* as it has been conceptualized by William Reddy. The term describes a set of emotional norms that govern not only

the expression of emotions in a given society but also the feeling of emotions itself. For—in Reddy's view—feelings are altered when perceived and acted on by the feeling individual itself and are therefore deeply engrained by societal and cultural norms about emotions. However, there might be conflicting norms or groups adhering to emotional regime(s) different from the dominant one—at least to a certain extent. Thus, there is a certain flexibility encompassed in the concept of emotional regimes. See William R. Reddy, "Against Constructionism: The Historical Ethnography of Emotions," *Current Anthropology* 38.3 (1997): 327–51, and *The Navigation of Feeling: A Framework for the History of Emotions* (Cambridge: Cambridge University Press, 2001). See also Jan Plamper, "The History of Emotions: An Interview with William Reddy, Barbara Rosenwein, and Peter Stearns," *History and Theory* 49.2 (2010): 237–65, 237–49.

5. Thus, fear and, more specifically, the use of fear balanced by hope is a recurrent feature in almost all studies that deal with early detection campaigns. For the United States, see Aronowitz, *Unnatural History*; Barron H. Lerner, *The Breast Cancer Wars: Hope, Fear, and the Pursuit of a Cure in Twentieth-Century America* (Oxford: Oxford University Press, 2001); and James T. Petterson, *The Dread Disease: Cancer and Modern American Culture* (Cambridge, MA: Harvard University Press, 1987). For France, see Patrice Pinell, *The Fight against Cancer: France 1890–1940* (London: Routledge, 2002). For Germany, no comparable story has been written yet. However, Robert Proctor has given a comprehensive account of anticancer campaigns under National Socialism. See Robert N. Proctor, *The Nazi War on Cancer* (Princeton, NJ: Princeton University Press, 1999). Also, a small collection of posters stemming mainly from the German Hygiene Museum (Dresden) is available. Susanne Roeßiger, ed., *Rechtzeitig erkannt—heilbar: Krebsaufklärung im 20. Jahrhundert* (Dresden: Michael Sandstein, 2001). For Swiss cancer awareness campaigns, which resemble the German ones in some respects, see Daniel Kauz, *Vom Tabu zum Thema? 100 Jahre Krebsbekämpfung in der Schweiz 1910–2010* (Basel: Schwabe, 2010).
6. There is a vast literature on the breast cancer experience and breast cancer campaigns in the United States. See, among others, Ellen Leopold, *A Darker Ribbon: Breast Cancer, Women, and Their Doctors in the Twentieth Century* (Boston: Beacon, 1999); James S. Olson, *Bathsheba's Breast: Women, Cancer and History* (Baltimore: Johns Hopkins University Press, 2002); Kirsten E. Gardner, *Early Detection: Women, Cancer, and Awareness Campaigns in the Twentieth Century United States* (Chapel Hill: University of North Carolina Press, 2006); and Maren Klawiter, *The Biopolitics of Breast Cancer: Changing Cultures of Disease and Activism* (Minneapolis: University of Minnesota Press, 2008). On cervical cancer most recently, see Ilana Lowy, *A Woman's Disease: The History of Cervical Cancer* (Oxford: Oxford University Press, 2011). For Germany, both East and West, there are no comparable studies in the history of breast and cervical cancer available.

Considering what is known about cancer awareness campaigns in Dresden, one can assume that during most of the twentieth century women's cancer was not the issue that it was in the United States; e.g., the "cancer wars" during the Nazi era targeted mainly lung cancer, which was at that time considered a typical men's cancer. But, even then, women were often described as being more prone to fear cancer. See Proctor, *Nazi War on Cancer*, 29–32. Moreover, all cancer campaigns until the 1970s were not only gender but also color biased, targeting almost exclusively white women from the middle classes. For a more detailed analysis, see Keith Wailoo, *How Cancer Crossed the Color Line* (Oxford: Oxford University Press, 2011).

7. Lerner, *Breast Cancer Wars*, 41–68.
8. Senatskommission für das Krebsforschungszentrum an der Universität Heidelberg [DFG], Denkschrift betr. Anstalt für Geschwulstforschung und Geschwulstbehandlung an der Universität Heidelberg, gez. K. H. Bauer [o.D.], BArch B 142/3434, Bl. 186–96. Heinrich Martius wrote in 1957: "Mit Rücksicht auf die Krebsangst sollte dem Publikum nicht immer wieder gesagt werden, daß nur 16% aller Krebse geheilt werden, eine Zahl, die in vielen Veröffentlichungen wiederkehrt, besonders von denjenigen Autoren, die das Heil in irgendeiner besonderen Diät sehen, durch die sie den Krebs nicht nur vermeiden, sondern auch heilen wollen. Die Zahl 16% ist angreifbar und bedeutet insofern einen Mißbrauch der Statistik, als inkommensurable Kollektive zusammengeworfen und gemeinsam verrechnet werden." Heinrich Martius, "Probleme und Aufgaben der Krebsforschung," *Deutsche Medizinische Wochenschrift* 82.35/2 (September 6, 1957): 1500–1502.
9. This holds true for most but not all physicians treating cancer patients. For instance, dermatologists were much more optimistic since cancers of the skin had much better chances of being cured at the time. Donald Oken (Chicago), who in the early 1960s asked 219 physicians about "telling" cancer patients, reported that there had been a general feeling among physicians "that we can do little to save lives and not a great deal to prevent suffering." Donald Oken, "What to Tell Cancer Patients: A Study of Medical Attitudes," *Journal of the American Medical Association* 175.13 (1961): 1120–28, 1126.
10. Carly Seyfarth, *Der Arzt im Krankenhaus ("Ärzte-Knigge"): Über den Umgang mit Kranken und über Pflichten, Kunst und Dienst der Krankenhausärzte*, 5th ed. (Leipzig: Georg Thieme, 1946), 93. Seyfarth used the same words already in the first edition. See Carly Seyfarth, *Der "Ärzte-Knigge": Über den Umgang mit Kranken und über Pflichten, Kunst und Dienst der Krankenhausärzte* (Leipzig: Georg Thieme, 1935), 61.
11. Seyfarth, *Ärzte-Knigge* (1946), 14.
12. The wording *benign deception* is used in Verdi's *La traviata*. An influential German article discussing the issue of truth telling referred to this word-

ing. See Hans-Wolfgang Becker and Hans-Lothar Kölling, "'Schonendes Betrügen?' Ermittlungen zur Frage der Aufklärung bei Geschwulstleiden," *Nova Acta Leopoldina* 28.168 (1964): 1–18.

13. William T. Fitts and I. S. Ravdin, "What Philadelphia Physicians Tell Patients with Cancer," *Journal of the American Medical Association* 153.10 (1953): 901–4; Oken, "What to Tell," 1120–28.
14. Oken, "What to Tell," 1126.
15. Ibid., 1123–24.
16. Jay Katz, *The Silent World of Doctor and Patient* (New York: Free Press, 1984).
17. Aronowitz, *Unnatural History*, 127–33.
18. G. Schläger, "Aufklärung und Verschwiegenheit bei Krebsverdacht," *Monatsschrift für Krebsbekämpfung* 10 (1943): 150–55.
19. Gerald Grünwald, "Die Aufklärungspflicht des Arztes," *Strahlentherapie* 114 (1961): 173–74.
20. "Beschluss OLG Hamburg vom 24.02.1958 1 W 8; 58," *Versicherungsrecht*, 1958, 388; "22.05.1958 Urteil LG Frankfurt/M. 2/4 0 232/57 Umfang der Behandlung von Zungenkrebs, therapeutisches Privileg," *Versicherungsrecht*, 1958, 868–69; "Urteil BGH vom 16.01.1959," *Neue Juristische Wochenschrift* 18 (1959): 814–15; "22.02.1962 Urteil LG Dortmund 8 0 167/59 Umfang der ärztlichen Aufklärungspflicht vor Beginn der Strahlenbehandlung," *Versicherungsrecht*, 1963, 689–90.
21. The German original is as follows: "Die Patientinnen neigten dazu, selbst ernsthafte Symptome zu bagatellisieren, sie klagten wenig, waren freundlich, zuvorkommend und umsorgten häufig Mitpatientinnen auf rührende Weise. Aggressive Äußerungen waren selten, wenn einmal durchaus berechtigte Anschuldigungen vorgetragen wurden, baten mich die Beschwerdeführerinnen am Schluß doch lieber alles beim alten zu lassen, als Ärger zu erregen." Charlotte Neumann, "Psychische Besonderheiten bei Krebspatientinnen," *Zeitschrift für psychosomatische Medizin* 5 (1958/59): 91–101, 91.
22. One Philadelphia physician stated, e.g.: "If they don't know definitely—that is, so long as the doctor doesn't tell them the truth—they still have a ray of hope that it might be something else, no matter how much they think they have cancer." Fitts and Raydin, "Philadelphia Physicians," 903.
23. See, e.g., Bolech, "Die seelsorgerliche Betreuung," 355–56.
24. Peter N. Stearns and Timothy Haggerty, "The Role of Fear: Transitions in American Emotional Standards for Children, 1850–1950," *American Historical Review* 96.1 (1991): 63–94; Holger Nehring, "Angst, Gewalterfahrungen und das Ende des Pazifismus: Die britischen und westdeutschen Proteste gegen Atomwaffen, 1957–1964," in *Angst im Kalten Krieg*, ed. Bernd Greiner, Christian Th. Müller, and Dierk Walter (Hamburg: Hamburger Edition, 2009), 436–64; Frank Biess, "'Everybody Has a Chance': Civil Defense, Nuclear Angst, and the History of Emotions in Postwar Germany," *German History* 27.2 (2009): 215–43.

25. "Fear of Fear" has been chosen as the title of a chapter in Patrick Bormann, Thomas Freiberger, and Judith Michel, eds., *Angst in den Internationalen Beziehungen* (Göttingen: Vandenhoeck & Ruprecht, 2010).
26. On the complex relationship of hope and fear in postwar Germany, see Frank Biess, "Feelings in the Aftermath: Toward a History of Postwar Emotions," in *Histories of the Aftermath: The Legacies of the Second World War in Europe*, ed. Frank Biess and Robert G. Moeller (New York: Berghahn, 2010), 31–48, 37–40. On the distinction between fear as an irrational emotion and *ratio* in the West German and British peace movements, see Nehring, "Angst," 451–52.
27. On the notion of security in West Germany, see Eckart Conze, "Security as a Culture: Reflections on a 'Modern Political History' of the Federal Republic," *German Historical Institute London Bulletin* 28.1 (2006): 5–34.
28. Biess, "Feelings in the Aftermath," 234–37. For the origins and implications of an American emotional culture of anti-intensity that emerged during the first half of the twentieth century, see Peter N. Stearns, *American Cool: Constructing a Twentieth-Century Emotional Style* (New York: New York University Press, 1994).
29. Ruth R. Faden and Tom L. Beauchamp, *A History and Theory of Informed Consent* (New York: Oxford University Press, 1986), 125; Walter J. Friedlander, "The Evolution of Informed Consent in American Medicine," *Perspectives in Biology and Medicine* 38.3 (1995): 498–510.
30. See Gere, chapter 5 in this volume.
31. On *Natanson v. Kline*, see Ellen Leopold, *Under the Radar: Cancer and the Cold War* (New Brunswick, NJ: Rutgers University Press, 2008), 42–58.
32. For the American context, see the discussion in Samuel Standard and Helmuth Nathan, eds., *Should the Patient Know the Truth? A Response of Physicians, Nurses, Clergymen, and Lawyers* (New York: Springer, 1955). For West Germany, see Grünwald, "Die Aufklärungspflicht des Arztes"; and, as a historical overview from a juridicial perspective, Alexander P. F. Ehlers, *Die ärztliche Aufklärung vor medizinischen Eingriffen: Bestandsaufnahme und Kritik* (Cologne: Carl Heymanns, 1987). From a history-of-medicine perspective, see D. von Engelhardt, "Wahrheit am Krankenbett im geschichtlichen Überblick," *Schweizerische Rundschau für Medizin (Praxis)* 85 (1996): 432–39.
33. In the *Journal of the American Medical Association* up to the 1980s, one can find articles and letters to the editor debating to what extent severely ill patients, namely, cancer patients, should be told the truth.
34. Paul S. Rhoads, "Management of the Patient with Terminal Illness," *Journal of the American Medical Association* 192.8 (1965): 661–65, 662; John S. Stehlin and Kenneth H. Beach, "Psychological Aspects of Cancer Therapy," *Journal of the American Medical Association* 197.2 (1966): 100–104, 102.

35. It was Donald Oken who asked for the doctor's "emotional reeducation." Oken, "What to Tell," 1128. See also Michael Balint, *The Doctor, His Patient and the Illness* (London: Tavistock, 1957). For a short overview of the Balint movement in the United States, see Alan H. Johnson, "The Balint Movement in America," *Family Medicine* 33.3 (2001): 174–77. Balint also discussed fear/thrill as a "mixed" feeling. See Michael Balint, *Thrills and Regression* (New York: International Universities Press, 1959).
36. Rhoads, "Management of the Patient," 661; Philip D. Woodbridge, "Telling the Truth about the Necessary End," *Journal of the American Medical Association* 194.3 (1965): 311–12 (letter to the editor).
37. Stehlin and Beach, "Psychological Aspects," 102.
38. Ibid., 102, 103.
39. See, e.g., Howard Brody, "Hope," *Journal of the American Medical Association* 246.13 (1981): 1411–12.
40. For the United States, see Dennis H. Novack et al., "Changes in Physicians' Attitudes toward Telling the Cancer Patient," *Journal of the American Medical Association* 241.9 (1979): 897–900. For Germany, see Engelhardt, "Wahrheit am Krankenbett," 438.
41. Samuel J. Kowal, "Emotions as a Cause of Cancer: 18th and 19th Century Contributions," *Psychoanalytic Review* 42.3 (1955): 217–27. This article tried to establish a long-standing tradition of psychosomatic cancer medicine.
42. For an emphasis on gendering, see Patricia Jasen, "Malignant Histories: Psychosomatic Medicine and the Female Cancer in the Postwar Era," *Canadian Bulletin of Medical History* 20.2 (2003): 265–97.
43. The most prominent example from the first half of the twentieth century is the Jungian Elida Evans's *A Psychological Study of Cancer* (New York: Dodd, Mead, 1926).
44. Milton Tarlau and Irwin Smalheiser, "Personality Patterns in Patients with Malignant Tumors of the Breast and Cervix," *Psychosomatic Medicine* 13.2 (1951): 117–21, 120.
45. For an affirmative but somewhat critical assessment of the Tarlau and Smalheiser study, see John L. Wheeler and Bettye McDonald Caldwell, "Psychological Evaluation of Women with Cancer of the Breast and of the Cervix," *Psychosomatic Medicine* 17.4 (1955): 256–68.
46. Viktor von Weizsäcker, *Der Gestaltkreis: Theorie der Einheit von Wahrnehmen und Bewegen* (Stuttgart: Thieme, 1940), 34.
47. Udo Benzenhöfer, *Der Arztphilosoph Viktor von Weizsäcker: Leben und Werk im Überblick* (Göttingen: Vandenhoeck & Ruprecht, 2007), 203–8; Tobias Freimüller, *Alexander Mitscherlich: Gesellschaftsdiagnosen und Psychoanalyse nach Hitler* (Göttingen: Wallstein, 2007), 152–60.
48. Wilhelm Kütemeyer, "Anthropologische Medizin in der inneren Klinik," in *Viktor von Weizsäcker: Arzt im Irrsal der Zeit; Eine Freundesgabe zum 70. Geburtstag*, ed. Paul Vogel (Göttingen: Vandenhoeck & Ruprecht, 1956),

243–65, "Psychosocial Aspects of Cancer" (paper presented at the Fourth International Conference on Psychosomatic Aspects of Neoplastic Disease, Turin, June 1965), and *Die Krankheit Europas: Beiträge zu einer Morphologie* (Frankfurt: Suhrkamp, 1951). Frank Biess has argued that the psychosomatic cancer etiologies of the 1970s were construed on pathologizing a kind of outward normal and emotionally retracted appearance deviating from the then dominant norm of emotional expressivity. As such, the cancer personality was marked by what was regarded as fascistic behavior patterns. Reading Kütemeyer, it is evident that this kind of parallelism goes back to late 1940s psychosomatic cancer etiologies. See Frank Biess, "Die Sensibilisierung des Subjekts: Angst und 'Neue Subjektivität' in den 1970er Jahren," in *WerkstattGeschichte* 49 (2008): 51–71, 65.

49. For the complex relationship between Weizsäcker and Mitscherlich, especially with regard to the establishment of the Heidelberg Psychosomatic Clinic and Mitscherlich's reorientation toward psychoanalysis, see Thomas Henkelmann, "Zur Geschichte der Psychosomatik," *Psychotherapie, Psychosomatik, medizinische Psychologie* 42 (1992): 175–86; Benzenhöfer, *Arztphilosoph*, 179–202; Martin Dehli, *Leben als Konflikt: Zur Biographie Alexander Mitscherlichs* (Göttingen: Wallstein, 2007), 176–234; and Freimüller, *Mitscherlich*, 134–205.
50. The term *reimport* is used in Freimüller, *Mitscherlich*, 160.
51. Claus Bahne Bahnson was a professor at Jefferson Medical College, Philadelphia (1963–69), the director of the Department of Behavioral Sciences, Research and Training of the Eastern Pennsylvania Psychiatric Institute (1969–80), and the president of the National Institute for the Seriously Ill and the Dying (1977–87). He wrote the chapter about cancer in one of the German classics of psychosomatic medicine: Claus Bahne Bahnson, "Das Krebsproblem in psychosomatischer Dimension," in *Lehrbuch der Psychosomatischen Medizin*, ed. Thure von Uexküll (Munich: Urban & Schwarzenberg, 1979), 685–98.
52. Fritz Zorn, *Mars*, trans. Robert Kimber and Rita Kimber (New York: Knopf, 1981). The book appeared in German as *Mars: "Ich bin jung und reich und gebildet; und ich bin unglücklich, neurotisch und allein . . ."* (1977; Frankfurt a.M.: Fischer, 1994). For German media coverage, see Hellmuth Karasek, "'Mars': Ein dreißigjähriger Krieg in Frieden," *Der Spiegel* 15 (1977): 219; and Hugo Leber, "Der Zorn des Fritz Zorn," *Die Zeit*, June 3, 1977, http://www.zeit.de/1977/23/der-zorn-des-fritz-zorn.
53. See, e.g., Zorn, *Mars*, 206–7.
54. Biess, "Sensibilisierung des Subjekts," 54, 62–67.
55. A second German edition of *Mars* was published in 1979. Today, the twenty-fifth edition is available. In 1978, *Mars* was translated into Swedish, Dutch, and Danish and, in 1980, into French.
56. The psychologist Lawrence LeShan was one of the first to popularize the psychosomatic perception of cancer through a self-help book. See Law-

rence LeShan, *You Can Fight for Your Life: Emotional Factors in the Causation of Cancer* (New York: M. Evans, 1977). Interestingly, he later changed the subtitle to the much less ambitious *Emotional Factors in the Treatment of Cancer* (1980). His book was translated into German in 1982 (*Psychotherapie gegen den Krebs* [Stuttgart: Klett-Cotta, 1982]), and and a tenth edition appeared in 2008. Equally influential and widely published was Carl Simonton, Stephanie Matthews Simonton, and James Creighton, *Getting Well Again* (Los Angeles: J. P. T. Tarcher, 1978), translated into German as *Wieder gesund warden: Eine Anleitung zur Aktivierung der Selbstheilungskräfte für Krebspatienten und ihre Angehörigen* (Reinbek: Rowohlt, 1982). A tenth edition in German was published in 2011. See also Jasen, "Malignant Histories," 289–90.

57. Susan Sontag, *Illness as Metaphor* (New York: Farrar, Straus & Giroux, 1978).
58. Rose Kushner, *Breast Cancer: A Personal History and an Investigative Report* (New York: Harcourt Brace Jovanovich, 1975); Betty Rollin, *First, You Cry* (New York: J. B. Lippincott, 1976); Audre Lord, *The Cancer Journals* (San Francisco: Aunt Lute, 1980).
59. Rachel Carson's *Silent Spring* (Boston: Houghton Mifflin, 1962) is often cited. It was initially published in three parts in the June 16, 23, and 30, 1962, issues of the *New Yorker.*
60. See two of the three cultures of action analyzed in Klawiter, *The Biopolitics of Breast Cancer,* 163–225.
61. Hildegard Knef, *Das Urteil oder Der Gegenmensch* (Vienna: Molden, 1975). The very same year, a first American edition was published: *The Verdict* (New York: Farrar, Straus & Giroux, 1975). Three more editions were published in the following two years.
62. For the history of psycho-oncology, see Jimmie C. Holland, "History of Psycho-Oncology: Overcoming Attitudinal and Conceptual Barriers," *Psychosomatic Medicine* 64 (2002): 206–21. For the hospice movement, see David Clark, "A Special Relationship: Cicely Saunders, the United States, and the Early Foundations of the Modern Hospice Movement," *Illness, Crisis and Loss* 9.1 (2001): 15–30.
63. A very striking example is the the notion of the five stages of grief, initially introduced by the Swiss-American psychiatrist Elisabeth Kübler-Ross in *On Death and Dying* (New York: Macmillan, 1969), first translated into German as *Interviews mit Sterbenden* (Berlin: Evangelische, 1971). Following Kübler-Ross, most dying patients go through the stages of denial, anger, bargaining, depression, and, ultimately, acceptance.
64. For the beginnings of psychoneuroimmunology, see Ronald Glaser, ed., *Handbook of Human Stress and Immunity* (San Diego: Academic, 1994).
65. Jimmie Holland and Sheldon Lewis, *The Human Side of Cancer: Living with Hope, Coping with Uncertainty* (New York: HarperCollins, 2000), esp. 13–25 (chap. 2, "The Tyranny of Positive Thinking").

66. For a critique of the American culture of positive thinking, see Barbara Ehrenreich, *Smile or Die: How Positive Thinking Fooled America and the World* (London: Granta, 2009). For a view more focused on the positive value of fear in dealing with cancer, see the German director Christoph Schlingensief's Internet-based project initiated at the Biennale in Venice in 2003: http://www.church-of-fear.net.

PART THREE

Psychiatry

SEVEN

The Concept of Panic: Military Psychiatry and Emotional Preparation for Nuclear War in Postwar West Germany

FRANK BIESS

Introduction

In June 1963, the Cold War became hot at the demarcation line between East and West in Germany: conventional forces of the Warsaw Pact crossed the inner-German border and sought to occupy the Central German region around Kassel in order to secure their further advance toward the Rhine. In response, NATO forces quickly resorted to nuclear weapons and detonated four nuclear bombs on West German territory with a combined destructive force of almost three times the Hiroshima bomb. Other West German forces attempted to halt the Warsaw Pact invasion by placing nuclear land mines on major streets and bridges. A Soviet conventional attack thus had quickly turned into nuclear war on the territory of the Federal Republic.

Fortunately, this scenario became reality only in the military planning exercise "Morning Greetings" of the Second West German Tank Grenadier Division in the summer of 1963.[1] Yet it exemplified the central security dilemma of the Federal Republic during the Cold War: because of the clear Soviet superiority in conventional weapons, the military defense of West Germany rested ultimately on the first and early use of nuclear weapons on German soil.[2] While military planners sought to cope with this scenario and also prepared for a limited and possibly nonnuclear war in Central Europe, they were ultimately faced with a paradox: any attempt to defend the Federal Republic against a potential attack from the East might very well imply its self-destruction. This situation changed only gradually from the late 1960s on when the NATO doctrine of "flexible response" moved away from the threat of massive retaliation and allowed for a more graduated response to a potential Soviet attack.

In light of the escalating tensions at the height of the Cold War as well as the Federal Republic's strategic dilemma, emotional and psychological preparation for nuclear war assumed a new significance during the late 1950s and early 1960s. How to contain extreme emotions such as fear and panic during a potential nuclear war was a problem that preoccupied West German civilian officials and military planners alike.[3] They echoed similar (and earlier) concerns in the United States, where civil defense experts and public commentators pondered—with widely varying findings—the likelihood that Americans would succumb to panic in the case of a nuclear attack.[4] All these considerations, to be sure, remained highly speculative since there was—apart from Hiroshima and Nagasaki—very little evidence of the impact of nuclear weapons on civilian and military morale. Their hypothetical nature notwithstanding, such contingency planning for military emergencies was central to the Cold War, which, after all, took place primarily in the realm of the imagination.[5] Panic discourses were part and parcel of the history of future scenarios during the Cold War.[6] Military planners and their civilian counterparts viewed the ability to withstand and master the potential psychological and emotional challenges of nuclear war as critical to the ability to fight and survive a possible nuclear war. Containing fear and panic thus became a central element in Cold War military planning. Military discourses on fear and panic pointed to the significance of emotional and psychological preparedness, and they revealed nagging self-doubts and persistent anxieties regarding the emotional and moral steadfastness of Western societies in the Cold War.

In its consideration of the genesis and nature of panic discourses in postwar West Germany, this chapter analyzes one aspect of the emotional preparation for military conflict during the Cold War. It focuses on the production and reception of two military service manuals produced for officer training in the West German military, the Bundeswehr. The first manual, "Panic: Recognition-Prevention-Overcoming," was published in 1962. It was followed by a successor, "Psychological Crisis in Nuclear War," which specifically addressed the issue of fear and panic in a potential nuclear confrontation. Yet, after a comprehensive internal review, the successor was never published and did not advance, for reasons to be discussed below, beyond the stage of various drafts.

By providing a genealogy of two military service manuals at the height of the Cold War, this chapter traces shifting normative conceptions of fear and panic. It does so by identifying two formative influences on conceptions of panic in postwar West Germany. The first consisted of diagnostic and conceptual continuities within West German military psychiatry dating back to the 1930s and even to the First World War. Following a series of other studies, the chapter emphasizes the central role of military psychiatry in defining the boundaries of legitimate emotions within the military.[7] Two psychiatrists in particular, Max Mikorey and Julius Deussen, embodied these continuities and exerted a decisive influence on official conceptions of panic in Cold War Germany. Both texts thus serve as concrete examples of a historically specific nexus between science and emotions—that is, of the ways in which scientific insights regarding the nature of specific emotions were transferred into the different institutional context of the West German military. Second, both manuals also exhibited an increasing transatlantic influence of US conceptions of panic. In particular, they incorporated American understandings of fear and panic as they were articulated not primarily by psychiatrists but rather in popular military histories or sociological analyses. Partly owing to these influences, West German military psychiatrists eventually began to revise their own normative conceptions of fear and panic. The real moment of discontinuity, however, did not occur until the early 1980s, when the confluence of the psychiatric reform movement with the peace movement led to a frontal assault on traditional panic concepts as they had been formulated by Mikorey and Deussen. A combination of internal developments within postwar psychiatry and external forces thus contributed to shifting emotional norms regarding fear and panic and also challenged the institutional link between psychiatry and the West German military as it had been forged during the Cold War. The result was a destigmatization of panic, which was now seen less as a symptom of a pathological personality structure and increasingly as an adaptive response to an overwhelming threat. This development was part and parcel of a broader "emotional democratization" in postwar West Germany that complemented the general process of postwar democratization by widening the range of legitimate emotions and increasingly tolerating their expression in public.[8]

Max Mikorey, Julius Deussen, and Panic in the Bundeswehr

The concept of panic first appeared in the Bundeswehr service regulations *HDv 100/2* of April 1961, which outlined guidelines for soldierly behavior in the case of a nuclear war.[9] The 1962 manual "Panic: Recognizing-Preventing-Overcoming" then addressed the issue more comprehensively and was included in the publication series *Innere Führung*. Alternatively translated as "internal guidance," "moral leadership," or "moral education," *Innere Führung* was the central concept of military reform in West Germany and aimed at creating a new type of democratic soldier, a "citizen in uniform" fully committed to the democratic state.[10] While this project challenged some of the traditional authoritarianism within German military culture, it also sought to instill a strong ideological commitment among individual soldiers. In what Count Wolf von Baudissin, the main proponent of *Innere Führung*, described as a global civil war between East and West, "calmness" and "inner strength" were particularly important characteristics for the modern soldier and especially for officers. Faced with the enormous destructive power of nuclear weapons, soldiers' psychological and emotional strength became an even more necessary component of military preparedness.[11]

Since the publication series *Innere Führung* was designed for the training of soldiers and officers, the 1962 panic manual represents a good source for discerning prevailing emotional norms within the West German military. The manual defined *panic* as an excessive emotional state that manifested itself either by an "affective explosion" or by its complete contrast, a complete absence of emotions or "panic freeze."[12] The former was conceived of as a phenomenon of larger crowds, whereas the latter was believed to occur primarily among individuals. Panic thus appeared as the pathological other to the dominant emotional norm of emotional restraint and anti-intensity in the postwar period.[13] The manual considered panic to be a regression to the "animal stages of the personality" in which "reason was switched off" and individuals were driven solely by their "primitive drives." Panic also appeared in distinctly gendered forms that contrasted an activist, male "panic storm" deriving from prehistoric man's instinctual reaction to an external threat with a passive "panic freeze" originating from women's allegedly inborn reflex to play dead in moments of danger.[14]

This recourse to evolutionary biology in defining panic drew on the cultural anthropology of the Bonn philosopher Erich Rothacker. Rothacker had developed his theory of the "layers of the personality" in a 1938 monograph, which reappeared in several new editions until 1966. Since Rothacker extended his notion of hierarchically ordered layers of the personality also to larger collectives and assigned, for example, "Slavs" and "Orientals" collective characteristics on this basis, his theory proved to be highly compatible with Nazi racial ideology.[15] The invocation of his theory in the context of the Bundeswehr panic manual reflected an all-too-common case of academic continuity in postwar West Germany. It also illustrated the particular forms of pathologization shaping conceptions of panic within West German military culture. Panic was associated with evolutionarily older, primitive, and inferior aspects of one's personality that had escaped the regulatory functions of higher and superior cognitive abilities.

Conceptual continuities within military psychiatry further compounded this pathologizing view of panic. The 1962 Bundeswehr manual drew heavily on the psychiatrist Max Mikorey's expert testimony on panic, commissioned by the Ministry of Defense in 1958. While it is not entirely clear why Mikorey was selected to provide such testimony at that particular moment, it is likely that the official interest in panic derived from the larger debate over the nuclearization of the West German military during the late 1950s.[16] Mikorey had published and lectured on the subject since the 1930s and was able to apply his expertise in the new context of the Cold War. Because he eventually exerted a strong influence on official panic conceptions within the Bundeswehr, some biographical information is in order here.

Born in 1898, Mikorey had studied medicine and psychiatry in Munich and became an early supporter of National Socialism, joining the party in May 1933. As an expert in forensic psychiatry, he became a member of the Academy of German Law, which aimed at the development of a specifically Nazi conception of law. In 1936, he delivered the lecture "The Jews in Criminal Psychology," in which he advocated major tenets of Nazi anti-Semitism.[17] With the onset of war in 1939, Mikorey served the German army as one of sixty consulting psychiatrists whose main task was to prevent the numbers of psychiatric casualties among German soldiers that had occurred in World War I.[18] In the course of their service, these consultants adopted the prevailing view of the etiology of "war neurosis" denying any causal link between the exposure to violence and psychic reactions. Mikorey defined *war neurosis* as a form of "not wanting" in the guise of "not being able to." To him, the experience of the First World War confirmed the basis for the "hereditary biological thinking" and "the racial hygienic legislation" by disproving the significance of external factors for mental illness among soldiers. As a result, he advocated stern treatment of "war neurotics," who should be concentrated in special units with the "odium of inferiority" while more active forms of psychopathic behavior such as desertion or self-wounding were to be dealt with swiftly and sternly.[19]

Mikorey's view of war neurosis constituted the conceptual starting point for his concept of panic. The main reason for isolating war neurotics from the troops was the possibility of contagion and hence of panic. The "psychopathological behavior" of a few individuals, as he argued in his 1939 lecture, was always at the root of "panic and mutiny" among troops.[20] In the final stages of World War II, with enemy armies advancing on all fronts, Mikorey's psychiatric expertise on panic assumed an even more heightened significance. In September 1944, he gave an invited lecture, "The Significance of Panic in War," during a visit to Hans Frank, the head of the General Government in Poland and a major Nazi perpetrator.[21] Shortly thereafter, in the spring of 1945, Mikorey disappeared only to reappear again in July 1946 in Salzburg. His biographer, Andreas Weidmann, casts doubt on Mikorey's claim to have ended up in Soviet captivity and suggests that he might have gone into hiding with friends to avoid prosecution in light of his compromised Nazi past.[22] He subsequently managed to delay his de-Nazification until 1948 and eventually submitted a long statement of justification in which he portrayed his life in Nazi Germany as "twelve years of resistance and sabotage."[23] He was eventually classified as a "follower" and—after having paid a small fine—reassumed his position in the Psychiatry Department of the University of Munich, where he was promoted to professor in 1952.

The onset of the Cold War and West German rearmament brought renewed interest in expertise on panic. Mikorey began to give public lectures on the subject as early as 1954, initially to civil defense experts who were keenly interested in containing fear and panic among civilians during a potential nuclear war.[24] He continued to lecture on the subject to high-profile military audiences throughout the 1960s, including, in February 1960, the high command (*Führungsstab*) of the Bundeswehr.[25] Widespread interest in the issue of panic thus allowed him to capitalize on his previous expertise without having to interrogate his basic assumptions and conceptual categories. His was a typical case of academic transformation and reinvention after 1945, a process that allowed for significant biographic and conceptual continuity from the Third Reich to the Federal Republic in many academic disciplines.[26]

Mikorey's 1958 expert testimony on panic was a sprawling document that, according to Andreas Weidmann, revealed the "basic lack of substance in [his] medical and academic works."[27] While this assessment is correct, what matters here is not so much the academic quality of Mikorey's testimony but rather its significance in defining conceptions of panic within the Bundeswehr. As indicated above, his testimony constituted the most important source for the Bundeswehr panic manual of 1962. The manual not only quoted his testimony verbatim at several places; it also drew on some of the same sources that he had employed. For example, like Mikorey, the manual quoted a 1958 article by the French general Pierre Gallois in the German military periodical *Wehrkunde* predicting the likelihood of panic in a nuclear war, and it reiterated the "view of a German scientists" (i.e., Mikorey) that previous military doctrine and military psychology had largely neglected the issue of panic.[28]

The manual largely adopted Mikorey's definition of *panic* as the "collective explosion of cowardice."[29] Similar to war neurosis, which he described as a "chronic and thinned out version of panic," panic constituted for Mikorey an unmistakable sign of individual and collective psychic weakness.[30] Following the work of his colleague Friedrich Panse, Mikorey asserted a disproportionate relationship between external force and panic. Whereas massive external force such as the aerial bombardments of World War II had not produced panic, minor incidents or even hallucinations had often done so.[31] Even in the nuclear age, panic was always unjustified and never the product of an overwhelming external threat. Rather, it was, as Mikorey stated in another lecture, a "harmful atavism" that transformed an initially functional evolutionary response to danger in the animal kingdom into its complete opposite, thus rendering humans "stupid, disoriented, uncontrollable, and disrespectful."[32] Panic resulted not from external forces but was rather the product of individual constitutional weakness. As a result, pathological personalities, such as the "alcoholic, weak, unstable, crazy, or godless," were more prone to succumb to panics.[33] As such, the Bundeswehr conception of panic echoed the classic endogenous interpretation of war neurosis that originated in World War I and located psychological symptoms among soldiers not in external forces but rather in "constitutional weakness" or "hereditary deficits" of the individual soldier.[34]

Reflecting the classic concerns of military psychology with discipline and desertion, the panic manual also largely followed Mikorey's assessment of the highly infectious nature of panic. Panic occurred, Mikorey argued, when individual "psychic infection" entered the "healthy body of the troops" and hence undermined traditional deterrents against cowardice such as the military justice system. It ultimately led to "open mutiny and revolution" and was therefore responsible for most military defeats in history, as he sought to demonstrate with an idiosyncratic analysis of military history from the Celts in 273 BC to World War II.[35] Along the same lines, the manual cited a plethora of historical examples—from ancient history to Stalingrad—that were supposed to demonstrate the nature of panics and their significance for deciding military battles. Citing Mikorey as scientific authority, it asserted that panic occurred only when the psychophysical solidarity of soldiers was reversed and the troops were transformed into a "flight pack" or "civilians who shiver because of every danger." For both Mikorey and the Bundeswehr manual, greater susceptibility to panics confirmed the inferiority of civilian crowds. Finally, the manual also followed prevailing pessimistic cultural sentiments by linking the susceptibility to panics to a preexisting vague anxiety without a clear object, as postulated by philosophers such as Kierkegaard and Heidegger. Such "fearful fantasy," moreover, was aggravated by a general loss of religious beliefs, which rendered individuals less capable of resisting panic.[36]

Both Mikorey's testimony and the Bundeswehr panic manual fell short in offering effective countermeasures to panic. For Mikorey, only the exemplary behavior and courage of individual military officers was ultimately capable of preventing or containing panic. Otherwise, those parts of an army caught up in a wave of panic should simply be abandoned—similarly, as he put it, to a "high amputation" of an infectious leg.[37] The panic manual elaborated on these ideas and also placed a large degree of responsibility on military leaders, who should familiarize themselves with the psychic nature of their subordinates and therefore be able to anticipate "who can be relied on and who might go crazy." Only the courageous behavior of individual military leaders was supposed to contain the potentially infectious nature of "anxiety, fear, and horror."[38] Clear acoustic and visual signals also played an essential role in correcting the sensory misperceptions that lay at the roots of emerging panic. These suggestions again pointed to the understanding of panic as a form of individual psychic malfunctioning that could be corrected by external stimuli. Similarly to the prevailing therapies for war neurosis such as hypnosis or electroshock, the individual was supposed to be deceived or shocked back into psychic normalcy.[39]

Its extensive influence notwithstanding, Mikorey's expert testimony did not constitute the only source for the Bundeswehr panic manual. Instead, the manual's authors made a concerted effort to integrate a broader international discourse on panic. Its opening paragraph cited Soviet and American military service regulations concerning panic. In particular, West German conceptions of panic were also influenced by US military psychiatry. The panic manual drew explicitly on US publications such as the 1947 book by the military historian S. L. A. Marshall, *Men against Fire*.[40] Incorporating American insights into West German definitions of *panic*, however, was not entirely unproblematic. American military psychiatry had in World War II followed a fundamentally different path than German military psychiatry. In particular, the mass of neuropsychiatric casualties especially during the early military campaigns in North Africa led to the ascendancy of exogenous etiologies for psychiatric breakdowns among soldiers emphasizing external factors rather than, as was common in the German tradition, the constitutional weakness of individual soldiers.[41] The newly emerging diagnosis of "combat fatigue" reflected a modernization of American military psychiatry, which also included a basic recognition of fear as a natural reaction in combat.[42] Nonacademic authors like Marshall may not have incorporated all these diagnostic innovations in their analysis and were therefore better suited as a source for German military psychiatrists than actual American psychiatric experts. Yet some of Marshall's observations nevertheless stood in uneasy tension with the traditional panic conception as advocated by Mikorey and German military psychiatry. For example, in his analysis of a panic freeze among the participants in an amphibious landing operation (presumably during the Allied invasion of Normandy), Marshall appeared more willing to attribute panic to an "obvious danger without escape."[43] He explained panic not only as the result of a "fear fantasy" but also as the product of an overwhelming threat that exceeded the coping mechanisms of the individual. Finally, he highlighted the "knowledge and sympathetic understanding of each other" within the troops as the most important antidote to panic, which contrasted markedly with the more hierarchical emphasis on military leadership in West German publications.[44] To be sure, his analysis was not completely incompatible with

West German military psychiatry—his book gave several examples of panics that resulted from misunderstandings and misperceptions that were also quoted or paraphrased in the Bundeswehr manual. Yet, in line with the conventional wisdom of American military psychiatry, Marshall did not see panic primarily as an outgrowth of pathological personality structures and was more willing to consider potential external triggers for panic reactions among soldiers than was Mikorey.

The reception of the 1962 panic manual among West German officers appears to have been largely positive. As late as 1969, several Bundeswehr units requested thousands of additional copies for the purpose of instruction. These requests, however, also suggested the need for some changes, such as the inclusion of references to biological and chemical weapons as well as historical examples from more recent conflicts in Korea and Vietnam.[45] The most extensive suggestions for revisions came from the agency in charge of officer training in the army (Inspektion für Erziehung und Bildung im Heer). It proposed the inclusion of additional evidence in the manual, including examples from "more recent military history" drawn primarily from Marshall's publications. This agency was especially dissatisfied with the sections on the prevention and containment of panic and suggested the incorporation of findings by another military psychiatrist, Julius Deussen, that had been published in a three-part series on panic in the journal *Wehrkunde* in 1962–63.[46] While Deussen's article series appeared too late to be included in the 1962 manual, it featured prominently in its reception, and it was cited in the never published 1963 successor volume (discussed below).

Deussen's intellectual biography as well as his views on panic exhibited many similarities to Mikorey. Like Mikorey's, he was an early supporter of National Socialism and subsequently became even more implicated with the Nazi regime than Mikorey was. He served as a local party leader in Heidelberg from 1937 on and joined the German Institute for Psychiatric Research in 1939. In collaboration with its director and one of the main Nazi racial scientists, Ernst Rüdin, he actively participated and was personally implicated in the medical murder of mentally disabled children. During the Second World War, he served, like Mikorey, as a consulting psychiatrist for the German army. After 1945, he initially practiced as an ordinary physician. He also engaged in a retroactive normalization of his biography by systematically obscuring his institutional affiliations and personal links to major perpetrators of the Nazi euthanasia program. For example, he reframed his engagement with central areas of Nazi racial policies by describing his areas of expertise as laying in fields such as "psychopathology," "constitutional biology," or "mind/body medicine." In so doing, he recommended himself as a reliable expert in military psychiatry and actually joined the newly formed Bundeswehr in 1956.[47]

Like Mikorey, Deussen shared the preexisting conceptions of German military psychiatry that portrayed panic as an abnormal, sick reaction to an external threat.[48] For Deussen, panic occurred when an individual's ratio and willpower were completely overwhelmed by "affectivity." Its analogy was mental illness or a regression to the lowest evolutionary level. Reflecting the classic view of war neurosis among military psychiatrists, resistance to panic depended solely on the psychological constitution of the individual. While "only strong characters were capable of resisting fear," certain "neuro-psychopathological conditions such as "affective irritability," "lack of willpower," or an "emotional personality structure" predisposed individuals to succumb to panic.[49] In contrast to Mikorey, however, Deussen offered more detailed suggestions for preventing and containing panic. As a nonprofessional army relying on the draft, the Bundeswehr was susceptible, according to Deusen, to "silent panic," "general insecurity," and "defeatism" as he perceived it in West German society more broadly. The prevention and containment of panic thus depended on the psychological preparation of the entire society, not just the army. While the authority of the doctor, with his "affectionate strictness," constituted the main therapeutic instrument, Deussen conceded that the credibility of authority in general had been badly damaged as a result of the Hitler dictatorship. He therefore suggested that therapy should not just be based on punishment and discipline but also consider emotional factors. Rejecting Gustave Le Bon's model of mass psychology, he advocated an approach that would make use of the new insights of group psychology, which, he argued, was more conducive to a democratic order. In particular, he propagated a strengthening of the emotional bond between the troops and their leader as the most promising antidote to panic. These suggestions reflected at least an initial effort to adjust his theory of the diagnosis and treatment of panic to the changed conditions of the postwar democratic state and to move away from the more authoritarian Wehrmacht model.[50] And it was precisely these insights on combating and preventing panics that later critics of the 1962 manual demanded be included in a revised edition.

Panic in Nuclear War

The 1962 panic manual also contained a short section on possible panics in nuclear war. Yet military planners must not have been completely satisfied with its cursory analysis. In the wake of the Cuban Missile Crisis in October 1962, when the world had come closer than ever to a nuclear escalation, the necessity of addressing the nuclear danger became even more apparent. As a result, the next issue in the series *Innere Führung* was supposed to focus solely on the issue of "psychological crises in nuclear war—possibilities for mastery." By approaching the topic not from the "military-psychiatric" or "legal" but rather from the "pedagogical-psychological" perspective, this manual sought to identify the psychological challenges of nuclear war as well as provide pragmatic guidelines for mastering them.[51]

"Psychological Crisis in Nuclear War" acknowledged the novelty and unprecedented destructive power of nuclear weapons while, at the same time, seeking to normalize them by placing them in a historical continuity of previous new weapons and their eventual mastery. Reflecting Mikorey's persistent influence on conceptions of panic, the historical precedents of Hannibal's elephants and the tanks of the First World War reappeared in this context as examples of initially new and frightening weapons that, however, eventually did not prove to be invincible. "Like the tanks, every new weapon has its weaknesses. And nuclear weapons are no exceptions to this rule," the manual concluded.[52] Nuclear weapons thus appeared not as qualitatively new but rather as an extension of a long history of military innovations—dating all the way back to elephants and the crossbow. Thus, the manual sought to square the circle by acknowledging the unprecedented destructive power of nuclear weapons while at the same time suggesting the possibility of their eventual mastery.

Illustrating a persistent belief in the possibility of mastering the threat of nuclear weapons, the manual asserted that individual psychological strength was the decisive factor in determining the outcome of a nuclear conflict. This is why psychological preparation and training for nuclear war—analogous to the training of astronauts for a space mission—was so essential. In seeking to prepare soldiers for nuclear war, the manual both drew on and transcended conventional psychiatric wisdom. While fear of radiation constituted, for example, one of the genuinely new and dangerous elements of nuclear war, the manual was excessively concerned with the possibility of a *simulated* radiation exposure. It followed German military psychology's traditional concern with war neurosis and argued that the symptoms of radiation poisoning were the same as the symptoms of fear—fatigue, headaches, nausea, thirst. Thus, nuclear war might provide an even stronger incentive for "escape into disease," which the manual equated with a form of desertion. Drawing on the previous discussion of panic, the manual made individual manifestations of fear appear to be deeply contagious and hence to have the potential to escalate into mass mutiny and desertion, or what Mikorey had called a *forward panic*.[53] It sought to counteract a controversial public statement by a high-ranking official in the West German Defense Ministry, Gerd Schmückle, who had asserted that, in a possible nuclear war, soldiers would basically engage in mass desertions and the "enemies would get out of their tanks and hug each other in their desperation."[54] Instead, the manual stipulated, the anticipation and overcoming of psychological crises was essential for waging and, ultimately, winning a nuclear war.

While this line of argument remained very much within the mainstream of West German military psychiatry, the manual also reflected some important shifts in West German military culture. Unlike the panic manual, it emphasized more strongly—rather in line with Deussen's thinking—trust in a military leader as an alternative to military coercion in seeking to avoid mutiny and desertion. Reflecting the new normative model of the citizen in uniform, it asserted that the individual soldier should be enlightened about the potential destructive impact of nuclear weapons and thus enabled to prevail in a potential nuclear conflict. In similar ways, the manual also began to reflect the gradual shifts in the psychiatric assessment of the psychological consequences of extreme violence. While traditional military psychiatry had denied any possible traumatic effect of war and violence and had asserted a "practically complete resistance of the human being toward psychological challenges," the manual now asserted that the possible challenges of nuclear war might exceed the available coping mechanisms of the individual.[55]

Like its predecessor, the 1963 manual also drew on a transatlantic discursive context, yet it did so in different ways. Unlike the 1962 publication, it did not rely on popular military history or military psychiatry but rather located the soldier's psychological crisis within a broader existential dilemma of modern man. The fighter in nuclear war was believed to experience an extreme version of modern "nervousness" that was also experienced by managers or schoolchildren in contemporary society.[56] The soldier's sense of psychological crisis in nuclear war appeared as part and parcel of a general crisis-ridden modernity that witnessed a quick succession of real and imagined crises. Reflecting the increasing popularity of psychoanalytic explanatory models in West German public discourse during the early 1960s, the manual asserted that the fear of nuclear war already "dominated consciously or unconsciously all of humanity" and, if repressed, was bound to reappear in the form of a "hectic yearning for life." A potential use of nuclear weapons would activate this preexisting, more generalized anxiety and "willingness to panic" that the invention of nuclear weapons had already introduced in modern society.[57] The manual thus ended up portraying the psychological situation of the soldier in nuclear war as paradigmatic for the predicament of modern man more generally.[58]

Paradoxically, it was precisely the rise of modern weapons of mass destruction that appeared to assign a new significance to the individual soldier. Under the conditions of nuclear war, he might be thrown back to a virtually primitive form of fighting. As such, the manual's proposed solution to the individual soldier's situation in nuclear war consisted in the appeal to his endurance and primitive fighting power. In so doing, the manual drew on a broader Western discourse that contrasted an alleged "softening" of Western soldiers in modern consumer society with the ostensible "absence of need" (*Bedürfnislosigkeit*) among "Asiatic soldiers."[59] This is why the strengthening of soldiers' emotional and psychological capabilities appeared essential for their capacity to prevail in a nuclear war. While these suggestions might have constituted, in the words of one historian, a "labored attempt to rescue military qualities from past chivalrous episodes in the age of mass destruction," they were also quite compatible with a very modern emphasis on the strengthening of the ego, vis-à-vis the anonymous forces of modern mass society, as it was propagated by psychoanalytic approaches that had been reimported from the United States to West Germany.[60] In this sense, the 1963 manual actually reflected the gradual dissolution of conventional wisdom in traditional military psychiatry and a renewed openness to external, especially American, influences in conceptualizing psychological crises in the modern age.

The topic of the manual was sufficiently new and controversial to warrant a broader external review. The Ministry of Defense sent a draft to the Protestant Church Office and the Catholic military bishop as well as to the high commands of the different military branches for their review.[61] The resulting evaluations were almost all very critical. And, even though a meeting of all the involved parties eventually concluded that, after significant revisions, the brochure should appear under the revised title "Psychological Crisis in Combat," it was ultimately not included in the series *Innere Führung* and never published. The volume that succeeded the panic manual focused on the inconspicuous topic "training made easy," and subsequent volumes never picked up the theme of the psychological challenges of nuclear war.

While it is not clear what exactly prevented the publication of the manual, commentators formulated a series of—sometimes contradictory—objections that, taken together, betrayed a considerable uncertainty regarding the possible psychological and emotional challenges of nuclear war. Some reviewers criticized one of the central premises of the brochure, namely, that past confrontations with new weapons offered guidelines for mastering the challenges of nuclear weapons.[62] These critics contested the essential historical continuity of nuclear weapons and asserted their incomparability to previous innovations in military technology. At the same time, however, reviewers also formulated the opposite criticism, namely, that the brochure's "bloody superrealism" might lead especially younger soldiers to the conclusion that there was no protection from or defense against nuclear weapons.[63]

Critics also charged that the manual did not live up to its promise of offering guidelines for mastering psychological crises in nuclear war. They challenged the content and the tone of prescriptive guidelines for individual behavior in the case of nuclear war. To the public relations division of the high command, the manual's suggestion to face the inferno with "conviction, courage, and belief in God" sounded too much "like past rhetoric when there was no real help anymore." In similar ways, the "language" and "style" of the brochure appeared outdated, expressing a "pathos one can no longer bear today."[64] Its rhetoric deviated from a "clear and calm soldierly language" and represented the opposite of the "sobriety, steadfastness, and calmness" that it sought to project.[65] This criticism reflected a broader suspicion of excessive emotionality and a more general propagation of sobriety in West German public discourse.

While the manual thus offered few useful guidelines for soldierly behavior in nuclear war, some reviewers even worried whether it "might not produce the opposite effect among the troops."[66] In particular, there was considerable unease regarding the extensive discussion of mutinies and desertions. Commentators, for example, did not agree with the inclusion of the above-cited statement that asserted desertion and fraternization among soldiers to be a likely consequence of a nuclear war.[67] The revised version therefore was supposed to treat these subjects only briefly.[68] Other critics suggested shifting the discussion away from the extreme circumstances of nuclear war to instead include "examples of daily life in peace time" in which individuals "have overcome their life anxiety [*Lebensangst*] and have withstood the threat of panic." Such examples promised to be more relevant than examples drawn from military history. But they also required a stronger emphasis on ideology since "courage and steadfastness" were possible only if individuals were convinced of their way of life. Reflecting a more general Western insecurity regarding the ideological commitments of its citizens, commentators worried that Soviet leaders had an easier job with respect to this issue since "communism represents for its followers the religion and the future of the world."[69]

Finally, reviewers criticized some of the sources on which the insights of the manual drew. In particular, American practices and experiences represented, at best, an ambivalent point of orientation. In light of the reference to American prisoners of war in Korea, observers argued, for example, that the Germans certainly have "very rich experiences in this area" and "do not need to borrow from the Americans." The same reviewers also highlighted the potential difference in mentalities between German and American soldiers.[70] On the other hand, the false assertion that US military ordinances had "not dared to approach this topic" of psychological crisis in nuclear war made the public relations division of the army high command wonder whether the subject was ripe for discussion within the West German military at all.[71] A representative of the Ministry of Defense noted that the "entire brochure neglected the significance of military psychiatrists," who "alone can decide over the disease character of a specific psychological reaction" and who are "solely qualified in eliminating weak human beings."[72] Contrary to the frequent references to popular US military history, this commentator sought to uphold the authority of military psychiatry over evaluating psychological responses to war and violence.

Continuity and Change in West German Conceptions of Panic

This defense of the traditional authority of military psychiatry failed to realize, however, that the field itself was undergoing important transformations. The classic rejection of exogenous explanations for war-related trauma, which formed the conceptual basis of traditional understandings of panic, was gradually dissolving by the late 1950s. This development resulted from the confrontation with—and increased receptivity toward—the prolonged psychological suffering of Holocaust survivors but also of former German soldiers and prisoners of war who in 1955–56 returned from ten years and more in Soviet captivity. In these cases, an increasing number of West German psychiatrists were willing to grant that extreme experiences of prolonged suffering and violence exceeded the normal coping mechanisms of individuals. Transatlantic influences from psychiatric diagnosis and practice in the United States were especially important in this process. As a result, West German psychiatrists began to diagnose Holocaust survivors and former prisoners of war with an "experience induced personality disorder" that transcended the traditional rejection of war-related trauma. Precisely at the time of the publication and planning of the two panic manuals discussed above, the conceptual ground of West German military psychiatry was shifting.[73]

This shift, however, did not encompass all aspects and did not affect, at least initially, conceptions of panic.[74] A 1973 synthetic study of fear by Walter von Baeyer, one of the younger and more conceptually innovative West German psychiatrists, provided a good indicator regarding the extent and limitations of the changes. Von Baeyer's understanding of war-related trauma and fear clearly reflected the new paradigm. Drawing on numerous Western European studies of Holocaust survivors, von Baeyer listed various forms of fear that resulted from what he called the experience of *annihilation*. In so doing, he explicitly rejected "classic German psychiatry" and its notion of an "imagined or desired traumatic neurosis." Citing the cases of Israeli Holocaust survivors, he insisted that their persistent psychological problems, especially nightmares, resulted not from any "psychopathological conditions" or an "abnormal reaction to the extreme experience of persecution" but rather from a deep and existential shock resulting from the prolonged threat of annihilation.[75]

While von Baeyer's conception of fear and trauma thus reflected a new and increasingly transnational orthodoxy, his remarks on panic continued to sound remarkably traditional. Like his predecessors in the 1950s and 1960s, von Baeyer characterized panic as a "collective, uncontrollable, all consuming fear reaction that spreads in moments of real or imagined life danger." While this definition still left open the possibility of panic as a response to a real threat, he depicted military panics as largely unjustified. They often derived, he argued, from "a general panic mood based on rumors" and were often prompted by "individuals who were disposed toward fear." He continued to invoke Friedrich Panse's classic study on the absence of panic among German civilians during Allied bombardment in World War II, thus highlighting its extraordinary and long-lasting impact on West German conceptions of panic. At the same time, he cited similar studies of English civilians during the Blitz that reaffirmed the traditional notion of the far-reaching adaptability of human beings to catastrophes while also signaling the opening up of West German psychiatry to the findings of international and Western research. In addition, he mobilized psychoanalytic insights from American studies among survivors of tornado catastrophes to support his general thesis. Only those individuals who had exhibited previous psychic maladjustment tended to display prolonged and intense fear reactions in these situations. Either they had denied any possible threat from natural disaster and subsequently needed to see their "infantile fantasies of omnipotence collapse," or they had exhibited significant fear before the catastrophe and had projected their own latent fears into the future. Nonneurotic individuals, by contrast, quickly developed a realistic perception of the catastrophe without abandoning their basic hope for survival and did not suffer any lasting damage to their psychic health.[76] Unlike Mikorey's, von Baeyer's conception of panic thus incorporated recent findings of transnational psychiatry, including psychoanalysis. Yet these findings served to support rather than undermine endogenous interpretations of panic as the expression of a pathological personality structure, not primarily as a natural response to an overwhelming external threat.

Traditional conceptions of panic as advocated by Mikorey also entered other institutional spaces. As part of the establishment of a West German civil defense system from the 1950s on, a "protective committee" was formed within the Ministry of the Interior. It brought together, in a series of subcommittees, renowned West German scientists from a variety of fields who were put in charge of harnessing scientific research for the protection of the civilian population against the impact of a possible conventional or nuclear war. As such, the protective committee was part and parcel of a virtual shadow state in the Cold War, that is, institutional structures that emerged parallel or next to the regular state structures and that served the sole purpose of preparing for military contingencies.[77] Mikorey had been a member of this protective committee from 1962 on, and, in 1971, he initiated the formation of a new subcommittee that was specifically dedicated to panic.[78] This "panic committee," as it was initially labeled, was eventually renamed the "committee on psychobiology" in order to denote that it was concerned not only with panic but also with a broader range of human behavior in extreme situations, including the attempt to prevent panics.

The committee also widened its purview from military psychiatry to more general research on stress as well as to the pharmacological development of psychiatric drugs that were supposed to stabilize "weak persons who are susceptible to panic." In so doing, it pointed forward to the still-ongoing biologization of pathological behavior in Western psychiatry.[79]

Still, by the early 1970s, traditional concepts of panic were in decline. In part, this was the case because the concept lost one of its main advocates when Mikorey passed away in November 1977. In a meeting of the subcommittee on psychobiology a few weeks later, the committee chairman, Hippius, declared: "Panic is not an issue of grave concern."[80] Even military officers moved away from the traditional and pathologizing concept as it had been advocated by Mikorey, at least on the level of rhetoric. At a 1978 meeting of the psychobiology committee, a Bundeswehr representative, one Oberst Schuh, for example, reiterated some of the basic aspects of the panic concept. Yet he felt compelled to preface his presentation with the assurance that "fear and anxiety are normal reactions to danger" and that "each human being needs to be entitled to his own feelings."[81] He therefore reflected broader changes in West German emotional culture that increasingly tended to depathologize feelings and their expression in public. When traditional Cold War conceptions of panic finally became subjected to vehement criticism in the early 1980s, critics were therefore aiming at a moving target.

Reform Psychiatry, the "Second Cold War," and Challenges to Cold War Panic

Despite gradual changes from the 1970s on, however, West German conceptualizations of panic exhibited a remarkable longevity throughout World War II and the early Cold War era. The discontinuous moment finally occurred as a result of the confluence of a psychiatric reform movement in the 1970s, on the one hand, and the renewed escalation of Cold War tensions in the early 1980s during the "Second Cold War" on the other. An increasingly critical history now denounced military psychiatrists as "machine guns behind the front" who had deployed considerable force and violence against their patients in order to uphold military discipline.[82] At the same time, the prospect and the possibility of nuclear war in Central Europe reemerged with the NATO double-track decision and the controversy over the deployment of US nuclear missiles in Central Europe to counter a perceived Soviet superiority. In West Germany, these developments prompted the largest peace movement in postwar history, a movement deeply informed by a new emotional regime that validated the expression of strong emotions in public as a sign of healthy subjectivity. Fear in particular moved to the center of the emotional economy of the peace movement and began to assume the status of a "new form of rationality."[83] In this context, prevailing conceptions of panic came under attack.

A pamphlet of the German Society for Social Psychiatry can serve as a case in point here.[84] The main target of the pamphlet was another military psychiatrist, Rudolf Brickenstein, who, like Mikorey and Deussen, had served as a consulting psychiatrist in the Wehrmacht and then became the director of the neuropsychiatric division of the Bundeswehr hospital in Hamburg, a position he held until 1980.[85] But the pamphlet made clear that Brickenstein's publications on panic derived from the "principal writings of the psychiatrists Mikorey, Panse and Deussen, which had been written already in the 1950s and early 1960s."[86] What this critique missed, however, was the fact that these panic conceptions had been undergoing revision already since the 1970s and no longer reflected the conventional psychiatric wisdom. Still, the pamphlet vehemently denounced these publications, as well as the 1962 Bundeswehr manual on panic, as propagating a repressive political strategy that sought to prepare West German society for a possible nuclear war. Drawing on a significantly revised understanding of fear, the authors of the pamphlet identified the "collective fear" of an impending nuclear war as entirely justified. According to them, the stigmatization of such normal and natural emotional reactions to the imminent threat of annihilation as panic thus provided the conceptual ground for the violent repression of any dissenting voices or acts. For these critics, military psychiatrists' notion of panic constituted not an academic or a scientific but rather a political concept. The notion of the "panic person," which had informed conceptualizations of panic from the 1940s on, thus appeared as a barely disguised form of pathologizing any form of political dissent. Moreover, the theories of military psychiatry as well as the activities of psychiatrists in military advisory committees significantly contributed to the militarization of West German society already in peacetime. The pamphlet thus directly challenged the close institutional nexus between psychiatry, state institutions, and the military that had so decisively shaped the academic biographies of panic researchers such as Mikorey and Deussen from the 1930s into the 1970s.

The confluence of the psychiatric reform movement with new social movements such as the peace movement was part and parcel of a broader challenge to scientific authority in the 1970s and 1980s. In what the sociologist Ulrich Beck has termed the *risk society,* scientific expertise was more important than ever in discerning hidden and often invisible threats. Yet, at the same time, scientists lost the uncontested institutional authority that psychiatrists like Mikorey and Deussen had still claimed and enjoyed. Instead, scientific expertise proliferated and was increasingly mobilized for competing political positions.[87] This demonopolization and increasing plurality of scientific approaches reshaped the relationship between science and emotions. It contributed to a broadening of emotional norms, which legitimized a wider range of emotional experiences. The depathologization of panic thus was part and parcel of a broader process of emotional democratization in postwar West Germany that ultimately also brought West German panic conceptions into sync with prevailing Western assumptions. This meant, above all, the increasing biologization of panic as it characterized the definition of the phenomenon in the fourth edition of the *Diagnostic and Statistical Manual of Mental Disorders,* which was released in 1994. *Panic* was thus defined as a "discrete period of intense fear that is accompanied by at least 4 of 13 somatic (e.g. palpitations, shortness of breath) or cognitive (e.g. fear of dying) symptoms."[88] Recent psychiatric studies also seek to depathologize the condition. "A panic attack" thus constitutes a "normal and appropriate emotional response occurring at an inappropriate time or in an inappropriate situation." It "does not represent a biological dysfunction per se" and "if it occurs during a specific threat, it would be seen as adaptive and useful."[89] By defining *panic* as an entirely healthy response to an overwhelming threat such as nuclear war, this conceptualization is diametrically opposed to the pathologization of panic as advocated by Mikorey and Deussen throughout the Cold War.

Postwar conceptualizations of panic thus underwent an important destigmatization that coincided with larger processes of liberalization and Westernization in postwar West Germany. Yet it would be misleading to see the history of the West German discourse on panic as simply moving from repressive and traditional beginnings to an emancipated and enlightened present. Such a teleological view misses both historical contingency and the potentially repressive potential of any normative emotional construct. Cognitive psychologists, for example, challenge the current biologization of panic and point to preexisting "catastrophic cognitions" as preconditions for panic attacks. Such conceptualizations aim not to reintroduce West German military psychiatrists' stigmatizing notion of constitutional weakness into the etiology of panic but instead to highlight the significance of specific historical and cultural contexts for the individual and collective susceptibility to panics. Rather than representing merely a functional and adaptive biological response of the organism, panics thus vary across cultures and historical periods depending on which "sensations are viewed as potentially catastrophic by members of a society or social group."[90] As such, contemporary panic conceptions again introduce an element of historical variability and cultural contingency that rejects biologizing and universal conceptions of panic. In so doing, they underline the historicity of emotional norms more generally and hence point to the historical variability and cultural contingency of the evolving relationship between science and emotions.

NOTES

My thanks to Daniel Gross, Ulrike Strasser, and Rebecca Plant for their helpful comments on an earlier version of this chapter.

1. This scenario is cited in Klaus Naumann, "Machtassymetrie und Sicherheitsdilemma: Ein Rückblick auf die Bundeswehr im Kalten Krieg," *Mittelweg 36* 14.6 (2005–6): 13–28. See also Helmut Hammerich, "Der Fall 'MORGENGRUSS': Die 2. Panzergrenadier Division und die Abwehr eines überraschenden Feindangriffs westlich der Fulda 1963," in *Die Bundeswehr, 1955 bis 2005: Rückblenden, Einsichten, Perspektiven*, ed. Frank Nägler (Munich: Oldenbourg, 2007), 297–312.
2. This military dilemma had already become apparent shortly after the Federal Republic joined NATO in May 1955 in exchange for full sovereignty. NATO military exercises such as "Carte Blanche" (June 1955) and "Lion Noir" (March 1957) projected the massive use of nuclear weapons on West German territory, which, in the first case, would result in civilian casualties up to 1.6 million. See, in general, Christian Greiner, "Die militärische Eingliederung der Bundesrepublik Deutschland in die WEU und die NATO," in *Anfänge westdeutscher Sicherheitspolitik*, vol. 3, *Die NATO-Option*, ed. Militärgeschichtliches Forschungsamt (Munich: Oldenbourg, 1993), 707–50. Mark Cioc, *Pax Atomica: The Nuclear Defense Debate in West Germany during the Adenauer Era* (New York: Columbia University Press, 1988); Bruno Thoss, *NATO-Strategie und nationale Verteididungsstrategie: Planung und Aufbau der Bundeswehr under den Bedingungen einer massiven atomaren Vergeltungsstrategie 1952 bis 1960* (Munich: Oldenbourg, 2006); and Marc Trachtenberg, *A Constructed Peace: The Making of the European Settlement, 1945–63* (Princeton, NJ: Princeton University Press, 1999).

3. On these attempts at emotion management with respect to civil defense, see my "'Everybody Has a Chance': Civil Defense, Nuclear *Angst*, and the History of Emotions in Postwar Germany," *German History* 27.2 (2009): 215–43.
4. See, e.g., Philip Wylie, "Panic, Psychology, and the Bomb," *Bulletin of the Atomic Scientists* 10.2 (1954): 37–40, 63. Postwar discussions of panic in the United States often invoked the 1938 panic in response to the Orson Welles *War of the Worlds* radio broadcast. See Joanna Bourke, *Fear: A Cultural History* (Emeryville, CA: Shoemaker Hoard, 2005), 178–84; see also Jackie Orr, *Panic Diaries: A Genealogy of Panic Disorder* (Durham, NC: Duke University Press, 2006), 79–164.
5. Jacques Derrida, "No Apocalypse, Not Now (Full Speed Ahead, Seven Missiles, Seven Missives)," *Diacritics* 14.2 (1984): 20–31.
6. Matthew Connelly et al., "'Generals, I Have Fought Just as Many Nuclear Wars as You Have': Forecasts, Future Scenarios, and the Politics of Armageddon," *American Historical Review* 117.12 (2012): 1431–60.
7. Jan Plamper, "Fear: Soldiers and Emotions in Early Twentieth Century Russian Military Psychology," *Slavic Review* 68.2 (2009): 259–83; Susanne Michl and Jan Plamper, "Soldatische Angst im Ersten Weltkrieg: Die Karriere eines Gefühls in der Kriegspsychiatry Deutschlands, Frankreichs und Russlands," *Geschichte und Gesellschaft* 35.2 (2009): 209–48.
8. I borrow the notion of emotional democratization from Daniel M. Gross, *The Secret History of Emotions: From Aristotle's "Rhetoric" to Modern Brain Science* (Chicago: University of Chicago Press, 2006), 5.
9. Georg Meyer, "Zur inneren Entwicklung der Bundeswehr bis 1960/61," in *Anfänge westdeutscher Sicherheitspolitik*, vol. 3, *Die NATO-Option*, ed. Militärgeschichtliches Forschungsamt (Munich: Oldenbourg, 1993), 851–1162, 914–15.
10. David Clay Large, *Germans to the Front: West German Rearmament in the Adenauer Era* (Chapel Hill: University of North Carolina Press, 1995), 176–204.
11. Meyer, "Zur inneren Entwicklung der Bundeswehr bis 1960/61," 912–13.
12. "Panik: Erkennen—Verhüten—Bekämpfen" (hereafter cited as "Panik"), Schriftenreihe Innere Führung, Reihe Erziehung—Heft 8, hrsg. vom Bundesministerium der Verteidigung, 1962, BM-BWD7-35-8-62, 9–10.
13. See Frank Biess, "Feelings in the Aftermath: Toward a Postwar History of Emotions," in *Histories of the Aftermath: The Legacies of the Second World War in Europe*, ed. Frank Biess and Robert Moeller (New York: Berghahn, 2010), 30–48; and, in general, Peter Stearns, *American Cool: Constructing a Twentieth-Century Emotional Style* (New York: New York University Press, 1997).
14. "Panik," 11.
15. Erich Rothacker, *Die Schichten der Persönlichkeit* (1938), 7th ed. (Bonn: Bouvier, 1966), 80–83. See, in general, Volker Böhnigk, *Kulturanthropologie als Rassenlehre: Nationalsozialistische Kulturphilosophie aus der Sicht des Philosophen Erich Rothacker* (Würzburg: Königshausen & Neumann, 2002).

Rothacker explicitly stated in a later edition of his book that he saw "no reason to change the theories as developed in [his] text." Rothacker, *Die Schichten der Persönlichkeit*, vi. The degree of Rothacker's personal involvement with National Socialism remains controversial. See Ralph Stöwer, *Erich Rothacker: Sein Leben und seine Wissenschaft vom Menschen* (Göttingen: Vandenhoeck & Ruprecht, 2011), 13–17.

16. The Federal Republic had renounced the production but not the possession of nuclear weapons in the 1955 Paris Treaties. For an analysis of the larger debate over nuclear weapons, see Cioc, *Pax Atomica*.
17. On Mikorey's biography, see Andreas Michael Weidmann, "Professor Dr.med. Max Mikorey (1899–1977): Leben und Werk eines Psychiaters an der Psychiatrischen und Nervenklinik der Ludwig-Maximilians-Universität München" (Ph.D. diss, University of Munich, 2007). I am grateful to Professor Weidmann for sharing with me copies from the personal papers of Max Mikorey. The treatment of psychiatric casualties in the German army appears to have been initially somewhat less severe in World War II than in World War I. Yet, from 1943 on, it became possible to subject soldiers to electroshock therapy without their consent. See Svenja Goltermann, *Die Gesellschaft der Überlebenden: Deutsche Kriegsheimkehrer und ihre Gewalterfahrungen im Zweiten Weltkrieg* (Munich: DVA, 2009), 176–91.
18. On the consulting psychiatrists, see Georg Berger, *Die beratenden Psychiater des Deutschen Heeres, 1939 bis 1945* (Frankfurt: Peter Lang, 1988).
19. Max Mikorey, "Das Psychopathenproblem im modernen Krieg," Public Lecture, Beratender Psychiater, München Wehrkreis VII, box 48, Nachlass Mikorey, Forensische Abteilung, Klinik für Psychiatrie, University of Munich. See also Weidmann, "Max Mikorey," 171–74.
20. Mikorey, "Psychopathenproblem."
21. Frank was a personal friend of Mikorey's family. He was sentenced to death at the Nuremberg trial of major war criminals and executed in October 1946. On Mikorey's visit to Frank, see Weidmann, "Max Mikorey," 178. Ironically, it was Frank who testified in favor of Mikorey's opposition to the Law for the Prevention of Hereditary Disease only five days before his execution. Ibid., 181–82.
22. Ibid., 192–94.
23. Ibid., 201–18.
24. Max Mikorey, "Die Atombombe und das Problem der Panik," box 8, Nachlass Mikorey; Weidmann, "Max Mikorey," 251–52. On the larger context of civil defense, see Biess, "'Everybody Has a Chance.'"
25. Weidmann, "Max Mikorey," 238–39.
26. Bernd Weisbrod, "The Moratorium of the Mandarins and the Self-Denazification of the German Academe: A View from Göttingen," *Contemporary European History* 12 (2003): 47–69; Bernd Weisbrod, ed., *Akademische Vergangenheitspolitik* (Göttingen: Wallstein, 2002).
27. Weidmann, "Max Mikorey," 258.

28. "Panik," 5–7.
29. Max Mikorey, "Grundsätzliches zur Paniksituation: Gutachten im Auftrag des Bundesverteidigungsministeriums" (1958), 6, box 102, Nachlass Mikorey.
30. Ibid., 38.
31. Friedrich Panse, *Angst und Schreck in klinisch-psychologischer und sozial-medizinischer Sicht* (Stuttgart: Thieme, 1952). On Panse, see Frank Biess, *Homecomings: Returning POWs and the Legacies of Defeat in Postwar Germany* (Princeton, NJ: Princeton University Press, 2006), 78–81.
32. Mikorey, "Die Atombombe und das Problem der Panik."
33. "Panik," 31.
34. On this history, see esp. Paul Lerner and Mark Micale, "Trauma, Psychiatry, and History," in *Traumatic Pasts: History, Psychiatry, and Trauma in the Modern Age, 1870–1930*, ed. Paul Lerner and Mark Micale (Cambridge: Cambridge University Press, 2001), 1–30; and Paul Lerner, *Hysterical Men: War Psychiatry and the Politics of Trauma in Germany, 1890–1930* (Ithaca, NY: Cornell University Press, 2003).
35. Mikorey, "Grundsätzliches zur Paniksituation," 37.
36. "Panik," 13–16, 35–37, 22.
37. Mikorey, "Grundsätzliches zur Paniksituation," 271.
38. "Panik," 25–26.
39. On these therapies, see also Ruth Klocke, Heinz Peter Schmiedebach, and Stefan Priebe, "Psychological Injury in the Two World Wars: Changing Concepts and Terms in Germany Psychiatry," *History of Psychiatry* 16.1 (2005): 43–60.
40. S. L. A. Marshall, *Men against Fire: The Problem of Battle Command in Future War* (New York: William Morrow, 1947), 145–50.
41. Hans Pols, "Die Militäroperation in Tunesien und die Neuorientierung der US-amerikanischen Militärpsychiatrie," in *Krieg und Psychiatrie*, ed. Phillip Rauh, Babette Quinkert, and Ulrike Winkler (Göttingen: Wallstein, 2010), 129–52; Gerald Grob, "Der Zweite Weltkrieg und die US-amerikanische Psychiatrie," in ibid., 153–64.
42. See Ellen Herman, *The Romance of American Psychology: Political Culture in the Age of Experts* (Berkeley and Los Angeles: University of California Press, 1995), 95–112; on similar discussions of panic as a natural and adaptive reaction in American social psychology, see Orr, *Panic Diaries*, 90–91. On divergences between German and American military psychiatry during World War II, see also Plant, chapter 8 in this volume.
43. Cited in "Panik," 20.
44. Marshall, *Men against Fire*, 150.
45. I. Korps to Bundesminister der Verteidigung, March 26, 1969 (1,930 copies), and Luftwaffengrupe Nord to Bundesminister der Verteidigung, March 19, 1969 (1,700 copies), BA-M, BW 3939.

46. Inspektion Erziehungs- und Bildungswesen im Heer, March 14, 1969, BA-M, BW 3939.
47. Volker Roelcke, "Kontinuierliche Umdeutungen: Biographische Repräsentationen am Beispiel der Curricula vitae des Psychiaters Julius Deussen (1906–1974)," in *Concertino: Ensemble aus Kultur- und Medizingeschichte: Festschrift zum 65. Geburtstag von Prof. Dr. Gerhahrd Aumüller*, ed. Kornelia Grundmann and Irmtraut Sahmland (Marburg: Universitätsbibliothek Marburg, 2008), 221–32.
48. J. Deussen, "Über Entstehung und Bekämpfung der Panik, I.–III. Teil," *Wehrkunde* 11.12 (1962): 665–71, 12.2 (1963): 84, and 12.3 (1963): 144–58; Willi Rothe, "Die Panik im Kriege: Ihre Ursachen und die Möglichkeiten ihrer Bekämpfung," *Wehrkunde* 11.9 (1962): 490–96.
49. Deussen, "Über Entstehung und Bekämpfung der Panik, I. Teil," 669.
50. Deussen, "Über Entstehung und Bekämpfung der Panik, III. Teil."
51. Entwurf, "Seelische Krisen im Atomkrieg—Möglichkeiten ihrer Meisterung" (hereafter cited as "Seelische Krisen"), Schriftenreihe Innere Führung—Reihe Erziehung, Heft 9, hrsg. vom Bundesministerium für Verteidigung, 1963, BA-M, BW 2-3939. According to Orr, *Panic Diaries*, 172, interest in collective panic began to wane in the United States at exactly the same time in the early 1960s.
52. Ibid., 23.
53. Ibid., 35–39, 62–73.
54. Gerd Schmückle, "Die Wandlung der Apokalypse: Eine Betrachtung über das Kriegsbild in Europa," *Christ und Welt* 15.4 (1962): 3.
55. "Seelische Krisen," 55.
56. Ibid., 6.
57. Ibid., 44. On the reception and popularization of psychoanalysis in postwar West Germany, see Tobias Freimüller, *Alexander Mitscherlich: Gesellschaftsdiagnose und Psychoanalyse nach Hitler* (Göttingen: Wallstein, 2007).
58. "Seelische Krisen," 13.
59. Ibid., 56.
60. Meyer, "Zur inneren Entwicklung der Bundeswehr bis 1960/61," 915. See also Freimüller, *Alexander Mitscherlich*.
61. Bundesminister der Verteidigung, July 1, 1963, BA-M, BW 2–3939.
62. Fü B I 6 to Fü B I 4, July 30, 1963, BW 2-3939. See also Katholisches Militärbischofsamt to Bundesminister der Verteidigung, August 1, 1963, and Fü B VII to Fü B I 4, November 7, 1963, BWI-313474.
63. Fü B VII 1 to Fü B I 4, July 6, 1963, BW 2-3939.
64. Ibid.
65. Katholisches Militärbischofsamt to Bundesminister der Verteidigung, August 1, 1963, BWI-313474.
66. Fü B VIII 1 to Fü B I 4, July 6, 1963, BW 2-3939.
67. Fü B I 3 to Fü B 1 4, July 19, 1963, and Fü B VII 1 to Fü B I 4, July 6, 1963, BW 2-3939.

68. Fü B I 4, January 24, 1964, BW 2-3939.
69. Bundesminister der Verteidigung to BMVtg, Fü B 1, November 29, 1963.
70. Fü B VII 1 to Fü B I 4, July 6, 1963, BW 2-3939.
71. Ibid.
72. Bundesminister der Verteidigung to BMVtg, Fü B 1, November 29, 1963.
73. On these developments, see Biess, *Homecomings*, 81–87, 212–14; and Goltermann, *Gesellschaft der Überlebenden*, 273–341.
74. In this sense, panic discourses exhibited a greater degree of discontinuity between psychiatric reform in the 1950s and the antipsychiatry movement in the 1960s and 1970s than did other areas of psychiatric theory and practice. See Catherine Fussinger, "'Therapeutic Community,' Psychiatry's Reformers and Antipsychiatrists: Reconsidering Changes in the Field of Psychiatry after World War II," *History of Psychiatry* 22.2 (2011): 146–63.
75. Walter von Baeyer and Wanda von Baeyer-Katte, *Angst* (Frankfurt: Suhrkamp, 1971), 175, 185, 91.
76. Ibid., 172, 175, 169–75.
77. The history of this protective committee—and of this wider shadow state—is virtually unknown and deserves further exploration.
78. Prof. Dr. H. Langendorff to Prof. Dr. M. Mikorey, July 14, 1971, box 57 "Schutzkomission 1971/72," Nachlass Mikorey; Curriculum Vitae Mikorey in BAK, B106/733374.
79. H. Hippius, "Psychobiologie (Verhalten in Belastungssituationen)," Wilhelm Janke, "Über einige Grundprobleme der experimentellen Stressforschung," and Erlo Lehmann, "Untersuchung zur Beeinflussung von Stress-Reaktionen durch verschiedene Dosierungen eines Tranquilizers," BAK, B106/54520. On the rise of psychopharmacology, see also Orr, *Panic Diaries*, 197–209.
80. Ergebnisbericht über die Sitzung des Fachausschusses VIII der Schutzkomission beim BMI am 3.12.1977 in München, BAK, B 106–73388.
81. Sitzung des Fachausschusses VIII der Schutzkomission beim BMI am 25. Juni 1977, August 15, 1977, "Merkblatt zum Verhalten in Katastrophensituationen." The speaker ultimately reiterated a more traditional panic concept in his presentation.
82. Peter Riedesser and Axel Verderber, *Maschinengewehre hinter der Front: Zur Geschichte der deutschen Militärpsychiatrie* (Frankfurt: Fischer, 1996).
83. Susanne Schregel, "Konjunktur der Angst: 'Politik der Subjektivität' und 'neue Friedensbewegung,' 1979–1983," in *Die Politik der Angst im Kalten Krieg*, ed. Bernd Greiner, Christian Th. Müller, and Dierk Walter (Hamburg: Hamburger Edition, 2009), 495–520; and Frank Biess, "Die Sensibilisierung des Subjekts: Angst und 'Neue Subjektivität' in den 1970er Jahren," *WerkstattGeschichte* 49 (2008): 51–72.
84. "Panikpersonen sofort eliminieren!" *Deutsche Gesellschaft für soziale Psychiatrie* (Hamburg, 1983), Hamburger Institut für Sozialforschung (HIS), SBe544, box 4.

85. Klee, *Das Personenlexikon zum Dritten Reich* (Frankfurt: Fischer, 2003), 75.
86. Ibid., 8. See also Rudolf Brickenstein, "Individualreaktionen, Summationsphänomene und Kollektivreaktionen in Katastrophen," *Münchner Medizinische Wochenschrift* 122.42 (1980): 1459–62.
87. Ulrich Beck, *Risikogesellschaft: Auf dem Weg in eine andere Moderne* (Frankfurt: Suhrkamp, 1986), 254–99.
88. Devon E. Hinton and Byron J. Good, "Introduction: Panic Disorder in Cross-Cultural and Historical Perspective," in *Culture and Panic Disorder*, ed. Devon E. Hinton and Byron J. Good (Stanford, CA: Stanford University Press, 2009), 4–12. Quotation from Georg W. Alpers, Stefan G. Hofmann, and Paul Pauli, "Phenomenology and Epidemology of Anxiety Disorders," in *Oxford Handbook of Anxiety and Related Disorders*, ed. Martin M. Antony and Murray B. Stein (Oxford: Oxford University Press, 2009), 34–46, 34.
89. Bruce F. Chorpita, David H. Barlow, and Julia Turosvsky, "Fear, Panic, Anxiety and Disorders of Emotion," in *Perspectives on Anxiety, Panic, and Fear* (Nebraska Symposium on Motivation), ed. Debra A. Hope (Lincoln: University of Nebraska Press, 1996), 251–328, 295. See also the discussion in Richard McNally, "Fear, Anxiety, and Their Disorders," in *Fear: Across the Disciplines*, ed. Jan Plamper and Benjamin Lazier (Pittsburgh: University of Pittsburgh Press, 2012), 15–34.
90. Hinton and Good, "Panic Disorder in Cross-Cultural and Historical Perspective," 18.

EIGHT

Preventing the Inevitable: John Appel and the Problem of Psychiatric Casualties in the US Army during World War II

REBECCA JO PLANT

On October 28, 1944, Major Morris J. Leslie spoke to medical officers taking a crash course in psychiatry at the U.S. Army's School of Military Neuropsychiatry on Long Island in preparation for assignment overseas. Earlier that year, Leslie had served as regimental surgeon with an infantry regiment that endured hard fighting in Italy. During the question-and-answer session following his speech, he offered the neophytes in his audience some stern advice: "Psychiatric cases are a very serious problem in combat. In this connection, gentlemen, you have got to remember that a medical officer in combat is not a humanitarian. Your primary function is to maintain the effective strength of your unit as far as possible, and from that point of view, you have got to stop being a doctor. We don't need doctors out there, we need medical officers."[1]

It is hard to imagine such a statement being made about any other medical condition. Soldiers who suffered from the myriad bodily assaults and ailments to which they fell victim—from combat wounds to malaria to syphilis—clearly needed doctors, not just discipline.

But the situation was far murkier when it came to "psychoneurosis" or "combat exhaustion," precisely because few military men regarded these conditions as strictly medical, even though they fell under the province of military medicine.[2] In treating such cases, medical officers, especially those at the front line, were therefore expected to adopt a skeptical stance. Citing his own experiences, Leslie described how he transformed himself into "a son of a bitch," evacuating only a minority of manifestly "hysterical" cases. "You will have five who think they can't take it for every one like that."[3]

Today, few Americans associate war trauma with the so-called good war and the greatest generation that fought it. But the proportion of men discharged from the U.S. military on neuropsychiatric grounds was in fact significantly higher in World War II than in World War I, Korea, or Vietnam.[4] By the time Leslie delivered his remarks, military command and the public more broadly had come to view psychiatric casualties as a serious problem that undermined the war effort and potentially threatened the nation's long-term security. Concerns over the high incidence of neuropsychiatric disorders grew so pronounced that, in 1946, Congress passed the National Mental Health Act, which for the first time appropriated federal funds to support research on the prevention and treatment of mental illness.[5]

Historians have thus rightly viewed World War II as a major boon to American psychiatry. They have shown how the war fueled a marked increase in the number of trained psychiatrists while dramatically accelerating the reorientation of the profession away from its prior focus on severe mental illness toward a greater concern with less disabling disorders. They have also analyzed how the war spurred the ascendance of psychoanalytic and psychodynamic approaches and the destigmatization and normalization of mental disorders, setting the stage for the flourishing of postwar therapeutic culture. In general, the historical literature conveys an image of psychiatrists (and psychological experts more generally) as emerging from World War II infused with new energy and self-confidence owing to wartime experiences that allowed them to expand the realm of their expertise.[6]

But links between the American psychiatric war effort and subsequent postwar developments appear less seamless when examined in light of the way in which military psychiatry was actually practiced—not stateside, or in rear hospitals far removed from the fighting, but at or near the front. Unlike the steady stream of government pronouncements and popular advice literature meant for public consumption, unpublished (and at the time classified) government documents reveal widespread resistance to attempts to normalize psychoneuroses and frame servicemen's neuropsychiatric disorders in nonjudgmental, medical terms. In particular, the experience of treating combat soldiers appears to have led some psychiatrists and medical officers to temporarily abandon a psychoanalytic or psychodynamic perspective. They instead reverted to traditional notions of masculine courage and cowardice even as they also emphasized the importance of external or environmental factors (the extreme stress of combat). In other words, the relationship between psychiatry and the military was by no means unidirectional. The rising influence of psychiatric authority no doubt altered the military's handling of men, but so, too, did exposure to military culture and the realities of warfare affect psychiatrists' own thinking and practice.

To better understand this dynamic, it is useful to employ the historian Barbara Rosenwein's concept of "emotional communities." Divergent views of war trauma cannot be attributed entirely to the endemic conflict between military and medical authority, for they also reflected the radically different social circumstances, from foxholes to DC desk jobs, in which individuals found themselves. As Rosenwein points out, throughout history people have constantly shifted between distinctive emotional communities, such as the church and the tavern, while adjusting their behavior and judgments accordingly. She urges fellow historians to seek to "uncover the systems of feeling" that characterized emotional communities—"what these communities (and the individuals within them) define and assess as valuable or harmful to them; the evaluations that they make about others' emotions; the nature of the affective bonds between people that they recognize; and the modes of emotional expression that they expect, tolerate, and deplore."[7] This is helpful advice for understanding the competing ideals and expectations of masculine behavior during World War II—a period when age-old notions of martial manhood stood in tension with newer, less stringent conceptions of masculinity. Historians have long recognized that conflicting ideals can coexist at any given moment in time, but most works on the history of manhood and masculinity have nonetheless offered essentially linear narratives that emphasize how newer gender ideals supplanted older ones over time.[8] Yet, when "the assumption is one of either change or continuity in masculinities," the complexity of how gender norms actually operate in social life is often obscured.[9] It becomes more difficult to perceive how contingent and provisional such standards can actually be and how rapidly they can shift in response to the demands of particular situations or emotional communities.

American psychiatrists did not articulate a uniform vision of what constituted honorable manhood, nor did they endorse a single view of servicemen's mental problems. Generally speaking, however, the closer psychiatrists and other medical officers came to the actual fighting, and the more intimate their association with combat troops and those who commanded them, the more moralistic they became in their assessments of men who quickly buckled under pressure owing to emotional or psychological difficulties. At the same time, they tended to be highly sympathetic toward those men who cracked after having endured an extended period of extreme stress and hardship. In other words, medical officers and psychiatrists who entered into the besieged emotional community at the front lines tended to adopt that community's commonsense view of emotional frailty and stamina. This reflected their heightened sensitivity to the needs of the combat unit as a whole, as distinct from the individual soldiers they treated; it also bespoke their deepened understanding of the difficulty of maintaining discipline and conserving manpower at the front.

In what follows, I explore this dynamic by analyzing the published and personal writings of the psychiatrist and psychoanalyst John W. Appel, who spent six weeks in the summer of 1944 studying the problem of psychiatric casualties near the front in Italy. Appel's trip coincided with a period of intense combat, during which the Fifth Army suffered its highest casualty rates of the war.[10] By the time he returned to the United States, Appel had concluded that "psychiatric casualties are as inevitable as gunshot and shrapnel wounds" and that efforts at prevention should be directed toward managing manpower losses rather than strengthening individual psyches.[11] In December 1944, the Office of the Surgeon General issued a special bulletin, "Prevention of Loss of Manpower from Psychiatric Disorders," based largely on the report that Appel filed on his return, that came to be widely read and circulated among the army command. Two years later, when the *Journal of the American Medical Association* published an extended and revised version, it introduced the paper as "without doubt one of the most important psychiatric documents" to emerge from the war.[12] Though Appel's central recommendation—that the army should adopt a policy of relieving men once they had spent a fixed number of days in combat—came too late to have a major impact on policy during World War II, it would prove highly consequential in Korea and Vietnam.

I argue that Appel's report bears traces of his emotional experiences within and his observations of the tightly knit communities that sustained men during wartime. His highly pragmatic and epidemiologic approach to the problem of psychiatric casualties, his sympathetic yet far-from-idealized depiction of the typical GI, and his willingness to retain terms like *goldbrick* and *coward* set his paper apart from most other psychiatric literature published in American medical journals during the 1940s.[13] For that reason, it appealed to commanding officers, even those who typically viewed psychiatry with great skepticism. Advocating a limited combat tour—he suggested between 200 and 240 days—Appel crafted a kind of compromise between martial and medical understandings of soldiers' psychiatric disorders.[14] On the one hand, he thoroughly normalized war trauma by insisting that every soldier not wounded or killed would sooner or later break down.[15] Yet he also established a punishingly high standard as to what the typical American man should be able to endure prior to that inevitable breaking point.[16] Thus, even as his report appeared to deal a blow to the embattled ideal of martial manhood, in practical terms it retained, and indeed codified anew, the dichotomy between real men and their failed counterparts.

The View from the Home Front

Like most of his colleagues, John Appel initially believed that screening examinations at induction centers would prove effective in keeping psychiatric casualties to a minimum. Well in advance of the US intervention, the nation's psychiatrists had successfully lobbied for a comprehensive screening program that would weed out men who were psychologically unfit for military service.[17] Examiners sought to flag not only men afflicted with any kind of neuropsychiatric disorder but even those *predisposed* to developing such disorders. Among other common characteristics and behaviors, "sulkiness . . . lonesomeness . . . shyness, sleeplessness, lack of initiative and ambition, and personal uncleanliness" were identified as grounds for rejection. As a result, the rejection rate was strikingly high: the Selective Service turned away a full 12 percent of potential recruits on neuropsychiatric grounds.[18] By way of contrast, during World War I, psychiatric screening had occurred only on a referral basis, with a mere 2 percent of men rejected on such grounds.[19]

Despite the program's ambitious scope, psychiatric screening proved a dismal failure: it did not prevent neuropsychiatric casualties, and it ultimately devolved into a public relations disaster.[20] Appel began to harbor doubts about the efficacy of screening as early as 1942. Helping screen potential recruits in Pennsylvania, he grew increasingly concerned with what he believed to be widespread apathy and resignation. Reasoning that soldiers would better withstand the mental strain of military life if they truly understood the threat posed by the Axis powers, Appel concluded that the army needed to implement a program in preventive psychiatry that would use political education to enhance motivation. Through personal connections, he secured an appointment with the director of neuropsychiatry in the Office of the Surgeon General, Roy Halloran, and persuaded him of the need for a mental hygiene branch.[21] Commissioned at the rank of first lieutenant in March 1943, Appel assumed directorship of the newly established branch. His plan, he later recalled, was to use the mass media to expose "Hitler's and Japan's intent—and ability—to conquer the United States along with the rest of the world." One of the first jobs he undertook was that of teaching basic Freudian concepts to the writers and producers of the director Frank Capra's famous "Why We Fight" film series.[22]

Yet, like the screening program, efforts to foster morale proved disappointing when it came to staving off mental disorders within military ranks. By early 1943, soon after the United States began its first large-scale engagement in North Africa, the army and the army air force faced a veritable flood of neuropsychiatric casualties.[23] At the year's end, *Fortune* magazine reported that nearly ten thousand psychiatric casualties were being discharged every month—a number so shocking that it prompted the War Department to issue a blackout on all information concerning military psychiatry.[24] Appel later described this period as one of "dark pathetic days . . . when it looked as though the whole psychiatric problem was going to tumble down over our ears."[25]

Appel and many others would later credit William C. Menninger, a well-known psychiatrist and psychoanalyst who took over as director of the Neuropsychiatry Division in December 1943, with turning the situation around.[26] Menninger moved quickly to implement damage control and to shift the focus away from screening and toward more effective prevention and treatment. It was as part of this broader effort that he ordered Appel to Monte Cassino, Italy, where the Fifth Army had been evacuating large numbers of men as neuropsychiatric casualties. Appel's mission was to study the problem on the ground and make recommendations that would help stem the tide.

Just days before his departure, Appel attended the 1944 annual convention of the American Psychiatric Association, where he presented a paper with the psychiatrist Malcolm Farrell that illustrates the state of his thinking prior to his fact-finding mission. The paper discussed the limitations of screening and the program's underlying fallacy—that only the predisposed would break down.[27] It also referred to a study conducted in an unidentified combat zone that had yielded two important findings: first, that neuropsychiatric casualties were higher among veteran combat troops than among green replacements and, second, that the rate of psychiatric disorders rose in direct proportion to the intensity of combat.[28] Thus, even before he set foot in Italy, Appel clearly recognized the critical role that external stress played in causing breakdowns, and he had already begun to approach the problem in the aggregate, zeroing in on the fact that combat troops suffered exponentially higher rates of mental disorder than soldiers serving outside the war zone.[29]

But the solutions that Farrell and Appel advanced did not anticipate the groundbreaking proposals that the latter would soon formulate. Reflecting Appel's belief in the efficacy of psychologically astute, morale-building propaganda, the authors' recommendations focused mainly on how to enhance soldiers' motivation. They advised that men should be given a "clearer understanding of the issues at stake in the war," that job assignments and reassignments should be handled with greater care, and that more effort should be made to disseminate psychiatric information to line officers, ideally through small group discussions. Had someone stationed at the front line read these suggestions, he would likely have dismissed them as the work of a clueless Washington bureaucrat. But, a few months later, Appel would write a far more pragmatic and hard-hitting report that suggests how a mere six weeks in a war zone could school a man.

In the Mediterranean Theater

Throughout his trip, Appel kept a detailed journal that is best described as a hybrid in terms of genre.[30] For the most part, the journal reads less like a personal diary than a record of his activities, conversations, and ruminations on the subject of war trauma—one that he may have written with the intention of sharing with his colleagues back in Washington. Many of the entries are essentially records of interviews he conducted with division psychiatrists, battalion surgeons, commanding officers, and personnel, often interspersed with or followed by his thoughts and reactions. Less frequently, the journal echoes other travel writings in what might be called the *American-abroad* tradition, characterized by a recognizable blend of ethnocentrism, voyeurism, and moralism.[31] And, occasionally, it reads like an evocative and dramatic war diary, as when Appel detailed his personal experiences under fire or sought to capture the euphoric chaos of postliberation Rome. Taken as a whole, the journal allows us to track Appel's intellectual and emotional journey as he traveled first to North Africa and then on to Italy.

From the outset, Appel encountered numerous line and medical officers who regarded psychiatrists and psychiatric diagnoses with deep skepticism. Several of the men he interviewed at NATOUSA (North African Theater of Operations, US Army) headquarters complained that military psychiatrists retained the outlook and approach of civilian practitioners, which led to higher "NP" (neuropsychiatric) casualty rates. Appel paraphrased the thoughts of one army medical inspector who "said there are still psychiatrists who haven't learned there is a war going on. Still retain their state hospital civilian viewpoint—baby the patient too much. The men just love to take advantage of a man like that and it's contagious." Soon thereafter, Appel spoke with a colonel who blamed the high incidence of neuropsychiatric disorders among American troops on the "cock-eyed ideas" of psychiatrists in the Surgeon General's Office. "Battle is something no one prefers, anybody wants to get out of it," he explained. "To plant in the minds of the line commander that N-P breaks are abnormal [was a serious mistake]. . . . They are not abnormal. They are normal. . . . The minute you get wavering leadership then you get a mass effect of N-P." These men viewed war neuroses more as a problem of weak leadership than of weak psyches: combat taxed everyone's nerves, and all men would become psychiatric casualties if they could. As the colonel put it: "You can't understand N-P cases just by seeing the cases, you must understand the stress situation."[32]

What initially perplexed Appel was the fact that frontline military psychiatrists to a large extent shared these views. On May 29, he flew in a weapons carrier from Tunisia to Italy, where he noted evidence of recent fighting—the wreckage of cars and trucks, mangled railroad tracks, enormous shell craters, and blown-up bridges as well as "forlorn, dirty-looked" Italians. He viewed this ravaged landscape en route to the 601st Medical Clearing Company (161st Medical Battalion), a special psychiatric field facility or "exhaustion center" where he would spend the majority of his six weeks working as a ward officer. He would also use the 601st as a home base from which he made various brief forays.

The 601st provided a unique forum where psychiatrists in the Mediterranean Theater gathered to debate competing conceptions of war trauma and exchange ideas concerning diagnosis and treatment. According to Albert J. Glass, a military psychiatrist who later helped chronicle the history of the US psychiatric war effort, this exhaustion center also offered a space where the typically "isolated" psychiatrist could himself rest and rejuvenate. Here, he received the "emotional support" he needed "to maintain objectivity" while coping with his "proximity to the combat scene and his own emotional involvement in daily decisions of whether to evacuate or return to duty with its implications of life or death for the individual concerned."[33] By seeking support from one another, military psychiatrists could process their own emotional reactions and quell whatever doubts or guilt they harbored in relation to the GIs over whom they wielded such authority.

On his first night at the 601st, Appel had a four-hour-long conversation with a small group of psychiatrists and other officers, who hammered home one point above all others—the same point that Major Leslie had tried to impart in his lecture at the School of Military Neuropsychiatry. In order to become an effective medical officer, the military psychiatrist had to undergo a major transformation: "The point stressed very strongly by all of them . . . was the necessity of being an officer first and a doctor second—e.g. the first duty is to keep as many men fighting as possible. I don't know still quite what they mean. They said it represented a major change in viewpoints. All of them who change learned the 'hard way.' . . . You have to be tough with patients—tough and firm."

Two days later, Appel began to treat neuropsychiatric casualties himself, though not at the front line. (The 601st clearing station was typically located between six and thirty miles from the front.) The 601st reopened for operation on June 1; by the day's end, it had 280 exhaustion patients. "The thing that impressed me about them most, of course, was their extreme fatigue," Appel noted in his journal. "Without sedation . . . the entire ward consisted of horizontal figures sound asleep. A few were sobbing and a few wandering about. They were all unshaven, dirty, ragged, hungry. . . . Almost all the faces gave the impression of having seen horror." By June 4, Appel had begun to place much greater emphasis on the external forces that precipitated breakdowns, writing: "There is not the slightest doubt in my mind that these N-P cases are 'caused' by danger." By June 24, his shift toward an environmental view of war trauma seemed complete. "Any discussion of neuropsychiatric cases in combat must focus [on] the fact of the overwhelming importance of the danger threat," he wrote in his journal. "Everything else shrinks into unimportance besides it as a cause of neuropsychiatric disorder."[34]

Yet, even as he increasingly came to appreciate the significance of external trauma in precipitating breakdowns, Appel continued to puzzle over how to define the role of the military psychiatrist (or that of the battalion surgeon who treated psychiatric casualties) and how psychoneurosis differed from cowardice. On May 31, he traveled to the Anzio beachhead, where over seven thousand Allied troops had been killed in recent fighting. There he encountered another group of military psychiatrists who echoed the same sentiments he had heard two nights before. "All reiterated the need to be tough on these patients" and insisted that they—psychiatrists—were in fact "much tougher . . . on patients than anyone else," including line officers and other medical officers. "By being tough," Appel wrote, "I still don't know what they mean; possibility: (1) tough suspicious, accusing attitude; (2) sending patients back to duty in spite of anxiety symptoms. I think of course they mean the latter." Back at the 601st, Appel spent the evening of June 3 with Fred Hanson, a leading military psychiatrist in the Mediterranean Theater, discussing whether neuropsychiatric casualties should be considered cowards. "We finally agreed that the men we psychiatrists now see in the Army are the men who in the past world would have been called cowards," he wrote, before further musing: "The term 'coward' has now been discarded by us as being meaningless and misleading. The term 'brave' logically should also be discarded for the same reason . . . [since] moral concepts of human beings are meaningless to a psychiatrist—or at least cannot be adapted to what he sees when he looks at people (or thinks he sees)."[35] Here, the convoluted prose that immediately followed the assertion of a total disconnection between psychiatry and morality mirrors the impossible position of the military psychiatrist, for in reality the pressure of combat inexorably led to a tendency to collapse the two.

Appel's conversations with line officers suggest that they rarely perceived any real distinction between psychoneurosis and cowardice. One commanding officer, "about 33, very pleasant and competent," believed that "being yellow and being a psychoneurotic were the same thing" and insisted that one "couldn't show too much sympathy to the neuropsychiatric cases."[36] A platoon leader shared this view, describing psychoneurosis as "a bad case of being scairt [*sic*]." When pressed by Appel to clarify whether he perceived *any* "difference between psychoneurosis and being yellow," he initially said no. But then, on reflection, he related the case of an "old man" who had been "one of the best men in the outfit" but was "now no good, burnt out, cried, etc." The platoon leader seemed very reluctant to deem this particular individual a coward or indeed even a psychiatric casualty. He confided to Appel that he had appealed to the battalion surgeon to have the man evacuated through medical channels.[37]

These men articulated an understanding of the relationship between manhood and emotional stamina that was widely shared by combat troops during World War II—one that stressed a capacity for endurance over absolute self-control. Combat troops typically resisted the leveling effects of the controversial term *psychoneurosis*, for they regarded the man who broke down soon after coming under fire (or, even worse, before ever being exposed to real danger) in a very different light than the one who sweated it out for many months. Appel captured this distinction in his report when he wrote: "After a man has been in combat for several months and has fought well through several campaigns, he has proved to himself and others that he is neither a weakling nor a quitter. How he behaves after this point cannot disprove this."[38] Thus, an "old man" who broke down and wept with abandon would likely be viewed quite differently—and treated more sympathetically—than a new man who behaved in the same manner.[39] The shame lay less in the symptoms or the loss of emotional control per se than in succumbing to fear and exhaustion without a sustained fight.

Appel's views in this regard may well have been influenced by Raymond Sobel, a twenty-seven-year-old physician who served as a battalion surgeon in the Thirty-fourth Infantry prior to being named division psychiatrist.[40] A few months prior to Appel's visit, during a long stretch that saw little action, Sobel completed a paper on what he termed "Old Sergeant's Syndrome" since the "most striking cases were found among non-commissioned officers who were old in combat experience."[41] He characterized the syndrome as one that afflicted "well motivated, previously efficient soldiers" who experienced a "progressive breakdown of their normal defenses against anxiety in long periods of combat," generally between 180 and 240 combat days. The soldier's most crucial psychological defense—identification with the group—gradually crumbled as the group itself eroded owing to death, injury and sickness, and other causes of separation. Symptoms included anxiety, depression, and a general loss of confidence. According to Sobel, the evacuation of "old sergeants" on neuropsychiatric grounds became a source of grave concern for the army command—"from the company commander to the commanding general of the division"—for these men had often been "among the best and most effective of the trained and disciplined combat infantry soldiers." They were "the key men, the 'old reliables,'" who firmly believed that "a soldier must be manly and courageous and must exhibit endurance and fortitude." Notably, Sobel did not propose a therapeutic solution that could restore them to their former selves; he accepted their own self-assessment of being "washed up" so far as combat was concerned. Instead, he proposed reassigning such men to positions outside shellfire range yet close enough to the front that they could feel of use to the men up front. Because "old sergeant's syndrome" was primarily a "situational reaction," improvement depended on "altering the environment": getting men out of combat.[42]

Sobel clearly conveyed his views to Appel, who recorded them in his journal and appears to have incorporated them into his final report.[43] But some seventy years later what Sobel recalled most about his exchange with Appel was not their discussions concerning the nature of war trauma but rather how "scared" and "clueless" the psychiatrist appeared. "I spent most of the interview explaining, teaching him how to tell the difference between the shell that was coming in and landing and exploding and the shell that was being shot outwards against the Germans," he noted. "I think the whole idea of [Appel's] being in a combat zone had come in from Washington. It was quite upsetting for him, I imagine."[44]

Appel's most serious brush with danger came in early July when he found himself caught up in heavy fighting during a visit to a battalion aid station near Cecina. On July 2, under heavy shelling, he sought refuge in a farmhouse outside the village of Babbino along with twenty GIs and fifteen Italian civilians. Even as they hunkered down, Appel continued writing in his journal; the jagged, erratic script is unlike any of the previous or subsequent journal entries. Decades later, he would recall the intensity of his fear that day and how the shared experience of danger had instantly (if fleetingly) altered his relationship to the men alongside him:

> The sense of danger hit me at once. . . . Constant fear of death accompanied by a feeling of immediate deep bonding with the other men in the unit. Today, more than fifty years later, the faces of the men in my outfit are still vivid in my mind. If, for example, Blackie, a sergeant, should show up at my front door . . . I would recognize him immediately, embrace him, and invite him to stay for as long as he wished. This is true despite the fact that I was with the unit for only three days. I

noticed, too, that once I received orders to return to the rear, the feeling of bonding disappeared immediately. I became a stranger in their midst.[45]

Soon after returning to safer ground, Appel reflected in his journal: "Certainly the main thing I learned . . . was how terrible it is in the front lines, how much the men suffer, and what an awesome thing it is that they keep on as long as they can."[46] The real question, he now believed, was not why so many men eventually broke down but how they could endure for as long as they did.

Yet, if Appel acquired a more visceral understanding of fear and mutual dependence in battle, he also experienced the exhilarating adrenaline rush that can follow in the wake of life-threatening experiences. On July 4, the day after the incident described above, Appel wrote an unusually self-referential entry: "As I walked around I felt tougher, felt perhaps I was a tougher person than I had ever been before in my life. I was freer, more sure of myself, more able to do what I wanted to. But I didn't feel like doing any work, writing any notes or seeing any patients."[47] Having survived a harrowing ordeal, Appel preferred, at least for the moment, to bask in a newfound feeling of masculine assuredness than to concern himself with those who had lost this feeling entirely.

Appel's Report to the Surgeon General

In the report he wrote immediately on his return to the United States, Appel spelled out a series of recommendations for reducing psychiatric casualties. He argued that infantrymen, who endured the most severe privations and faced the greatest danger of death and injury, should receive greater recognition and be granted special privileges. He suggested that the command should do a better job of communicating objectives to the troops and that evacuation be limited to legitimate medical cases. He urged that training be retooled to emphasize mental stamina and self-reliance as well as technical preparedness and proposed that the selection of officers should be more concerned with their "personality structure." He also argued that commanders should learn to regard the rate of psychiatric casualties as "an important source of information" concerning a unit's morale and the quality of its leadership. And he insisted that the replacement system was badly flawed and that new men should always join combat units in small, tight-knit groups rather than as isolated individuals. But, above all, he emphasized the need to establish an incentive for fighting by capping the number of days men were kept in combat.[48]

Appel's tumultuous experiences in the Italian theater shaped his report in fundamental ways. For starters, Appel essentially gave up on the idea that GIs would prove more resilient if only they could be educated to understand the nefarious intent of the Axis powers. Early on during his visit, he had recorded the comments of one combat officer who explained that frontline troops harbored "a distinct feeling of antagonism against any speaker—no matter how expert—if he speaks on solving combat morale problems when he himself knows of these problems only second hand."[49] It would seem that Appel took this and other comments to heart. Though his report briefly discusses how to make "the goal of winning the war a more meaningful one to the combat soldier," this section seems almost pro forma; what he had previously seen as the most effective means of stemming neuropsychiatric casualties became just one of a series of secondary recommendations. Now, he frankly acknowledged that, while it "would be convenient if the soldiers were more concerned with winning the war," victory was "unimportant to the average American soldier" except insofar as it would allow him to return home. The Russians, the French, and the British were naturally more driven to fight because they faced an enemy who had threatened or occupied their lands. In contrast, the typical GI in the European Theater fought "for his buddies and his self respect" but mostly because he had no choice.[50]

This unvarnished depiction of the nation's combat troops was not, however, lacking in sympathy. In contrast to his previous work, Appel's report demonstrated a real grasp of the outlook and grievances of the average combat soldier. "He feels that no one at home has the slightest conception of the danger his job entails or of the courage and guts required to do one hour of it," Appel wrote. "He feels that command does not distinguish between him and the base area soldier and is actually less concerned with his welfare." Although infantrymen contributed and sacrificed far more than others, they felt that their efforts went unrewarded, for they could gain relief only after they had become worthless as fighters. This spelled a future of grim and limited prospects—"death, mutilation or psychiatric breakdown." What the GI on the front line really needed, Appel concluded, was a concrete incentive that truly mattered to him, and what every combat soldier wanted "above everything else except his self respect" was "to get out of combat." Hence the idea of a limited tour of duty, a concept premised on the notion that patriotism and a sense of duty could carry a soldier only so far. For the long haul, he needed a more personal and immediate incentive.[51]

To arrive at this point, Appel had to relinquish his belief that antifascist sentiments would emotionally sustain men. Yet his report is refreshingly free of moralism in its attitude toward tested soldiers who yearned for a break. In marked contrast, he ultimately adopted a hard line toward those men who failed miserably as soldiers. A passage that appears in the draft of his report but that ended up being cut from the version distributed by the Surgeon General's Office reveals how fully he had embraced the line officers' disgust with such men: "As a corollary of increasing the rewards for performance in combat duty, the possibility might be considered of increasing the penalties for failure to perform. Two types of failure are involved: Those who fail through choice; the AWOL's, the malingerers, the goldbricks and men with self-inflicted wounds, and those who fail through weakness; the cowards, the psychoneurotics. It is believed that measures could be taken to increase the stigma of social disapproval attached to the former group by increasing the court martial sentence and providing greater publicity to the individual and his offense."[52] If Appel conceived of the limited tour of duty as the proverbial carrot, he also suggested that harsher sentencing and shaming be used as the stick. The passage is remarkable in that it reveals how harshly he had come to view not only those who failed "through choice" (and hence deserved to be punished and stigmatized) but even those who failed "through weakness." These "psychoneurotics" were not to be confused with the potential "exhaustion" cases with whom Appel sympathized—the "old men" he hoped to shore up with the promise of future respite. His wording appeared to lump or even equate the former with "cowards"—a point not lost on Arthur Ruggles, the chairman of the Committee of Neuropsychiatric Civilian Consultants and one of the many psychiatrists who responded to the initial draft of Appel's report. "I hate to see Psycho-neurotics coming right after cowards and before AWOL and malingerers," he wrote. "I know what you mean but am afraid some of the medical officers and many line officers would immediately put all psychoneurotics in the general classification of cowards and malingerers."[53] In short, what Appel had done was to reproduce the widespread association between psychoneurosis and cowardice even as his own office back in DC was seeking to pry them apart. This eyebrow-raising deviation from the Neuropsychiatry Division's party line reflected his exposure to a very different emotional community—one that could not risk the consequences that would likely follow should psychoneuroses be truly destigmatized.

In theory at least, the long-standing ideal of martial manhood that mandated courage was incompatible with psychodynamic constructions of the self that postulated unconscious drives over which men had little or no control. But, in the end, Appel and many other military psychiatrists—especially those who treated men near the front—frequently resorted to such concepts, in part because they witnessed how ineffective soldiers could adversely affect the unit as a whole. As Appel asserted in his report: "If an inexperienced battalion surgeon or an overly sympathetic psychiatrist permits goldbricks, cowards, and poorly motivated soldiers to escape [through medical channels], the morale of the entire unit is undermined and genuine psychiatric disorders develop."[54] When the *Journal of the American Medical Association* published a revised version of the bulletin in 1946, Appel felt compelled to explain in a note that he had used these admittedly "unscientific" terms "because they had a definite though not precise meaning for the nonmedical officer who would read the report and because they referred to important concepts for which no satisfactory scientific terms are available."[55] But, in truth, the problem was not really a lack of "satisfactory scientific terms." What Appel's recourse to this older, moralistic discourse reveals is simply how difficult it was to hold on to a psychodynamic or psychoanalytic sensibility in the unforgiving and lethal context of war.

A final way in which the initial draft of Appel's report differed from later versions is that it included passages in which he abandoned a stance of detached, scientific objectivity and asserted the importance of personal experience as a means of acquiring knowledge. This is evident in the opening paragraph of the section headed "Findings," reproduced below. The italicized text signifies the material that was ultimately cut and does not appear in the version distributed by the Surgeon General's Office:

The key to an understanding of the psychiatric problem is the simple fact that the danger of being killed or maimed imposes a strain so great that it causes men to break down. [*This fact is frequently not appreciated and cannot be fully understood until one has either seen psychiatric cases just out of the line or himself has actually been exposed to bombing, shell and mortar fire. One look at the shrunken apathetic faces of psychiatric patients as they come stumbling into the medical station, sobbing, trembling, referring shudderingly to "them shells" and to buddies mutilated or dead is enough to convince most observers of this fact. Anyone entering the combat zone undergoes a profound emotional change which cannot be described. To one who has been "up there" it is obvious that*] there is no getting used to combat. Each moment of it imposes a strain so great men will break down in direct relation to the intensity and duration of their exposure. Thus, psychiatric casualties are just as inevitable as gunshot and shrapnel wounds in warfare.[56]

Appel believed that being in the war zone had transformed his views in ways that no other experience could have—a fact that, at least initially, he did not attempt to conceal. As the deleted material clearly shows, somewhere along the way someone—perhaps Appel himself, perhaps William Menninger, or perhaps Surgeon General Norman Kirk—decided to prune the text in ways that expurgated its emphasis on the centrality of subjective experience. It is possible that the passage in question was deleted because of the bulletin's intended audience; after all, the military hardly needed the Surgeon General's Office informing it of the "profound emotional change" that combat induced in men. Yet two years later, when the *Journal of the American Medical Association* published its version of Appel's report, only one of the deleted sentences made its way back into the published text.[57] In this case, references to subjective emotional experiences may have appeared too far at odds with the ideal of detached, scientific objectivity to which medical journals aspired. But, whatever lay behind the excisions, the effect was to cover Appel's tracks, transforming a report that had at least hinted at its author's emotional journey into a less extraordinary document—part internal critique of army policy, part epidemiological study.

Appel's Legacy

Appel could not have foreseen the response his report would garner: no previous work by a psychiatrist had ever been as broadly circulated or as widely read by the military command. He first submitted it to William Menninger, who sent it in memorandum form to Brigadier General Hugh Morgan, the chief medical consultant to the surgeon general. Morgan pronounced it "superb" and suggested that it should not only be forwarded to the surgeon general but also "accompany the recommendations to the line." In September 1944, Surgeon General Kirk sent the memorandum, now under his own signature, to two high-ranking military authorities; he also published it in his monthly report on the army's overall health, where it drew the attention of Army Chief of Staff General George Marshall. Marshall personally sent copies to Generals Eisenhower, Douglas MacArthur, and Mark Clark, commander of the Fifth Army in Italy. Similarly impressed, Eisenhower ordered copies distributed to all commanding officers down to the regimental level in the European Theater. The report played a critical role in the army's decision in late May 1945 to establish a limit of 120 aggregate combat days. (Note that this was a period far shorter than that which Appel had proposed.) But, by then, the war in Europe was over, and, soon thereafter, V-J Day obviated plans to implement the policy in the Pacific.[58]

Looking back at his wartime experiences many years later, Appel claimed that his most important contribution had been to establish the fact that "every man has a breaking point." "This had not been known previously to psychiatrists or to anyone," he wrote.[59] In fact, it was a view held by innumerable GIs and many of those who commanded and treated them. "Is there any reason why four divisions in Italy have to be kept on the fighting for so long—from 60 to 90 [consecutive] days?" asked one group of medical evacuees from the Italian campaign. "There wouldn't be so many NP casualties if the men could get a little rest, warmth, good food, a bath and some sleep. Don't the higher-ups know that even a machine has to be oiled and greased? Is a human being any different?"[60] Appel's real contribution was to provide evidence that verified this widely known truth by tracking the rate of attrition according to the number of days in combat. In other words, he compiled data that statistically verified the breaking-point theory and articulated that theory in terms that resonated with both military officials and medical personnel.

Though Appel was by no means the only person, or even the only psychiatrist, who called for limited tours of duty for infantrymen, his report can nonetheless be credited with having a major impact on military policy in the postwar period. When the United States sent troops to Korea in 1950, it implemented a rotation system in which men were relieved after serving either nine months in combat (270 days) or thirteen months in support units.[61] And in Vietnam, as the historian Ben Shephard notes, "one of Appel's ideas—that everyone wears out in the end—would prove very influential, providing part of the justification for the one-year tour of duty by American soldiers."[62] In both Korea and Vietnam, limited tours of duty were widely credited with keeping psychiatric casualty rates low—although, with regard to the latter conflict, numerous critics would argue that the military's statistics grossly underestimated the war's real psychological impact.

If Appel's report significantly affected military policies, however, his views also came in for criticism as unduly "defeatist and fatalistic," in the words of Colonel Albert J. Glass, the man who led the psychiatric war effort in Korea. (Recall that, during World War II, Glass had served in the Mediterranean and had been interviewed by Appel.) Like virtually all his colleagues, Glass embraced the concept of "forward treatment," meaning that psychiatric casualties should be treated as close to the front line as possible. Ever since World War I, experience had shown that, when men were treated far back from the front, their symptoms became more fixed and the likelihood that they could be restored to duty declined. Psychiatrists attributed this phenomenon to the "secondary gain" that patients experienced: their symptoms proved valuable in preventing them from being sent back to an intolerable situation. But Glass ultimately concluded that another factor—the "emotional reactions and attitude" of the medical officer who treated the neuropsychiatric casualty—was also crucial in determining whether symptoms could be quickly resolved.[63]

In a 1954 article incongruously titled "Psychotherapy in the Combat Zone," Glass explained that the typical young military psychiatrist experienced insecurity and guilt when treating patients in hospitals in the rear but that his attitude underwent a significant "alteration" when he moved forward to the front. There, he not only observed men who successfully "adjusted" to life on the front lines; he also "decreased" his own "feelings of guilt by participation." As a result:

> An inevitable emotional reorientation occurred, namely, the division psychiatrist became identified with the welfare of the group rather than the wishes of the individual. With this change the psychiatrist lost anxiety and guilt when making decisions because he became convinced that it is for the best interest of the individual to rejoin his combat unit, for in no other way can the patient regain confidence and mastery of the situation and prevent chronic tension and guilt. This attitude of the division psychiatrist, stemming from participation with the combat group, makes it possible for him to assume the traditional role as an exponent of reality which insists that the individual continue functioning despite anxiety rather than allowing withdrawal or a disabling neurotic compromise.[64]

According to Glass, the vast majority of acute neuropsychiatric casualties could in "reality" carry on, and it was the psychiatrist's job to be "an exponent of reality"—to make that reality real. But the psychiatrist could accomplish this only through an emotional or psychological readjustment of his own. He had to transfer his identification and loyalty from the individual to the group and assume a disciplinary role within the combat unit—a tightly bound and highly distressed emotional community.

Here, in a nutshell, is what the men whom Appel interviewed had meant when they insisted that, in treating neuropsychiatric disorders, medical officers had to undergo a "major change" that would enable them to act as officers first and doctors second. It was a more complete version of the emotional transformation that Appel himself had undergone during his time in North Africa and Italy. Because he was never part of a unit, and because he and those with whom he interacted knew he would soon return to the States, his evolution was no doubt less complete than that which Glass hoped to see in military psychiatrists and medical officers serving in Korea. But it was consequential nonetheless. While only faint traces of his emotional journey are evident in the published record, its trajectory is clearly discernible in his journal, where he recorded not only his thoughts but also the varying emotions—fear and exhilaration, bewilderment and determination, contempt and admiration—that so profoundly informed his conclusions.

Conclusion

The largely laudatory accounts of the psychiatric war effort that psychiatrists penned in the postwar period typically emphasized how, once they acknowledged the failures of the screening program, they succeeded in dramatically lowering neuropsychiatric discharge rates by employing effective treatment methods that quickly returned men to their units. But what did such methods actually look like as practiced on the front line? Above all, the effective battalion surgeon refused to admit men who reported with symptoms not observable to the naked eye. He quickly learned to "ignore most of what the patient says and evaluate the degree of disability almost exclusively on objective findings," as Appel explained.[65] Or as Sobel put it: "If a guy could walk around, back to duty."[66] When confronted with a soldier who truly could no longer function, the standard protocol was to allow for a short rest, possibly aided by sedation, followed by a strong dose of persuasion, which typically meant a heavy-handed reminder that good soldiers never abandoned their comrades. Obviously, this did not resemble the image of psychiatric care depicted in 1940s cinema and popular culture, which showed talk therapy, hypnosis, or narcosynthesis occurring within a hospital or an office setting. To the extent that a battalion surgeon succeeded in restoring men to the line, he largely did so by steering clear of approaches that encouraged introspection or sought to probe the depths of the unconscious mind.

In the end, it is perhaps not surprising that those physicians who treated men fresh from combat often came away from their experiences more impressed with the profession's limitations than with its allegedly limitless possibilities. "That the tenets of Psychiatry and the philosophy of the military, under conditions of combat, are diametrically opposed has become obvious to me as a result of experiences as Division Neuropsychiatrist," wrote David I. Weintrob in December 1944 after months serving with a division that landed in Normandy. "It is my contention that one has to accept blindly either one view or the other to clarify problems that are constantly arising. . . . [T]here is no 'happy medium.'"[67]

This was not a lesson to take home. As medical officers once again became doctors who embarked on further training, entered into private practice, or assumed university teaching positions, what did they carry with them of their wartime experiences? It is the same question asked more broadly of World War II veterans, who, in contrast to their Vietnam successors, are generally seen as having reintegrated to civilian life to a remarkable degree. In Vietnam, the official psychiatric casualty rate may have been exceedingly low, but public attention to the problem of war trauma gathered force in the war's aftermath, ultimately resulting in a campaign for a new diagnosis that would acknowledge its ongoing traumatic impact. In contrast, during World War II, the official psychiatric casualty rate was exceedingly high, but public attention to veterans' psychological scars dissipated to a vanishing point in subsequent years. The end result was a kind of amnesia surrounding what, during and immediately after the war, had been identified as a problem of massive proportions.

NOTES

I would like to thank Frank Biess, Daniel M. Gross, Frances M. Clarke, Stefan Ludwig-Hoffmann, Rachel Klein, Ruth Leys, John Armenta, and the students in my graduate research seminar on US history for reading and commenting on earlier drafts of this chapter. Thanks, too, to Hilary Coulson for her able research assistance and to Dr. Raymond Sobel for sharing his recollections with me.

1. Major Morris J. Leslie, transcript of question-and-answer session, October 28, 1944, folder "Speech Delivered by Maj. Morris J. Leslie, MC,

Former Regimental Surgeon," box 1302, RG 112, Office of the Surgeon General/Army, National Archives and Records Administration II (hereafter NARAII), College Park, MD.

2. The psychiatric terminology employed by the army medical corps during World War II is an important topic in its own right. At first, medical officers diagnosed both seasoned soldiers and men who broke down in training as suffering from *psychoneurosis*, the most commonly used psychiatric diagnosis. But, in 1943, the War Department instructed battalion surgeons to use the term *combat exhaustion* or simply *exhaustion* on the medical tags of combat troops. Officials wanted to protect men who had stuck it out for many months from the stigma of psychiatric diagnosis; they also feared that men who read their own medical tags might end up developing more severe symptoms. William C. Menninger, "Diagnostic Labels," in *Psychiatry in a Troubled World: Yesterday's War and Today's Challenge* (New York: Macmillan, 1948), chap. 18.
3. Leslie, transcript of question-and-answer session.
4. Statistics on the incidence of wartime psychiatric disorders are notoriously problematic. Although strict diagnostic criteria help explain the comparatively low incidence of neuropsychiatric casualties in World War I, it is nonetheless striking that by conservative estimates the rate of psychiatric casualties is at least two or three times higher during World War II. William C. Menninger, "Psychiatry in World War I and World War II," in *Psychiatry in a Troubled World*, 338–47. In World War II the psychiatric casualty rate was over 10 percent, in Korea it was 4–5 percent, and in Vietnam it was only 1 percent. In World War II, about one-third of all medical discharges were for neuropsychiatric reasons, as compared to 24 percent in Korea and 14 percent in Vietnam. Jack McCallum, "Medicine, Military," in *Encyclopedia of the Vietnam War: A Political, Social and Military History* (2nd ed.), ed. Spencer C. Tucker (Santa Barbara, CA: ABC-CLIO, 2011), 729–33. See also Edgar Jones and Simon Wessely, "Psychiatric Casualties: An Intra- and Interwar Comparison," *British Journal of Psychiatry* 178 (2001): 242–47.
5. Jeanne L. Brand, "The National Mental Health Act of 1946: A Retrospect," *Bulletin of the History of Medicine* 39 (May–June 1965): 231–45.
6. See esp. Ellen Herman, *The Romance of American Psychology: Political Culture in the Age of the Experts* (Berkeley and Los Angeles: University of California Press, 1995), chap. 4; Gerald Grob, *From Asylum to Community: Mental Health Policy in Modern America* (Princeton, NJ: Princeton University Press, 1991); Hans Pols, "War Neurosis, Adjustment Problems in Veterans, and an Ill Nation: The Disciplinary Project of American Psychiatry during and after World War II," *Osiris* 22 (2007): 72–92; Nathan G. Hale Jr., *The Rise and Crisis of Psychoanalysis in the United States: Freud and the Americans, 1917–1985* (Oxford: Oxford University Press, 1995), chap. 11; and Eva S. Moskowitz, *In Therapy We Trust: America's Obsession with Self-*

Fulfillment (Baltimore: Johns Hopkins University Press, 2001). Other important sources on military psychiatry in World War II include Hans Pols, "The Tunisian Campaign, War Neuroses, and the Reorientation of American Psychiatry during World War II," *Harvard Review of Psychiatry* 19.6 (2011): 313–20; Josephine Callisen Bresnahan, "Danger in Paradise: The Battle against Combat Fatigue in the Pacific War" (Ph.D. diss., Harvard University, 1999); Ellen Dwyer, "Psychiatry and Race during World War II," *Journal of the History of Medicine and Allied Sciences* 61.2 (January 2005): 117–43; Ben Shephard, *A War of Nerves: Soldiers and Psychiatrists in the Twentieth Century* (Cambridge, MA: Harvard University Press, 2001), chaps. 14–22; Ruth Leys, *Trauma: A Genealogy* (Chicago: University of Chicago Press, 2000), chap. 6; Rebecca Schwartz Greene, "The Role of the Psychiatrist in World War II" (Ph.D. diss., Columbia University, 1977); Christina S. Jarvis, "'If He Comes Home Nervous': U.S. World War II Neuropsychiatric Casualties and Postwar Masculinities," *Journal of Men's Studies* 17.2 (2009): 97–115; and Alan Bérubé, *Coming Out under Fire: The History of Gay Men and Women in World War II* (New York: Free Press, 1990), chap. 6. The official military history is also a very valuable source: U.S. Army Medical Department, *Neuropsychiatry in World War II*, ed. Albert J. Glass, 2 vols. (Washington, DC: Office of the Surgeon General, Department of the Army, 1966–73).

7. Barbara H. Rosenwein, "Worrying about Emotion in History," *American Historical Review* 107.3 (June 2002): 821–45, 842. William Reddy employs the concept of emotional regimes, which he defines as "the set of normative emotions and official rituals, practices, and emotives that express and inculcate them; a necessary underpinning of any stable political regime." William R. Reddy, *The Navigation of Feeling: A Framework for the History of Emotions* (New York: Cambridge University Press, 2001), 129. For my purposes, Rosenwein's notion of multiple, coexisting, and often competing emotional communities is more germane.

    In recent years, a number of scholars have approached the subject of war trauma from perspectives informed by the history of emotions. See, e.g., Joanna Bourke, *An Intimate History of Killing: Face-to-Face Killing in Twentieth-Century Warfare* (New York: Basic, 2000), and *Fear: A Cultural History* (London: Virago, 2005); and Jan Plamper, "Fear: Soldiers and Emotion in Early Twentieth-Century Russian Military Psychology," *Slavic Review* 68:2 (2009): 259–83.

8. See, e.g., Gail Bederman, *Manliness and Civilization: A Cultural History of Gender and Race in the United States, 1880–1917* (Chicago: University of Chicago Press, 1995).

9. Michael Roper, "Between Manliness and Masculinity: The 'War Generation' and the Psychology of Fear in Britain, 1914–1950," *Journal of British Studies* 44.2 (April 2005): 343–62, 360. Roper also warns that, if we reduce mascu-

linity to "a set of abstract codes," we end up focusing only on the social, losing sight of the individual psyche. He proposes a view of masculinity as "a process in which social scripts are negotiated, one on another, within the self." Ibid., 345, 360. While this advice is well worth heeding, my focus here is less on how individuals negotiated gendered scripts internally and more on how individuals had to negotiate different scripts according to the communities or subcultures in which they found themselves.

10. Appel was in the Mediterranean Theater from May 17 to July 29, 1944. The Allies seized Rome on June 4, two days before D-Day. See Rick Atkinson, *The Day of Battle: The War in Sicily and Italy* (New York: Henry Holt, 2007); and Matthew Parker, *Monte Cassino: The Hardest Fought Battle of World War II* (New York: Anchor, 1945).
11. John W. Appel, "Prevention of Loss of Manpower from Psychiatric Disorders: A Report of the Surgeon General," Special Technical Bulletin no. 3, December 1, 1944, file 147, box 1310, RG 112, NARAII. For the earlier (unpublished) draft of the report, see John W. Appel, "Prevention of Manpower Loss from Psychiatric Disorders," file 146, box 1310, RG 112, NARAII.
12. John W. Appel and Gilbert W. Beebe, "Preventive Psychiatry: An Epidemiologic Approach," *Journal of the American Medical Association* 131.18 (August 31, 1946): 1469–75, 1469. Appel actually authored the report; Gilbert Beebe, a statistician, analyzed data concerning rates of attrition to arrive at a combat tour length that would reduce the number of psychiatric casualties without seriously compromising the need for manpower.
13. The term *goldbrick* dates from the 1880s and originally referred to an item, like a gilded brick, that appeared extremely valuable but was in fact largely worthless. During World War I, people began using the term to refer to soldiers who shirked their duties and failed to pull their weight.
14. In the initial draft of his report, Appel recommended: "Upon completion of 200 (or 240) aggregate days of combat an infantryman be relieved from combat duty for a period of six months and given the option of serving this period in the United States." The bulletin distributed by the Surgeon General's Office does not make such an explicit proposal. Instead, it notes that limiting combat tours to 240 days would lead to a mere 5 percent increase in the cost of replacement troops. The bulletin also omitted a statement that appears in the initial draft and in the *Journal of the American Medical Association* version: "Actually, many of the line officers were emphatic in stating that the limit of the average soldier was considerably less than two hundred or two hundred and forty aggregate combat days. Most men, they stated, were ineffective after one hundred and eighty or even one hundred and forty days. The general consensus was that a man reached his peak of effectiveness in the first ninety days of combat, that after this his efficiency began to fall off, and that he became steadily less valuable thereafter until finally he was useless." Appel, "Prevention of Manpower Loss from Psychiatric Disorders" (draft), 3, 5, and "Prevention

of Loss of Manpower from Psychiatric Disorders: A Report of the Surgeon General," 5; Appel and Beebe, "Preventive Psychiatry," 1470.

15. For a discussion of the wartime normalization of psychoneuroses, which "altered the subjects and purposes of clinical work by reorienting theory and practice away from mental illness and toward mental health," see Herman, *Romance of American Psychology*, chap. 4 (quotation 83).
16. According to studies Appel consulted when writing his report, a regiment could expect that, after 120 days in combat, 50 percent of its men would be killed, wounded, captured, or missing in action; by the 200-day mark, close to 90 percent of the original men would be gone. John W. Appel, "Fighting Fear," *American Heritage* 50.6 (October 1999), http://www.americanheritage.com/content/fighting-fear.
17. In making their case, psychiatrists argued that the nation had paid a steep price for failing to institute such a program in 1917–18. Between 1925 and 1940, the cost of caring for neuropsychiatric veterans had reached almost a billion dollars, and veterans suffering from chronic war neuroses came to occupy nearly half of all beds in Veterans Administration hospitals. Herman, *Romance of American Psychology*, 86. Other sources reported that every veteran hospitalized for a chronic psychiatric condition cost the government between $30,000 and $35,000 per year. Willard Waller, *The Veteran Comes Back* (New York: Dryden, 1944), 166.
18. Greene, "The Role of the Psychiatrist in World War II," 64–65. See also Naoko Wake, "The Military, Psychiatry, and 'Unfit' Soldiers, 1939–1942," *Journal of the History of Medicine and Allied Sciences* 62.4 (2007): 462–94.
19. During World War I, men could be rejected on psychiatric grounds only if they had a "definite, corroborated history of mental disease that required hospitalization or treatment or observation." Moreover, the army surgeon general instructed examiners: "When there is doubt as to a registrant's physical or mental fitness for military service, he should be considered acceptable." Greene, "The Role of the Psychiatrist in World War II," 34–35.
20. The fact that many men felt stigmatized after being rejected on neuropsychiatric grounds created a backlash against the program. See, e.g., Henry C. Link, "The Errors of Psychiatry," *American Mercury* 59 (July 1944): 72–78. In addition, the disproportionate rejection rate for African Americans prompted black leaders to condemn psychiatric screening. Dwyer, "Psychiatry and Race during World War II."
21. Appel's father-in-law was Rear Admiral Alan G. Kirk, who would later serve as the senior US naval commander during the Normandy landings on D-Day.
22. Appel believed that the "Why We Fight" films would be more effective if the producers took into account Freud's tripartite conception of the mind. He emphasized: "The soldier's superego would tell him it was his duty to fight; his ego would convince him fighting was a good idea; and his id would arouse his emotions, his fear, and his anger." Appel, "Fighting Fear."
23. Pols, "The Tunisian Campaign."

24. "The Psychiatric Toll of Warfare," *Fortune*, December 2, 1943, 141–49. On the government's censorship of information related to the psychiatric war effort, see William C. Menninger, "Public Relations," in Glass, ed., *Neuropsychiatry in World War II*, 1:29–151.
25. John Appel to William Menninger, June 27, 1946, in a bound collection of congratulatory letters, William C. Menninger Papers, Menninger Family Archives, Kansas Historical Society, Topeka, KS.
26. For background on Menninger, see Lawrence J. Friedman, *Menninger: The Family and the Clinic* (New York: Knopf, 1990); W. Walter Menninger, "Contributions of William C. Menninger to Military Psychiatry," *Bulletin of the Menninger Clinic* 68.4 (Fall 2004): 277–96; and Rebecca Jo Plant, "William C. Menninger and American Psychoanalysis, 1946–48," *History of Psychiatry* 16.2 (2005): 181–202.
27. The question as to whether war trauma should be attributed to predisposition was by no means a new one; psychiatrists from various combatant nations debated the issue extensively during and after World War I. See, e.g., Paul Lerner, *Hysterical Men: War, Psychiatry, and the Politics of Trauma in Germany, 1890–1930* (Ithaca, NY: Cornell University Press, 2003). For a discussion of how the concept of predisposition in military psychiatry "waxed, waned, and then waxed again" over the course of the twentieth century, see Ben Shephard, "Risk Factors and PTSD: A Historian's Perspective," in *Posttraumatic Stress Disorder: Issues and Controversies*, ed. Gerald Rosen (Chichester: Wiley, 2004), 39–62.
28. This report was likely authored by the neurologist Colonel Frederick R. Hanson on the basis of the findings of a group of psychiatrists who worked under him in the Mediterranean Theater. Appel later described these men as "a particularly able group of psychiatrists" who "made the epidemiological and clinical studies which laid the groundwork for the development of preventive psychiatry in combat troops." John W. Appel, "Preventive Psychiatry," in Glass, ed., *Neuropsychiatry in World War II*, 1:373–414, 393. The papers of a symposium on military psychiatry that Hanson presided over in February 1943 appear in Colonel Frederick R. Hanson, ed., *Combat Psychiatry: Experiences in the North African and Mediterranean Theaters of Operation, American Ground Forces, World War II* (1949; Honolulu: University of the Pacific Press, 2005).
29. Appel's coauthor was the psychiatrist Malcolm Farrell, who served as assistant director of the Neuropsychiatry Division of the Office of the Surgeon General. Malcolm Farrell and John W. Appel, "Current Trends in Military Neuropsychiatry," *American Journal of Psychiatry* 101 (1944): 12–19. This paper was one of two test cases that the Neuropsychiatry Division used to challenge censorship restrictions. William Menninger described how "with great effort" the paper was "hand-carried through various echelons to obtain approval," which was finally granted on the condition that all figures be deleted. Menninger, "Public Relations," 141. The American Psy-

chiatric Association conference at which the paper was first read was held May 15–18, 1944, in Philadelphia.

30. Notebooks, file 154, box 1310, RG 112, NARAII. (These notebooks are mislabeled as having been written by William C. Menninger; however, the dates and places noted exactly match John Appel's itinerary.) There is also a typed transcript. See "Diary. Personal. Author Unknown," box 1301, RG 112, NARAII. (A handwritten note on this document states: "This report is supposed to be John W. Appel's.") Hereafter, I cite this source as Appel Diary.
31. His earliest entries, e.g., refer to the "whore houses" in the Cashbah, where the most "horny, repulsive set of female flesh I have ever seen" practiced its trade. Appel Diary, May 20, 1944.
32. Appel Diary, May 22, 1944. This informant also insisted that "N-P patients among the French and French Arabs" were virtually "unknown."
33. Edwin A. Weinstein, "The Fifth U.S. Army Neuropsychiatric Center—the '601st,'" in Glass, ed., *Neuropsychiatry in World War II*, 2:127–41, 141.
34. Appel Diary, May 29, 1944, June 1, 1944, and June 24, 1944.
35. Ibid., May 31, 1944, June 3, 1944. The psychiatrist, referred to as "Fred," was most likely Frederick Hanson.
36. Ibid., June 3, 1944. Similarly, a captain serving as a medical officer described psychoneurosis as the modern version of cowardice. According to Appel, he related with disgust: "There is no such thing as being yellow anymore—when the case comes up in court, the psychiatrist says he had a psychoneurosis and [he] gets let off." Ibid., July 2, 1944. This conversation appears to have found its way into Appel's report: "It is currently stated in Italy that there is no longer any such thing as cowardice in the U.S. Army, for any man who runs away from the enemy falls into the hands of a psychiatrist before he can be court-martialed and is thereupon declared not responsible for his acts on the grounds of psychoneurosis. Many line officers are very bitter about this." Appel, "Prevention of Loss of Manpower from Psychiatric Disorders: A Report of the Surgeon General," 6.
37. Appel Diary, July 2, 1944.
38. Appel and Beebe, "Preventive Psychiatry," 1471.
39. GIs and their officers routinely used the phrase *old man* to describe anyone who had been in combat for more than five to six months, as did a man Appel quoted in his report: "There's only two of us old men left, and they're no better off than I am." Appel and Beebe, "Preventive Psychiatry," 1470. The expression captured combat troops' feelings of inevitable, impending death as well as their understanding of how combat duty ground a man down and aged him at an astonishing rate.
40. Sobel had received no formal training in psychiatry prior to being named a division psychiatrist in March 1944. "They made me the division psychiatrist because I wrote a paper for the division medical newspaper on the battalion surgeon as psychiatrist," he explained. "That was my quali-

fication." Dr. Raymond Sobel, interview with author, October 27, 2012, Lebanon, VT.

41. Sobel developed his ideas in consultation with two other division psychiatrists, Captains Joseph R. Campbell and Douglas Kelling; they called themselves the Anzio Beachhead Psychiatric Society. The paper was later published as Raymond Sobel, "Anxiety-Depressive Reactions After Prolonged Combat Experience—the 'Old Sergeant Syndrome,'" *Psychiatry* 10 (1947): 315–21, and reprinted in Hanson, ed., *Combat Psychiatry*, 137–46. For background, see Menninger, *Psychiatry in a Troubled World*, 143; and Glass, ed., *Neuropsychiatry in World War II*, 2:48–52.
42. Sobel, "Anxiety-Depressive Reactions After Prolonged Combat Experience." Like the men he discussed, Sobel was by then an old man himself, having survived landings at Salerno (September 1943) and Anzio (January 1944) as well as heavy fighting at Monte Cassino. He had seen many of his buddies wounded or killed and had only narrowly escaped the same fate on several occasions. At age ninety-five, Sobel reflected, only half jokingly: "You asked why I wrote that article. I think I was developing Old Psychiatrist Syndrome. I had been in combat for a long time." Sobel, interview with author. Although Appel did not refer to Sobel by name in his report, he mentioned him several times in his journal. Moreover, in his retrospective account of his experiences, he noted: "The division psychiatrists drew my attention to what they referred to as the old sergeant syndrome." Appel, "Fighting Fear."
43. "Saw some more 'old sergeants' today," Appel wrote on June 14. "These cases are particularly interesting to me because although it is obvious there is something seriously wrong with these men, it is impossible for me to demonstrate any pathology—all that can be elicited is the simple statement 'I'm not good anymore under combat—my nerves are shot.'" Appel Diary, June 15, 1944.
44. Sobel, interview with author.
45. Appel, "Fighting Fear."
46. Appel Diary, July 10, 1944.
47. Ibid., July 4, 1944. On July 8, Appel sent a letter to his boss, William Menninger, in which he must have detailed the dangers he had recently endured, for it prompted the latter to caution him against "unnecessary" risks. "I want you to get all the experience you can," wrote Menninger, "but I don't want you to needlessly be experimenting just to find out how it feels, without considering the fact we haven't got any, and never could get any replacement for John Appel right here in an extremely important Army wide job." William C. Menninger to John W. Appel, July 24, 1944, October 2, 1944, file "Manpower," box 1309, RG 112, NARAII.
48. Appel, "Prevention of Loss of Manpower from Psychiatric Disorders: A Report of the Surgeon General."
49. Appel Diary, May 27, 1944.

50. Appel, "Prevention of Loss of Manpower from Psychiatric Disorders: A Report of the Surgeon General," 5, 3.
51. Ibid., 2–4.
52. Appel, "Prevention of Manpower Loss from Psychiatric Disorder" (draft), 8.
53. Arthur H. Ruggles to John Appel, October 2, 1944, file "Manpower," box 1309, RG 112, NARAII.
54. Appel, "Prevention of Loss of Manpower from Psychiatric Disorders: A Report of the Surgeon General," 6.
55. Appel and Beebe, "Preventive Psychiatry," 1473.
56. Appel, "Prevention of Manpower Loss from Psychiatric Disorders" (draft), 4, and "Prevention of Loss of Manpower from Psychiatric Disorder: A Report of the Surgeon General," 1.
57. The reinserted sentence is: "One look at the shrunken apathetic faces of psychiatric patients as they come stumbling into the medical station, sobbing, trembling, referring shudderingly to 'them shells' and to buddies mutilated or dead is enough to convince most observers of this fact." Appel, "Prevention of Manpower Loss from Psychiatric Disorders" (draft), 4; Appel and Beebe, "Preventive Psychiatry," 1469–70.
58. Appel discusses the history of the report in Appel and Beebe, "Preventive Psychiatry," 1475.
59. Appel, "Fighting Fear."
60. Untitled and undated document, quotation identified as "Observations from patients returning from overseas on the Hospital Ships Algonquin and Acadia, from Report No. 10," folder (Neuropsychiatry) Seventh Infantry Division, box 1333, RG 112, NARAII.
61. Whereas combat troops in Korea were required to serve longer than the 120 aggregate combat days approved at the end of World War II, fighting tended to be more intermittent, and men were granted more short breaks for R & R. Few scholars have written on the history of military psychiatry and neuropsychiatric casualties during the Korean War. See Elspeth Cameron Ritchie, "Psychiatry in the Korean War: Perils, PIES, and Prisoners of War," *Military Medicine* 167.11 (2002): 898–903; Edgar Jones and Simon Wessely, *Shell Shock to PTSD: Military Psychiatry from 1900 to the Gulf War* (New York: Psychology Press, 2006), 119–27; and Shephard, *War of Nerves*, 341–43.
62. Shephard, *War of Nerves*, 245. Shephard also points out, however, that Appel's recommendations concerning ways of improving the replacement system would be "disastrously ignored." Ibid.
63. Albert J. Glass, "Psychotherapy in the Combat Zone," *American Journal of Psychiatry* 110.11 (April 1, 1954): 725–31, 728, 730. Glass delineated the cardinal principles of military psychiatry as follows: *proximity* (treat close to the battlefield), *immediacy* (treat as quickly as possible), and *expectancy* (convey the expectation of a full recovery and prompt resumption

of duties)—or *PIE* in shorthand. Military psychiatrists still teach these principles. As Ruth Leys notes in her astute discussion of Glass's preferred treatment regimen, in rejecting techniques designed to produce abreaction (along with all other therapies that urged the patient to look inward), he "created new disciplinary requirements for patient and physician alike." Leys, *Trauma*, 221–22. For background on Glass, see Jane C. Morris, "Albert Julius Glass, 1908–1983," in *Builders of Trust: Biographical Profiles from the Medical Corps Coin*, ed. Sanders Marable (Detrick, MD: Borden Institute, 2011), 135–44.

64. Glass, "Psychotherapy in the Combat Zone," 730.
65. Appel and Beebe, "Preventive Psychiatry," 1473. As with many other passages in Appel's report, this one directly incorporated a remark by a frontline medical officer whom he interviewed, in this case, Captain Charles Mills, who advised: "Never believe a history—go on objective evidence only." Appel Diary, July 2, 1944.
66. Sobel, interview with author.
67. David I. Weintrob, "Limitations of Psychiatry in Combat," December 2, 1944, Folder, "Combat Exhaustion," Box 1339, NARAII.

NINE

Feeling for the Protest Faster: How the Self-Starving Body Influences Social Movements and Global Medical Ethics

NAYAN B. SHAH

As a medical ethicist and physician, Dr. George Annas observed: "The power of the hunger strike comes from the striker's sworn intent to die a slow death in public view unless those in power address the injustice or condition being protested."[1] In the twentieth and twenty-first centuries, hunger strikes and protest fasts are bracing forms of political protest that summon passionate responses by followers, the public, and authorities. Voluntary protest fasting as a political strategy has ranged from the protest fasts of Mohandas Gandhi and Cesar Chavez to the hunger strikes of the incarcerated British suffragettes in 1912 and 1913, Irish Republican prisoners in Northern Ireland in the 1980s, Soviet dissidents and refuseniks, immigration detainees in Australia and Britain, and political prisoners in Turkey and Guantanamo at the turn of the twenty-first century.[2]

Diverse publics have been riveted by the boldness and desperation of men and women denying themselves food for days and weeks on end. Protesters used extreme bodily distress to draw attention to the intolerable conditions of confinement or to broader social injustices. The protest faster induced political change by publicizing the process of "slow death"—the refusal to nourish the body, the body's ensuing deterioration, and attendant fears of imminent death.[3] The protest fast to the death was a tool of social movement mobilization and charismatic leadership that served to communicate a political ethos of nonviolent civil disobedience frequently infused with spiritual imagery and moral crisis. The act of refusing food focused attention on the passivity of the vulnerable body and drew scrutiny on state and elite aggression, violence, and injustice. This social justice vision of political change emphasized how public scrutiny would pressure aggressors to respond to and remedy the problem identified by the protester.

In the twentieth century, as both prison administrators and social movement allies enlisted medical professionals to monitor hunger strikers, physicians were confronted with an ethical dilemma. They had to decide whether to treat and feed the protest faster despite the protester's intent to continue to fast irrespective of bodily consequences. The hunger strikes carried out by those in the custody of the state produced the most intense dramas of life and death and catalyzed crises of medical ethics and human rights. The prison hunger strike drew attention to the state's tactics of unjust incarceration and detention, political oppression and suppression. It also revalued those incarcerated as political prisoners and denied any state claims that justified incarceration by social criminality. The intractability of the standoff between state officials and prisoners forced physicians to confront ethical dilemmas of care for patients and complicity with state power.

The public scrutiny of the suffering body of the protest faster in the twentieth century reveals the formation of new political communities and emotional frameworks that involve the engagement of social movement publics, the intervention of "politically neutral" medical expertise, and the radicalization of justice movements in liberal democratic societies. As the historians Barbara Rosenwein and William Reddy have argued, the formation of emotional community and emotional style is critical to how political communities understand, measure, and identify emotional suffering. For Reddy, the "self-management" and "self-exploration" of emotions sustain a community's appreciation, understanding, and expression of "emotional styles" with a human utterance, gesture, and disposition.[4]

The politics of responding to hunger strikers marshaled three dimensions of a repertoire of emotional responses and action to the body and psyche of the protester. The first dimension was the social movement mobilization of the sympathy of followers and the broader public, which focused on the suffering body and amplified, within an emotional community, feelings of outrage, sadness, or anxiety for the leader's health and well-being and the larger political cause. The second dimension draws attention to how the self-management and regulation of emotions is critical to the protest faster's use of mental discipline to project resolve and convey fearlessness in the face of pain and death, similarly to how Daniel M. Gross characterizes apathy as not the absence of "emotion" but "a node of special density" that accrues a "rhetorically constituted shadow economy against which a positive economy of emotion is fashioned."[5] In the shadow economy of emotions of frustration, anger, and despair, psychologists observed how charismatic leaders such as Gandhi and Chavez communicated their fearlessness through their ability to overcome pain and basic bodily urges. The third dimension examines the bundle of conflicting emotions—anger, frustration, helplessness—that physicians confronted both in coping with patients who were hostile to treatment and within incarceration regimes that undermined patient-physician trust and autonomous decisionmaking and that they sought to mask through professional distance and clinical objectivity stances. Physicians sought to make their outrage at state constraints, as well as their impatience with self-starvation strategies, visible by arguing that state pressure undermined the quality of care and patient trust as well as identifying a variety of obstacles for patients under confinement to have informed consent in refusing treatment.

Medical Ethics and Care of the Fasting Body

As states called on physicians to intervene and treat hunger strikers in their custody, public and professional debates erupted over the ethical medical intervention for the protesters who are hospitalized and require treatment. The ethical dilemma drew into relief the scientific understanding of nutrition and starvation, the psychological understanding of the human psyche under acute bodily distress, and psychological assessments of charismatic nonviolent leadership in social movement mobilization.

In the first half of the twentieth century, dramatic changes in nutrition science identified deficiencies in amino acids, vitamins, and carbohydrates as critical to fatal consequence of starvation. This diverged from the conventionally prescribed regimens of "beef, eggs, and brandy" that physicians and prison and asylum officers had employed in intensive refeeding and force-feeding in Britain, Ireland, and India.[6] In the late 1960s, Erik Erikson led new inquiries into the consciousness of the protest faster and the emotional charge of the fasting body on charismatic leadership and social movement mobilization. Numerous standoffs in prisons and detention centers in Europe, the United States, Africa, and Asia during the last forty years and the harrowing deaths in 1981 of ten hunger strikers in Maze Prison in Belfast after forty-five to sixty-one days of fasting weighed heavily on the conscience of medical professionals and jolted physicians and human rights advocates to revise protocols for the medical treatment of and intervention in the cases of hunger strikers.[7] At the turn of the twenty-first century, debates in medical ethics and therapy intensified over the physiological and psychological understanding of the condition, treatment, and decisionmaking capacity of hunger strikers in the custody of the state.

In the early 1970s, professional medical associations in European and North American nations and international medical associations addressed the ethical dilemmas confronting physicians treating incarcerated hunger strikers. In 1975, British Medical Association Ethics Committee determined that physicians have final decisionmaking power in intervening in prison hunger strikes and that even a competent adult does not have overriding authority to refuse medical treatment.[8] In 1975, the World Medical Association (WMA) issued the Tokyo Declaration, which prohibited any physician from participating in torture, whether actively or through passive complicity. Under this concern about torture, the declaration advised physicians: "Where a prisoner refuses nourishment and is considered by the physician to be capable of forming an unimpaired and rational judgment concerning the consequences of voluntary refusal of nourishment, he or she shall not be fed artificially."[9] It further stipulated that doctors were obligated to approach prisoners as patients and that they would therefore undertake renourishment only with the patient's express consent.[10] Both these ethical pronouncements retained medical autonomy from the state and attempted to protect the independence of medical judgment as well as the sanctity of the doctor-patient relationship from state or other political and societal pressure. However, the deliberations did not effectively resolve how to determine patient consent and how to understand the physiological and psychological impacts on the hunger striker. The possibility that patient consent could be overridden or that the rehabilitation of a patient could make him or her available for further incarceration and possible torture continued to haunt physicians who confronted the care of hunger strikers.

Medical guidance regarding the observed impact of voluntary self-starvation and the protocols of renourishment had a remarkably limited evidence base and relied on the physiological and psychological measurements and conclusions of a clinical study conducted at the University of Minnesota between November 1944 and December 1945, referred to as the "Minnesota Starvation Experiment." Professor Ancel Keys, the founder of the Laboratory of Physiological Hygiene, designed the study of thirty-six young adult white men, selected from a pool of conscientious objector volunteers, to investigate the physiological and psychological effects of human starvation. The study's results were intended to increase the effectiveness of Allied relief assistance for and dietary rehabilitation of famine victims in Europe and Asia at the end of World War II. Keys and his research team's 1950 *Biology of Human Starvation* provided one of the few clinical studies of the physiological and psychological effects of prolonged fasting and the nutritional and psychological assistance necessary for recovery. The study participants did not undertake a total fast; rather, during the six-month semistarvation period, the subjects' dietary intake was cut to roughly 1,560 calories per day and their meals approximated foods available during the latter stages of war in Europe. Physiologically, the study observed marked declines in basal metabolic rates and an increase in risk of kidney failure, cardiac arrest, and swelling of the limbs.

Prolonged semistarvation also significantly increased symptoms of depression, hysteria, and severe emotional distress and depression, signs of social withdrawal and isolation.[11] The participants reported a decline in concentration, comprehension, and judgment capabilities, although the standardized tests administered showed no actual signs of diminished capacity. Clinicians emphasized the participants' mental stamina, focus, and resilience and their ability to regulate emotional responses to pain and discomfort. The scientists who managed the study had a vested interest in affirming both the autonomy and the stability of the majority (with the exception of two who withdrew because they were unable to comply with the exacting regimen) of the participants' full consent and their own unassailable objective clinical observation and judgment. When concerns about the participants' emotional disruption or psychological impairment arose, they were swiftly settled in favor of an appraisal of the courage and adaptability of the human subjects.

Keys observed that induced semistarvation approximated the physical effects of undernutrition. Drawing from the conclusions of the study and research on eating disorders, researchers speculated about the similarity of psychological effects in patients diagnosed with anorexia nervosa and bulimia nervosa. The Keys study was intended to inform medical protocols for replenishing the body in instances of famine or endemic malnutrition and reinforced informal medical guidance of those engaged in voluntary fasting with a protocol of the careful management of a regimen of gradually increasing fluids, food, and nutritional supplements.[12] The management of bodily regimens, however, does not account for the complexity of the management and exploration of the emotions of the protest fasters and their community.

Creating the Emotional Community of Fearlessness

Fearlessness is an approach that catalyzes intense emotions of anger, distress, and despair and transmutes them into a capacity to endure for the sake of justice physical pain, deprivation, and suffering without capitulating and abandoning deeply held principles. As the psychologist Erik Erikson addresses how Gandhi develops a practice and cultural vocabulary of nonviolent resistance, he taps into an understanding of the emotional community of Gandhi's supporters, who themselves are able to express intense emotions elicited by Gandhi's protest fast. In his 1969 *Gandhi's Truth*, which won the Pulitzer Prize and a US National Book Award, Erikson argues that individual emotional and psychological development can best be understood within the formation of culturally specific and national political communities.[13]

Erikson examines Gandhi's use of a hunger strike during the 1918 textile mill labor protests over wages in Ahmedabad in conveying suffering, mobilizing spiritual symbolism and purification, motivating political allies, and influencing political outcomes. In charting the cultural specificity of Gandhi's milieu, the "emotives" of sadness, outrage, anxiety, and guilt that supporters feel for Gandhi's physical deterioration are a collective and intelligible response of sympathy within a spiritual economy of self-denial and purification. Gandhi's emotional style of patience and disciplining of intense emotions and urges is viewed as a pedagogic model of "striving to conquer the flesh" and "curbing animal passion."[14]

Gandhi adapts the emotional styles of a merchant class and professional aspirational community in his work both in South Africa and in the specific emotional community in Ahmedabad, Gujurat, to which he returned during World War I. Specifically, in central Gujarat and in Bombay, he drew supporters from among high school– and college-educated young men from the upper castes and merchant communities. These young men's communities straddled specific devotional Hindu, Jain, as well as Muslim merchant communities and were central to commercial agriculture, industry, and trade in Gujarat Peninsula and Bombay.[15]

The use of the hunger strike was both a creation of individual human resolve and relationally connected to the response of followers. For this reason, Erikson's approach attends to the reception of fearlessness and the cultural resonance of Gandhi's charismatic appeal among students and young householders because, as Erikson argued, this relationship was framed socially and culturally in Indian (Hindu) society as the transmission of spiritual learning that was channeled between men in their forties and fifties and young men in their twenties. For Erikson, these young men were the most amenable to exhibiting the capacity for spiritual discipline and the ability to undertake nonviolent resistance. On his own terms, Gandhi adopts the protest fast as part of a larger struggle for purification and clarification of the mind, body, and spirit that he pursued prior to assuming leadership of the nationwide civil disobedience movement. Erikson developed a contextual framework for understanding how emotion was mobilized and organized in the service of a revolutionary political movement. He explored Gandhi's process of modeling and providing instruction in self-discipline to his followers and inculcating in them how distress, sorrow, and pain could be converted into spiritual strength and fortitude. As Gandhi conveys in his autobiography and subsequent writing, experimentation with the ability to control and direct emotions was integral to the successful communication and inculcation of a mass social movement based on nonviolent discipline. The protest fast and its severity were intended to make manifest the sacrifice to death—"never start what you have not clearly circumscribed in your mind or what are not ready to suffer for to the very end."[16]

However, Gandhi did recognize that the management of fearlessness did not emerge from the absence of emotion. In his autobiography, he writes about transmuting his own anger, frustration, disappointment, and "distress" at his students' behavior at his ashram, Tolstoy Farm, in South Africa into the imposition of a seven-day fast. He used the fast to conquer his own unmanageable feelings and was so "greatly relieved" once he undertook it that the "heavy load on his mind" had eased and the "anger against the guilty parties subsided and gave place to the purest pity for them." He recognized that he had transmuted his anger and disappointment into pain for those who observed his penance, and he found that the empathy he and his students shared throughout the process "bound [him] to the boys and girls" even more.[17]

Gandhi was well attuned to the psychological impact of his actions on his followers. Several years later, at the millworkers' strike in Ahmedabad, he initiated a fast because after two weeks of "great courage and self-restraint" he saw that the frustration of the strike had made the millworkers "irascible" and prone to respond with violence to the abusive mill guards and police. Because of his fear of "rowdyism," and to stoke the workers' flagging conviction and attendance, he declared that he would fast. Committed workers and followers felt guilty and distressed and wanted to share in the fast or substitute themselves as the fasters, but he dissuaded them. He acknowledged that the mill owners, who were friends and supporters in the nationalist independence struggle, expressed anger and sarcasm because they viewed the tactic as a high-handed attempt to force them to settle an industrial labor dispute. He also acknowledged the grave physical toll that the protest fasts exacted in terms of the muscular pain that resulted from suppressing his urge for self-preservation and self-care. With the rising urgency of hunger, he used spiritual recitation as a means to acknowledge his emotions of anger, hatred, and selfishness and his urges to nourish himself and stimulate his sensory desire and palate. He developed over time and with the frequent use of short- and long-term fasting a set of learned responses to food deprivation and an appreciation of the need for regular water intake irrespective of how "nauseating and distasteful" consuming large quantities of water without food would make him feel.[18]

For Gandhi, the critical process for cultivating fearlessness involved mental control over the emotions and passions as a practice of creating the strength to endure physical suffering. Gandhi explained that the capacity of fearlessness involved many dimensions of detachment: "Fearlessness should connote absence of all kinds of fear—fear of death, fear of bodily injury, fear of hunger, fear of insults, fear of public disapprobation, fear of ghosts and evil spirits, fear of anyone's anger. Freedom from all these and other such fears constitutes fearlessness."[19] It was a spiritual practice that harnessed and replaced emotions, transferred intense emotions to followers, and required self-reflection on motivations—the absence of "arrogance and aggressiveness" and the overcoming of the reactions of the physical body in order to create the quality of calmness or equanimity, which is often perceived as the absence or suppression of emotional response. The practice of fasting requires "great stamina, discipline, courage, and a complete fearlessness from death."[20]

Fearlessness and the harnessing of familiar spiritual models also characterized the protest work of Cesar Chavez. His protest fasting in California, which was publicized across the United States, tapped into rituals and practices that resonated with a distinct emotional community familiar with the suffering of Christ as a model for the management of one's emotions. In his campaign to draw attention to the dire working and living conditions of Mexican and Filipino farmworkers in California's Central Valley, Chavez drew inspiration from Gandhi's nonviolent civil disobedience and the strategy of protest fasts.

On February 15, 1968, Chavez went on a twenty-five-day fast that within weeks garnered nationwide and even international attention. When he broke the fast on March 10, he took communion in front of eight thousand farmworkers and supporters and with Senator Robert Kennedy by his side, an event that was televised nationwide. Through a priest who read his statement, he employed Catholic spiritual and ritual messages to inform the world of the motivation behind his fast. He converted grief and empathetic pain "for the sufferings of farm workers" into a vehicle for expressing a political vision, a call to action in service of "non-violence," and "a call to sacrifice" to both those in the union and those who rallied to support the union's struggle. He promoted a recognition of suffering on behalf of aggrieved peoples as the "truest act of courage" and "unadulterated fearlessness" and explicitly gendered, and thus redefined, this framework of empathy and fortitude: "The strongest act of manliness is to sacrifice ourselves to others in a totally non-violent struggle for justice. To be a man is to suffer for others. God help us to be men."[21] Chavez's model of Christ's suffering for the sins of others resonated with the use of Catholic ritual at the end of the hunger strike. His own definition of manliness depended on channeling the suffering of others into his own burden of responsibility and righteous action. His rhetoric mobilized and infused his model of Christ-like suffering to make his convictions unassailable and unself-interested. His body's endurance symbolized the fusing of spiritual and political leadership and underlined the righteousness of his cause. Like Gandhi, Chavez also relied on medical monitoring during this first protest fast and subsequent ones. Dr. James McKnight advised him to end his fast after twenty-one days, and, while he did not, he did adopt a regimen of "liquids in order to prevent serious damage to [his] kidneys" and muscles.[22]

One of Erikson's critics characterized the communication strategies of charismatic leadership during hunger strikes as problematic. J. L. Masson, a self-described Sanskritist with a profound interest in "classical psychoanalysis," criticized Erikson's interpretation by casting the external responses to fasting as elements of an "enormously complex fantasy, involving both suicidal and homicidal impulses, oral aggression and satisfactions usually associated with infantile passivity." Masson linked Gandhi's practice of fasting with his "odd ideas on bodily intake, self-castigation and liberation from sexual need" and his "narcissistic self-sufficiency" and "subsequent fantasized omnipotence reinforced by the view that his fasting would mysteriously affect other people directly."[23]

Nevertheless, Erikson's and Masson's approaches were attuned to the individual's volition in undertaking voluntary protest fasting. Gandhi and Chavez explained their emotional state in undergoing their fasts as a fearlessness that necessitates the cultivation of spiritual conviction to fortify one's mental resolve when actions that are detrimental to one's own body can be understood to be for the collective good. Erikson and Masson debated whether the fast was a demonstration of individual heroism or individual pathology, but the focus of their concern was the impact on political followers.

South African Detainees' Hunger Strikes

Two decades later, the mass hunger strikes of detainees in South Africa also registered with both the national and the global emotional community. The hunger strikes of hundreds of anonymous men and youths had a profound impact on the medical ethics of treating hunger strikers, the development of international medical professional protocols, the medical understanding of human starvation, as well as debates within the professions of psychology and psychiatry about the nature of depression. Since 1986, the South African government had used emergency regulations to detain political activists. About thirty thousand people had been arrested under these regulations in sweeps in predominately black townships in Transvaal, Natal, and Cape Provinces. By January 1989 more than one thousand were still in detention with no trial or expectation of release in sight. Most of the detainees were black men, along with some Indian and colored men and a handful of women. Most were affiliated with trade unions and United Democratic Front–affiliated student congresses, youth organizations, and civic associations, but some were young men who had been swept up for being in proximity to protests and activities of organizations that the state security force considered subversive.

Although there were dozens of documented hunger strikes in the prisons as early as August 1986, most were individual or isolated instances in particular prisons, often related to prison conditions, the quality of food offered, and the treatment by prison guards and officials. However the hunger strikes, that began on January 23 with a group of twenty young men at Johannesburg's Diepkloof Prison, were indefinite, demanding the immediate and unconditional release of all detainees. It sparked a national wave of hunger strikes by detainees in Transvaal, Natal, and Cape Provinces and solidarity actions elsewhere in South Africa. By early February, the South African and international press began reporting on the hundreds of hunger strikers among the population of detainees. By February 10, a group of forty-two lawyers affiliated with Lawyers for Democracy and Lawyers for Human Rights, which had struggled to represent detainees that the South African state refused to put on trial, had announced a two-day sympathy fast in four cities to support about three hundred hunger strikers. The attorneys reported that, on January 23, seven detainees from among the original twenty hunger strikers at Diepkloof Prison had been hospitalized in serious condition.[24]

In St. Albans Prison near Port Elizabeth, the detainees' support group released a message to the press explaining why, after up to thirty-two months of continuous detention with no hearing or trial, detainees believed that they had "no alternative but to take [their] lives into [their] own hands" to bring attention to their unjust incarceration and demand "immediate release from this dehumanizing detention without trial." The hunger strike at St. Albans began on February 6 and enlisted 105 detainees to "embark on the only action open to [them]—[the] total end of all food consumption (liquid and solid)." This "final action" was necessary because every other avenue for redress, including complaints lodged with prison authorities, visiting judges, and the advocate general and even short-term hunger strikes targeted at "conditions in prison"—"diet, health, education, recreation, visits and communication"—had been unsuccessful in bringing about change and in addressing the reasons for detention without a trial.[25]

Human rights lawyers and detainees' support groups not only served as advocates for detainees and distributed information to the public; they also monitored the physical condition of the hunger strikers. The Pietermaritzburg Detainees Aid Committee, a human rights group responsible for providing "material and moral help to political detainees and their families, and for monitoring and publicizing the effect of detention on society," announced the beginning of a one-hundred-detainee strike in regional prisons and detention centers beginning on February 18. They reported that, within a week, forty detainees who had continued the total protest fast suffered from dizziness, headaches, problems with their joints, and difficulty urinating. Eight days into the hunger strike, participants were split up and held in different police stations in the region. By February 28, twenty detainees were dispatched to hospitals.[26] The scale of the strikes and intensity of the strikers and their national and international publicity finally forced the South African government to rethink its policy of detention without trial. Five days after the suspension of the strikes on March 8, the Ministry of Law and Order began to slowly release detainees. Within two weeks, about two hundred had been set free. Protests followed the limited releases, and, with the intensification of the hunger strikes by detainees, Minister Vlok was forced to accelerate releases. By the end of March, over seven hundred detainees had participated in hunger strike protests nationwide, and the human rights committee documented the release of five hundred detainees. The slow pace of release increasingly frustrated some detainees. Several escaped from hospitals and sought refuge at international embassies and consulates, further drawing attention to the plight of detention. The hunger strikes also spread to detainees who had been in solitary confinement.[27]

In a subsequent study, Drs. William Kalk and Yusuf Veriawa, reputed physicians and medical researchers at the University of Witwatersrand, documented the histories of thirty-three prisoners who had gone on a hunger strike to protest their conditions of detention. Once the prisoners were hospitalized, they were duly informed of Article 5 of the Tokyo Declaration and told that there would be no force-feeding. Moreover, Dr. Kalk considered that their detention without trial constituted a form of torture and refused to discharge them back into detention after they had recovered from the effects of the hunger strike, which became known in medical and human rights circles as "Kalk's refusal." The South African Medical Association (MASA) urged the WMA to review the ethical guidelines on voluntary protest fasting and its treatment.

In 1991, the WMA issued the Malta Declaration, which shielded physicians from participation in coercive state practices, explicated patient autonomy and intentions, and provided guidance to inform the actions of physicians facing the ethical dilemmas of responding to hunger strikes. The development of medical ethics protocols stemmed from the inadequacy of the 1975 Tokyo Declaration in addressing vexing ethical dilemmas of treating political prisoners. The Malta Declaration advised physicians to respect patient autonomy but "ascertain the individual's true intention, especially in collective strikes or situations where peer pressure" could inhibit an individual's ability to consent. In debates, physicians, human rights activists, and state officials argued about how one could discern whether the prisoner was fasting voluntarily or subject to coercion by other prisoners, political groups, or family members. The declaration counseled physicians to act for the benefit of their patients and left open the possibility that they might be able to intervene and "feed artificially" a hunger striker who had lapsed into a semicomatose state or whose ability to consent had been impaired.[28]

Kalk and Veriawa's clinical study of thirty-three South African political poisoners hospitalized after hunger strikes of up to twenty-eight days was the first to detail the physiology and psychology of hunger strikers and provide a more acute medical understanding of the bodily and temporal process of starvation.[29] It detailed rapid physiological changes. During the first few days of starvation, the body consumes its stores of glycogen in liver and muscle, leading to substantial weight loss. After the second week, the body enters "starvation mode" and then "mines" the muscles, bone marrow, and vital organs for nutrients, a process that can impair the functioning of muscles, bone, and organs. Within these two weeks, hunger strikers experience symptoms of "feeling faint and dizzy" and are often confined to bed. Many patients experienced a substantial reduction in effective thyroid function, often resulting in weakness and a sensation of feeling cold. Three-quarters of those studied experienced abdominal pain and dehydration. Under voluntary total fasting, individuals may lose their feelings of thirst and hunger. The study affirmed the vernacular knowledge about fasters' fluid intake and need for salt by advising that a fluid intake of about 1.5 liters per day, supplemented by half a teaspoon of salt, must be maintained. Working from assessments made by an independent psychiatrist, it also showed 77 percent of hunger strikers to be clinically depressed at the time of admission to hospital, although they also demonstrated features similar to those of posttraumatic stress syndrome.[30]

In 1997, Michael Peel, the senior medical examiner of the Medical Foundation for the Care of Victims of Torture in London, published an editorial in the *British Medical Journal* outlining the results of Kalk's study to inform physicians of the physiology of hunger strikers so that they could provide better medical advice to either people who plan to go on a hunger strike or physicians who find themselves treating those who have finished a prolonged fast/hunger strike. He focused on cases of "voluntary total fasting, usually by politically motivated prisoners or prisoners supporting a specific cause," who are willing to "go all the way" and accept the physiological consequences of a prolonged fast.[31]

Both the Tokyo and the Malta accords are based on the premise that the physician's primary duty is the effective and attentive care of "hunger-striking prisoners." Peel advised that the refusal to be force-fed must be respected and that doctors must use their clinical and moral judgment to do their best for their patients. He provided graphic details of prohibited "coercive feeding," such as "prisoners being tied down and intravenous drips or esophageal tubes being forced into them," and, following the WMA Malta Declaration, affirmed that these actions constituted "a form of torture" and that doctors must refuse participation in such measures on the pretext of "saving the hunger striker's life." However, he suggested that physicians could take action if it were for a patient's benefit, which might sometimes mean "disobeying" his express wishes and reviving him, if they were "convinced that the patient will ultimately be glad to be brought back from the verge of death." This slippery ethical quandary, however, also led to the alternative counsel that, if "a prisoner at an advanced stage of a hunger strike is restored to consciousness or to a physiological situation where there can be no doubt about his state of mind, and that prisoner clearly indicates disapproval of the doctor's action, then the doctor should be prepared to step back and not intervene again." In such cases, ensuring the patient's welfare means allowing fasting prisoners the "last possibility of freedom of action, and letting them at least die with dignity." Heeding the informed consent of a hunger striker, confirmed within the trust of the doctor-patient relationship, and respecting the intrinsic dignity of the fasting prisoner is certainly part of the doctor's duty in looking after the patient's welfare."[32]

Kalk, Peel, and the WMA relied on a model of the hunger striker's physical state as one of fearless courage and unwavering and conscious determination. However, in the last decade, medical scientists have applied clinical knowledge regarding anorexia nervosa and self-starvation and offered caution about what the hunger striker's exhibition of fearless determination might mask in terms of the pathology of hunger. The repeated experience of prolonged voluntary fasting might ultimately decrease the fear of pain and death. Self-starvation could induce significant psychological changes.[33]

Documentation from recent studies of "hyper-irritability and alarming levels of aggression" in obese patients on "crash" diets, in Keys's study of experimental starvation of normal subjects, and in studies of those diagnosed with anorexia nervosa has renewed a reconsideration of the psychological changes in hunger strikers, who also exhibit "increasing levels of aggressivity, impulsivity, and anger." The key argument in the analogizing of these similar psychological symptoms for voluntary total fasting, dieting strategies, and eating disorders is that an individual's competence levels can change with each successive day of fasting. It raises a clinical question about whether it is possible to continue to "judge hunger strikers as competent even as they become increasingly outraged at their oppressors, increasingly focused on their own successes, and increasingly indifferent to the possibility of their own deaths." Dr. Daniel Fesser advises that, when briefing hunger strikers about the physical consequences of their actions, physicians have an ethical responsibility to advise them of the sometimes "radical personality alterations that often accompany starvation." He counsels physicians to observe and ask hunger strikers "whether they are feeling angry, impulsive, or indifferent to the prospect of death above and beyond the responses elicited by their sociopolitical circumstances."[34]

Although psychiatric diagnosis was critical to judgments about the competency of hunger strikers and their ability to make a life-threatening decision to refuse nourishment to the death, neither the state nor the medical establishment recognized the impact and aftermath of detention. Grassroots community psychology and social work advocates challenged the notion that the crisis ceased with the end of the hunger strike and the return of the protester to home and community. In South Africa, detainees' aid committees provided detainees and their families with information and material support, including referrals to lawyers, doctors, and mental health workers, and also challenged the idea that the problems of sadness, anxiety, uncertainty, and fear that former detainees, their families, and their communities faced were their own fault or within their own control. Physicians responded to the immediate and individual crisis of the hospitalized hunger striker. However, grassroots mental health professionals addressed the structural oppression and the impact on the communities and families of those who struggled after the release of detainees. In a pamphlet distributed in townships with a concentration of detainees and also with a history of violence involving South Africa's security forces, physiological impacts were addressed by analogies to the consequences of state-sponsored physical violence: "If a rubber bullet hits the body, the pressure hurts. Emotional pressure on the mind and feelings of people hurts in the same way as pressure in the body hurts." The detainees' support groups offered training, guidance, and literature to help people and communities address the emotional stress they faced and provided self-help strategies for use after release from detention. The pamphlets used layperson's language to help people self-diagnose and treat members of their community by outlining symptoms of distress and depression, including feelings of sadness, "not being able to sleep well," "not being able to concentrate," and "feeling fearful and not being able to eat well." Their self-help strategy focused on context and the origin of the crisis: "When you have been in crisis, you may have been in a situation where other people or events have been in control of you. Self-help means that you are in control of your life. You are responsible for yourself."[35] Visualization, meditation, breathing techniques, and communication strategies with intimates and friends were part of the program for addressing and healing people so that they could function better and work better to create a new democratic society. The self-help literature directly addressed the range of emotions—sadness, anxiety, fear, and depression—that emerged in the aftermath of the crisis when the media focus on the fearlessness and bravery of the hunger strikers had been eclipsed.

However, in this prescriptive advice to medical professionals regarding ethical and diagnostic protocols in treating hunger strikers, the focus is on the emotion of the hunger strikers, their families, and members of their community members, but there is no reflection on the medical professionals' own emotional responses and the reverberations of those responses among hospital workers and other patients. This division in recognizing emotion underlines the class and racial divide. Characteristically, physicians and psychiatrists distance themselves from their emotions in order to maintain their reputation for clinical objectivity. Responding to hunger strikers took on the air of an abstract ethical quandary, even when a range of passions surfaced in each of the events. There was little mention in the medical literature and protocols of reckoning with the discomfort and horror of being implicated in systems of torture, punishment, and coercion. There are hints that physicians who are forced to take action in a political standoff and decide whether someone lives or dies will experience feelings of guilt, anger, or bewilderment—but hints only.

Both Gandhi and Chavez relied on personal physicians to monitor their condition while they were fasting and to address their debility during the slow process of rehabilitation. Where conflicts regarding medical therapy and patient response emerged, they were meditated by the patients personal relationships with their physicians and the ensemble of support and pressure from family, friends, and supporters. Gandhi and Chavez experimented with regimens of water and dietary intake both during and after fasting that helped them manage different stages of the fast. They were under the constant care of family and friends and were able to time their fasts to last from three days to no longer than twenty-five days. Gandhi's and Chavez's ability to consciously manage their care and to have trusted caretakers assist them contrasted sharply with the experience of incarcerated political prisoners, whose hunger strikes were interfered with by wardens, guards, and the physicians called on by the state to keep them alive.

Not only were the political prisoners isolated, desperate, and constrained, but so too were the physicians whose autonomy to make decisions was restricted. In these circumstances, physicians questioned not only the constraints of incarceration but also the volition of the protest faster. They worried that prolonged protest fasts would make individuals numb to pain, impervious to the condition of physiological deterioration, and pathologically incapable of self-preservation. They questioned the ability of the hunger striker to reason and consent to the refusal of treatment and underscored the importance of their own clinical objectivity. Nevertheless, the medical ethical literature draws attention to but does not critically examine the presumption of the physician's fearlessness in the wake of pressures and possible reprisals from the state, social justice movements, and physicians in their nations and worldwide. Clinical objectivity provided both a tool to determine medical diagnosis and intervention and a shield from scrutiny and reflection about the emptying of emotions in the physician's involvement.

The emotional arena of fearlessness shifts from understanding the affect and resolve of the hunger striker/protest faster to the estimation of the physician's psyche as he or she weighs clinical and ethical options. The physician's imperative to preserve life is at odds with the worldview of the recalcitrant and resistant patient, who may expressively override the value of self-preservation in favor of efforts to catalyze social and political change. The emotional framework of fearlessness here again spotlights the independent and heroic persona, but this time it is the physician who appears to both ignore and presume leadership over the nurses, attendants, medical professionals, guards, and nonmedical caregivers who attend, monitor, and respond to the hunger striker. In ethical guidelines, the WMA, physicians, and medical scholars bolster the fearlessness of the physician by offering support for his or her imperviousness to the pressures of the state, the public, human rights organizations, the media, the families and friends of the total voluntary faster, and the social movements and solidarities catalyzed by the protest. The presumption of independence and objective distance underline the physician's actions as uncompromised by their own fears, anxiety, and conflicted affect.

NOTES

1. George J. Annas, "Hunger Strikes," *British Medical Journal* 311 (October 28, 1995): 1114.
2. Kevin Grant, "The Transcolonial World of Hunger Strikes and Political Fasts, c. 1909–1935," in *Decentering Empire: Britain, India and the Transcolonial World*, ed. Durba Ghosh and Dane Kennedy (New Delhi: Orient Longman, 2006), 243–69; Tim Pratt and James Vernon, "'Appeal from This Fiery Bed . . .': The Colonial Politics of Gandhi's Fasts and Their Metropolitan Reception," *Journal of British Studies* 44.1 (2005): 92–114; Taylor C. Sherman, "State Practice, Nationalist Politics and the Hunger Strikes of the Lahore Conspiracy Case Prisoners, 1929–39," *Cultural and Social History* 5.4 (2008): 497–508; Neeti Nair, "Bhagat Singh as 'Satyagrahi': The Limits to Non-Violence in Late Colonial India," *Modern Asian Studies* 43 (2009): 649–81; Patrick Anderson, *So Much Wasted: Hunger, Performance and the Morbidity of Performance* (Durham, NC: Duke University Press, 2010); N. Oguz and S. Miles, "The Physician and Prison Hunger Strikes: Reflecting on the Experience in Turkey," *Journal of Medical Ethics* 31.3 (March 2005): 169–72; Oleg Khlevniuk, *A History of the Gulag* (New Haven, CT: Yale University Press, 2004); Steven Barnes, *Death and Redemption: The Gulag and the Shaping of Soviet Society* (Princeton, NJ: Princeton University Press, 2011).
3. Annas's expression, the "slow death" of the hunger striker, implies the willful self-withdrawal of sustenance toward death and the spectacle of individual human agency exercised to fatal extremes. Lauren Berlant has recently developed the concept of slow death to refer to the "physical wearing out of population and deterioration of people" that is an ordinary condition, thereby questioning the very grounds in terms of which agency and political sovereignty are defined. Lauren Berlant, "Slow Death (Sovereignty, Obesity, Lateral Agency)," *Critical Inquiry* 33.4 (Summer 2007): 754–80.
4. Jan Plamper, "The History of Emotions: An Interview with William Reddy, Barbara Rosenwein and Peter Stearns," *History and Theory* 49 (May 2010):

237–65, esp. 242–43. See also Barbara Rosenwein, "Worrying about Emotions in History," *American Historical Review* 107.3 (2002): 821–45, 43.

5. Daniel M. Gross, *The Secret History of Emotion: From Aristotle's "Rhetoric" to Modern Brain Science* (Chicago: University of Chicago Press, 2006), 55.
6. Kevin Grant, "Fearing the Danger Point: The Study and Treatment of Human Starvation in the United Kingdom and India, c. 1880–1974," in *Comparative Physiology of Food Limitation, Fasting, and Starvation*, ed. Marshall McCue (New York: Springer, 2012), 113–43.
7. David Beresford, *Ten Men Dead: The Story of the 1981 Irish Hunger Strike* (New York: Atlantic Monthly Press, 1987); Padraig O'Malley, *Biting at the Grave: The Irish Hunger Strikes and the Politics of Despair* (Boston: Beacon, 1990).
8. "Ethical Statement: Artificial Feeding of Prisoners," *British Medical Journal* 3 (July 6, 1974): 52–53.
9. World Medical Association, "Declaration of Tokyo—Guidelines for Physicians concerning Torture and Cruel, Inhuman or Degrading Treatment or Punishment in Relation to Detention and Imprisonment," adopted by the Twenty-ninth World Medical Assembly, 1975, http://www.wma.net/en/30publications/10policies/c18/.
10. Hernán Reyes, "Medical and Ethical Aspects of Hunger Strikes in Custody and the Issue of Torture," extract from the publication *Maltreatment and Torture* in the series Rechtsmedizinische Forschungsergebnisse (Research in legal medicine), vol. 19 (Lübeck: Schmidt-Romhild, 1998), http://www.icrc.org/eng/resources/documents/article/other/health-article-010198.htm.
11. Ancel Keys, Josef Brozek, and Austin Henschel, *The Biology of Human Starvation*, 2 vols. (Minneapolis: University of Minnesota Press, 1950), 1:123–25, 161.
12. Todd Tucker, *The Great Starvation Experiment: The Heroic Men Who Starved So That Millions Could Live* (New York: Free Press, 2006); Leah M. Kalm and Richard D. Semba, "They Starved So That Others Be Better Fed: Remembering Ancel Keys and the Minnesota Experiment," *Journal of Nutrition* 135 (June 2005): 1347–52; J. Alexander Palesty and Stanley J. Dudrick, "The Goldilocks Paradigm of Starvation and Refeeding," *Nutrition in Clinical Practice* 21.2 (April 2006): 147–54.
13. Erik Erikson, *Gandhi's Truth: On the Origins of Militant Nonviolence* (New York: Norton, 1969).
14. Mohandas K. Gandhi, *Autobiography: The Story of My Experiments with Truth* (1927, Boston: Beacon, 1993), 277. See also William M. Reddy, The *Navigation of Feeling: A Framework for the History of Emotion* (Cambridge: Cambridge University Press, 2001).
15. Maureen Swan, *Gandhi: The South African Experience* (Johannesburg: Raven, 1985); David Hardiman, *Gandhi in His Time and Ours: The Global Legacy of Ideas* (New York: Columbia University Press, 2004), and *Peasant*

Nationalists of Gujarat: Kheda District, 1917–1934 (New York: Oxford University Press, 1981); Lisa Trivedi, *Clothing Gandhi's Nation: Homespun and Modern India* (Bloomington: Indiana University Press, 2007).

16. Erikson, *Gandhi's Truth*, 84–89. See also Gandhi, *Autobiography*.
17. Gandhi, *Autobiography*, 287–89.
18. Gandhi, *Autobiography*, 358–59. See also Stanley Wolpert, *Gandhi's Passion: The Life and Legacy of Mahatma Gandhi* (Oxford: Oxford University Press, 2001); Judith M. Brown, *Gandhi: Prisoner of Hope* (New Haven, CT: Yale University Press, 1989); and Joseph Lelyveld, *Great Soul: Mahatma Gandhi and His Struggle with India* (New York: Random House, 2011).
19. Gandhi in *Bapu Ke Ashirwad* (Ahmedabad), November 26, 1944.
20. Gandhi in *Harijan* (Yerwada), November 3, 1946. See also Wolpert, *Gandhi's Passion*.
21. Quoted in John C. Hammerback, *The Rhetorical Career of Cesar Chavez* (College Station: Texas A&M University Press, 2003), 116.
22. "Statement by Cesar Chavez at the End of His 24-Day Fast for Justice," June 4, 1972, http://chavez.cde.ca.gov/ModelCurriculum/Teachers/Lessons/Resources/Documents/EXR1_Cesar_E_Chavez_Statements_on_Fasts.pdf.
23. Jeffrey Masson, "India and the Unconscious: Erik Erikson on Gandhi," *International Journal of Psycho-Analysis* 55 (1974): 519–26, 520.
24. Max Coleman, ed., *A Crime against Humanity: Analysing the Repression of the Apartheid State* (Johannesburg: David Philips, 1998), 140.
25. "St Albans Prison—Press Statement," February 1989, file PD—F04.2.1.4, South African History Archives (SAHA), University of Witwatersrand.
26. "Detention under Three Emergencies: A Report of the Natal Midlands, 1986–1989," published by the Detainees Aid Committee, Pietermaritzburg (August 1989), file F.04.1.0.5, Popular History Trust, SAHA.
27. Coleman, *A Crime against Humanity*, 140–41.
28. World Medical Association, "WMA Declaration of Malta on Hunger Strikers," adopted by the Forty-third World Medical Assembly, 1991, http://www.wma.net/en/30publications/10policies/h31/.
29. W. J. Kalk, M. Felix, E. R. Snoey, and Y. Veriawa, "Voluntary Total Fasting in Political Prisoners: Clinical and Biochemical Observations," *South Africa Medical Journal* 83 (1996): 391–94; G. R. Keeton, "Hunger Strikers: Ethical and Management Problems," *South African Medical Journal* 83.6 (June 1993): 380–81. For an earlier study, albeit one more about ethics than physiology, see J. C. de Villiers, "The Medical Care of Prisoners and Detainees," *South African Medical Journal* 64.4 (1983): 116–18.
30. Michael Peel, "Hunger Strikes: Understanding the Underlying Physiology Will Help the Doctors Provide Proper Advice," *British Medical Journal* 315 (1997): 829–30.
31. Ibid.

32. Ibid. See also Johannes Wier Stichting, *Assistance in Hunger Strikes: A Manual for Physicians Dealing with Hunger Strikes* (Amersfoort: Johannes Wier Foundation for Health and Human Rights, 1995); Kalk et al., "Voluntary Total Fasting"; and W. J. Kalk and Y. Veriawa, "Hospital Management of Voluntary Total Fasting among Political Prisoners," *Lance*t 337 (1991): 660–62.
33. Edward A. Selby, "Habitual Starvation and Provocative Behaviors: Two Potential Routes to Extreme Suicidal Behavior in Anorexia Nervosa," *Behaviour Research and Therapy* 48 (2010): 634–45.
34. Daniel M. T. Fessler, "The Implications of Starvation-Induced Psychological Changes for the Ethical Treatment of Hunger Strikers," *Journal of Medical Ethics* 29 (2003): 243–47, 244–45.
35. *Coping after Crisis: A Self-Help Manual*, OASSA-DACOM publication (printed in Pietermaritzsburg), file F.4.10, SAHA.

PART FOUR

Social Sciences

TEN

Across Different Cultures? Emotions in Science during the Early Twentieth Century

UFFA JENSEN

In 1935, the Polish-Jewish biologist and physician Ludwik Fleck published *Genesis and Development of a Scientific Fact*. For the following three decades, this pathbreaking sociological analysis of the production of scientific knowledge was nearly forgotten, until Thomas Kuhn excavated it in 1962 for *The Structure of Scientific Revolutions*, his study of the social preconditions for innovation in the natural sciences.[1] Fleck's concepts of *thought style* and *thought collective* proved very helpful to Kuhn. But Kuhn's rediscovery of Fleck was incomplete: he forgot the emotions. For Fleck, a thought style does not just entail collectively shared assumptions about scientific objects, methods, and approaches but also always constitutes a "certain mood," a "readiness both for selective feeling and for correspondingly directed action." Using the case of bacteriology, Fleck describes the situation of the researcher who makes a new observation: "The first, chaotically styled observation resembles a chaos of feeling: amazement, a searching for similarities, trial by experiment, retraction as well as hope and disappointment. Feeling, will, and intellect all function together as an indivisible unit. The research worker gropes but everything recedes, and nowhere is there a firm support. Everything seems to be an artificial effect inspired by his own personal will. Every formulation melts away at the next test. He looks for that resistance and thought constraint in the face of which he could feel passive."[2]

Thus, researchers who observe a phenomenon that contradicts the conventional wisdom in their area of expertise enter a state of emotional chaos. To overcome this state, Fleck tells us, they try to remain faithful to the common thought style and thought collective. Emotional tranquility can be reestablished only if they find an explanation for the observed phenomenon in accordance with the prevalent thought style. If this fails, the uncertainty and anxiety will persist until a new explanation in a novel thought style is found. From such a perspective, emotions like confusion, and even bewilderment, reassurance, confidence, and trust, prove to be an integral part of the research process in the natural sciences; in fact, these emotions are part of the very core of scientific observation and knowledge production. True, Fleck never developed a full-fledged theory of emotions in the natural sciences, but he had certainly prepared the ground for such a venture. Yet, when Kuhn picked up the pieces from Fleck, he described scientific procedure with words like *puzzle, anomaly,* or *crisis.*[3] His portrayal of scientific innovation resembles an exclusively cognitive process without the dynamic and troublesome personal dilemmas that Fleck took into consideration. Thus, he could only descriptively take note of certain phenomena that Fleck had actually explained, that is, the scientist's lasting loyalty to an established scientific paradigm in the face of mounting counterevidence.[4] Even when he quoted emotional reactions of scientists, he refrained from discussing the significance of their affective language.[5] In short, Fleck's insistence on the important role that emotions play in scientific processes of observation, explanation, and theory building was lost on Kuhn.

Kuhn's simplification of Fleck's account and perhaps even Fleck's reluctance to delve into a full-fledged assessment of emotions in science points to a more fundamental problem: psychological reductionism in the explanation of scientific processes. Placing too much emphasis on emotions in science runs the risk of reducing the complexities of scientific changes to mere biographical reasons. Many historians, sociologists, and philosophers of science share a hostility toward psychologizing approaches: a "horror of the psychological," as Lorraine Daston has put it.[6] Daston developed her idea of a "moral economy of science" in order to avoid this reductionism while not losing sight of values and emotions in science. In her definition, a moral economy is a "web of affect-saturated values" and as such it amounts to the collective mental states of scientists, a *Gefühls-* and a *Denkkollektiv,* as she phrased it in Fleckian terminology.[7] However, in her concrete readings, the role of emotions in moral economies gets short shrift, especially in comparison with morals and values, the foci of her analysis.[8] Taking Daston as a point of departure, it is high time to address the *emotional* economy of science.[9]

Before doing so, a historical question remains: Why was this dimension forgotten, even repressed, for such a long time, despite its early prominence in Fleck's foundational work? His 1935 book came at an interesting moment in the history of science.[10] The 1930s were a time when, in a broad array of (natural) sciences, a new form of objectivity, defined as trained judgment, was eclipsing ideas of mechanical objectivity.[11] This shift had a subterranean emotional current.[12] Specific forms of epistemology correspond with specific scientific selves, and the ideal of mechanical objectivity that emerged in the mid-nineteenth century relied on a subjective, active, and assertive self that had to exercise restraint in order to produce valid scientific observation and representation.[13] This "will to willessness" was undermined by the epistemology of trained judgment during the early twentieth century.[14] If in the earlier phase subjectivity and consequently also emotionality were perceived as threats to scientific objectivity, the scientific self was now expected to embrace its unconscious, intuitive side. Because this emotional quality was held to be decisive for scientific process, scientists now needed to train their intuition by learning how to see and to judge.

In his theory of scientific processes, Fleck highlights a similar emotional dimension and in so doing dovetails nicely with Daston and Galison's discussion of the new epistemology that became so influential during the early to mid-twentieth century. Yet this correspondence between Fleck and Daston and Galison also opens up a new question: If the intuitive side of the scientific self was increasingly valued in this period, what does it mean that Kuhn overlooked Fleck's argument about the emotions when he rediscovered *Genesis and Development of a Scientific Fact*? To put this question somewhat differently, why were emotions marginalized when scientific knowledge moved across the Atlantic—as was the case with Fleck, whose ideas traveled from his native Lwów to Kuhn's Berkeley? In what follows, I will examine two surprisingly similar examples for the marginalization of emotions in the history of transatlantic science: the legacies of Weber's sociology and Freud's psychoanalysis. In discussing them together with Fleck's and Kuhn's accounts of the natural sciences, I seek to challenge notions about the two (or three) science cultures, notions that pervade the practice, organization, and theory of science as well as its historiography to this day.[15]

I do this with a number of considerations in mind. Studying the emotional economy of science could help in linking the human and natural sciences in new and meaningful ways without falling into the well-known pitfalls of psychological reductionism.[16] Following Fleck, it is imperative to scrutinize emotional thought styles independently from their different theories about emotions. These styles may very well transcend the different disciplines and the infamous two (or three) cultures of science. The two fields I want to discuss in greater depth here, sociology and psychoanalysis, eo ipso transcend the taxonomy of the different science cultures, belonging, as hybrid disciplines (at least at the time of Weber and Freud), neither solely to the humanities nor exclusively to the natural sciences.[17]

Weber and Freud as Cartesians?

In recent years, a number of scholars have pointed to the—forgotten—centrality of emotions in the work of canonical late nineteenth- and early twentieth-century social science or humanities luminaries.[18] Indeed, many fin de siècle thinkers were preoccupied with analyzing emotions in ways that mid-twentieth-century scholars no longer were. Consider only Georg Simmel or Émile Durkheim in sociology, William James in psychology, Aby Warburg in art history, or the Annales school in historiography.[19] In this context, Max Weber and Sigmund Freud are particularly interesting cases. On the one hand, both have been described as original thinkers in the study of emotions. Weber is seen to have linked the history of capitalism to the emotional effects of religious history and to have made transparent what he famously called the *iron cage* of modern rationality.[20] Freud is seen by many to represent the intentionalist and hermeneutic tradition of understanding emotions, which today represents one of the important alternatives to a neurobiological study of affects.[21] On the other hand, Weber and Freud have also been portrayed as prime examples of the conventional view of emotions, as proponents of the modern contrast between rationality and emotion, the invention of which is usually ascribed to Descartes.[22] In this view, the two scholars stood at the threshold of the later, mid-twentieth-century period of the marginalization of emotions, to which they seemed to contribute in manifold ways. Their foundation of influential discursive fields was, at least in part, aided and solidified by transfers of their ideas from Europe to the United States, starting in the 1920s. It could be argued that these processes of transfer ended up downplaying the significance of the emotional dimension of their thought.

There is a lot of common ground between Weber and Freud and their understanding of emotions. Some residual Cartesianism shapes their work as both perceive emotions as affects and cast their descriptions in energetic metaphors.[23] This view of affects is clearly visible in Weber's discussions of politics in a mass society, and it also is evident in his description of Calvinist asceticism.[24] Energetic notions underpin Freud's basic theory of the drives, which he sees as putting immense pressure on the mental system.[25] In both Weber and Freud, emotions are conceptualized as energies that need to be contained, suppressed, or channeled—in Freud's case, through repression, diversion, or sublimation and, in Weber's case, with the help of ascetic denial or rational political force. In such concepts, human beings emerge as mere objects of emotions who possess, at best, a limited degree of control over them. They never really have agency in the production of their emotions even if those emotions are understood to originate in their inner self. Such is a widespread and, in many ways, understandable reading of Weber's and Freud's theories. But a closer reading reveals a much more complicated picture.

Weber's *Leidenschaften*

The early Weber of the *Protestant Ethic* indeed largely subscribed to the conventional Cartesian view, surmising that in Calvinist thought and practice emotions have to be controlled and even suppressed.[26] However, Weber also claimed that a Calvinist *hates* emotionality and irrationality. Thus, even a rationalist and ascetic style proved incomplete without an emotional foundation. What is more, Weber's later *Theory of Social and Economic Organization* and the concept of affectual action conform to the energetic view, with affectual behavior appearing in contrast to the rational forms of action (*zweckrational* and *wertrational*) and as a threat to rationality. The same perspective is manifest in his political writings, for example, in the concluding section of "Suffrage and Democracy in Germany," a text he published in 1917. Weber here asserts that the "irrational instincts of the masses" can exert "pressure" on a political system and that universal suffrage can undermine rational politics.[27] Weber consistently associated the masses with affectual action and granted them no or very little agency in their emotional experience and conduct.

Yet this reading of Weber is incomplete because it fails to take into account his ideas about emotional management, which can be found in his writings on *Berufsmenschentum*.[28] In his *Science as a Vocation* (1917) and *Politics as a Vocation* (1919), Weber puts forward the concept of passion (*Leidenschaften*) as a specific way of dealing with emotions.[29] Whereas he described affects as dimensions of perception and cognition that can force people to act in a certain way, passions can motivate people's actions more directly and internally: "Affects move us, but with our passions we can move things."[30] In Weber's account, politicians and scholars need to be passionate about what they do; otherwise they fail in their vocation: "We can say that three qualities, above all, are of decisive importance for a politician: passion, a sense of responsibility, and a sense of proportion. Passion in the sense of a *commitment to the matter in hand* [*Sachlichkeit*], that is, the passionate dedication to a 'cause' [*Sache*], to the God or demon that presides over it."[31] The control of one's affects (or those of others, of the masses) is possible only via passions. Consequently passions are the prerequisite for rational action. With politicians, this kind of emotional management can even take the shape of charismatic authority. Thus, in Weber's view it is an important question how to control or even suppress one's emotions. An increasingly rationalized and bureaucratic modern society dramatizes this question.

Freud's Angst

At first glance, Freud too seemed to subscribe to the conventional Cartesian view. From early on, he preferred the term *affect* and put a strong emphasis on the quantitative and energetic nature of affect.[32] A good example of this economic theory of affect is anxiety, which Freud initially took to be caused by frustrated sexual desire in the unconscious.[33] The transformation of desire into anxiety was a purely quantitative, not a qualitative, matter. But even this affect is, for Freud, worth considering only because it forms a disruptive energy that exerts pressure from below—pressure that the system needs to discharge in one way or another. Here, the limited capacity of the person to rationally control affects comes into play; it takes the forms of repression, diversion, and sublimation.

In this play of affects and energies, there is very little room for emotions, in particular because studying emotions implies, first and foremost, taking into account their qualitative differences.[34] Anxiety, hatred, and love differ in quality, not quantity. However, Freud, I believe, modified his views over the years. As I have argued elsewhere, this was partly due to the continuous importance of patients' emotional reactions during the therapeutic process.[35] But there were also conceptual reasons, considering that this shift occurred in the 1920s when Freud was in the midst of exploring further the implications of his new structural model of the psyche. He departed from his earlier quantitative view of affects, and, from the mid-1920s on, anxiety became *the* central emotion in his theory of mental development.[36] Starting with birth anxiety, and continuing with the later anxiety of losing the motherly object up to castration anxiety, this emotion structured every step in mental maturation. In the end, anxiety would even become internalized as an abstract form of conscience in the superego.[37] Here, the moral development of the individual person became unthinkable without the transformative experience of anxiety, and the same applied to sociability. Thus, virtually the entire mental development of a person depended on the experience of anxiety. Even though Freud never says this explicitly, it is clear that rationality, rather than overcoming emotions, in fact results from emotionality.

Weber and Freud Move across the Atlantic

Both Weber's and Freud's work could have served as a launching pad for theories of emotion. Yet this is not what the reception looked like once their ideas began to exert influence on the academic world of the United States. The development of sociological thought in the footsteps of Weber—primarily through Talcott Parsons's reinterpretation[38]—moved along the path of a very rationalistic theory that deliberately denied emotionality its role in social and cultural life. The Parsonized Weberian sociology and its substantial success in postwar America relied on a systematically rationalized Weber.[39] When the German sociologist Wilhelm Hennis remembered his time as a postgraduate student in the United States during the early 1950s, he described the protagonists of what he called, tongue firmly in cheek, the American *Weberei*.

Theirs was an attempt to create a veritable anti-Marx, and for this purpose they had to downplay Weber's German patriotism, his uncanny Nietzscheanism, and the passion (*Leidenschaftlichkeit*) of his thinking.[40] Freud's ideas were subjected to a similar reshaping in the name of science in the postwar United States. In these years, ego psychologists like Heinz Hartmann or David Rapaport tried to make psychoanalysis more "scientific," a project that corresponded to the strong medicalizing tendencies in American psychoanalysis.[41] As ego psychologists aspired to move away from the analysis of pathological mental states and instead sought to foreground normal mental functioning, they took particular interest in the functioning of the ego and its own strength for a healthy mental development. In this context, Hartmann's ideas about adaptation and the nonconflictual autonomous sphere of the ego can be described as the rationalist triumph of the ego over drives, sexuality, irrationality, and, of course, emotions.[42]

The recalibration that went hand in hand with transnational transfers amounted to a normalization and rationalization of Weber's and Freud's ideas. Admittedly, this goes first and foremost for the academic and science-oriented reception. In a more popular genre of sociological analysis, Freud's and Weber's thoughts were sometimes used differently. To give just one example, David Riesman's *The Lonely Crowd*, which he published together with Nathan Glazer and Reuel Denney in 1949, explicitly stood in a neo-Freudian tradition—Riesman was a student of Erich Fromm's—and also frequently referred to Weber's ideas. In Riesman et al.'s description of the social character types that prevailed in contemporary American society, emotions structured the much-debated distinction between inner directedness and other directedness: "The other-directed person wants to be loved rather than esteemed . . . ; he seeks less a snobbish status in the eyes of others than assurance of being emotionally in tune with them."[43]

But even the academic silencing of the emotional dimension in Freud's and Weber's work did not arise from a vacuum; as I have shown above, such tendencies were latent in their original theories. At the same time, it would make little sense to defend a "true" Weber or a "true" Freud against "heretical" interpretations of their ideas. Yet the question remains, What caused this neglect? Many factors were certainly involved, among them linguistic ones, given that a rich and complex German had to be rendered in English.[44] Obviously, the scientific, academic, and intellectual landscape for sociology and psychoanalysis was rather different in the United States at midcentury. In central Europe two decades earlier, Weber and Freud were preoccupied with establishing their respective disciplines, and much of their work transcended boundaries between the different disciplines, often between the two cultures.[45] As their disciples tried to solidify their heritage in the more stable institutional setting of the postwar United States, the nature of their boundary-crossing work also changed. Another factor was the fact that the emphasis on normalization and rationalization in Weber's and Freud's thought dovetailed with much broader concepts of society, economy, and politics in the kind of Fordist mass society that had emerged in the United States during the 1920s.[46] In this context, it also played a role that the political alternatives to Fordism, fascism, Nazism, and Soviet-style communism were seen as dangerously irrational and excessively emotional.[47] In this politically volatile climate, Weber's writings about politicians or scientists—the idols of emotion management via passionate vocation à la Weber—seemed less relevant and even more dangerously ambivalent than his comments on the dangers of affect-guided behavior. In the case of ego psychology, it made political sense to reshape Freud's emphasis on anxiety and—related to this—on the relative weakness of the ego in the face of demands from the id and the superego: now only a mentally strong ego could battle the irrational forces of political desires. Here, a distance from therapeutic work and a preoccupation with systematizing Freud's intellectual heritage further strengthened the tendency to neglect emotions.

Different Sciences and Epistemic Fear

Neither Weber's preoccupation with professional passions nor Freud's interest in his patients' anxiety, then, resonated with their American academic disciples. However, it is necessary to make a distinction between the history of emotions as objects of scientific inquiry and the history of emotional thought styles among scientists, for, in Fleck, Weber, Freud, and many other fin de siècle scholars, these two dimensions overlapped to some degree. These thinkers focused on emotions as scholarly topics because they became aware of the importance of emotions in their own ways of practicing scholarship. Later, these two dimensions fell apart again. What was marginalized in the history of emotions of the postwar sciences were emotions as objects of scientific inquiry. Yet emotions did continue to play a role in their thought styles. Against the backdrop of the varieties of political extremism of the interwar years, and during the highly emotional confrontation of the Cold War years, science was cast as rational for emotional reasons.

Passions in the Weberian sense may not have been talked about much among scientists, but they certainly still structured scientific work. What we are in fact witnessing in recent decades is a new merging of emotions as objects and as thought style in a variety of different scholarly agendas.

Daston and Galison conclude their study of objectivity with an interesting point about the emotional character of epistemology: "All epistemology begins in fear—fear that the world is too labyrinthine to be threaded by reason; fear that the senses are too feeble and the intellect too frail; fear that memory fades, even between adjacent steps of a mathematical demonstration; fear that authority and convention blind; fear that God may keep secrets or demons deceive."[48] From this perspective, it is primarily epistemic fear—and not political concerns or scholarly agendas—that is key in guarding the scientific self with an emotional thought style as well as in elaborating or ignoring emotions in science. Fleck's insights into the insecurities of the researcher confronted with the inexplicability of his raw data provide remarkable glimpses of this fundamental fear. Does the same apply to all fields of scholarship in the natural, social, and human sciences? If this were the case—and to me it looks like it is indeed true—it would become possible to transcend the different cultures of science and to write a history in which times of scientific interest in emotions alternate with times of relative silence about them. This would turn out to be a history in which objectivity and emotions are not opposites but support one other structurally in the production of scientific knowledge.

NOTES

1. Ludwik Fleck, *Genesis and Development of a Scientific Fact*, trans. Fred Bradley and Thaddeus J. Trenn (Chicago: University of Chicago Press, 1979); Thomas S. Kuhn, *The Structure of Scientific Revolutions* (1962), 2nd ed. (Chicago: University of Chicago Press, 1970).
2. Fleck, *Genesis and Development of a Scientific Fact*, 94, 99.
3. Kuhn, *Structure of Scientific Revolutions*, esp. 52–91.
4. Ibid., 78.
5. Ibid., 83–84.
6. Lorraine Daston, "The Moral Economy of Science," *Osiris* 10 (1995): 3–24, 4.
7. Ibid., 5. Daston's text is the earliest reference to this lost aspect of Fleck's work that I could find. Another rare example is Jack Barbalet, "Science and Emotions," in *Emotions and Sociology*, ed. Jack Barbalet (Oxford: Blackwell, 2002), 132–50.

8. To be sure, Daston claims—referring to Gaston Bachelard—that morals and emotions are very closely related. Daston, "Moral Economy of Science," 4. In my view, however, they should not be collapsed into a single category of analysis. An examination of the values that scientists attach to their work seems hardly the same as an investigation of their emotional states.
9. See "Emotional Economy of Science," special issue, *Isis* 100 (2009). Interestingly, these new attempts to integrate emotions into the study of science do not use Fleck as an inspiration.
10. Fleck's attention to emotions seems less extraordinary when we take into account the widespread discussions of emotions in various scientific disciplines at the turn to the twentieth century. Uffa Jensen and Daniel Morat, eds., *Rationalisierung des Gefühls: Zum Verhältnis von Wissenschaft und Emotionen, 1880–1930* (Munich: Wilhelm Finck, 2008).
11. On this, see Lorraine Daston and Peter Galison, *Objectivity* (New York: Zone, 2007).
12. This transpires from Daston and Galison's account, even if the role of fear in the history of objectivity makes only a short appearance. See ibid., 372–73. For a more extensive discussion of "epistemic fear," see Lorraine Daston and Peter Galison, "Objectivity and Its Critics," *Victorian Studies* 50 (2008): 666–77.
13. This is Daston and Galison's argument. It resonates with recent findings on the bourgeois concept of self in the nineteenth century. See, e.g., Philipp Sarasin, *Reizbare Maschinen: Eine Geschichte des Körpers, 1765–1914* (Frankfurt: Suhrkamp, 2001); and Andreas Reckwitz, *Das hybride Subjekt: Eine Theorie der Subjektkulturen von der bürgerlichen Moderne zur Postmoderne* (Weilerswist: Velbrück, 2006).
14. Daston and Galison, *Objectivity*, 203.
15. For the "two" or "three cultures," see, respectively, C. P. Snow, *The Two Cultures* (Cambridge: Cambridge University Press, 1993); and Wolf Lepenies, *Between Literature and Science: The Rise of Sociology*, trans. R. J. Hollingdale (Cambridge: Cambridge University Press, 1988). For a recent invocation, see Jerome Kagan, *The Three Cultures: Natural Sciences, Social Sciences, and the Humanities in the Twenty-first Century* (Cambridge: Cambridge University Press, 2009).
16. Of course, Weber as well as Freud (and, possibly, also Fleck, Kuhn, Parsons, and Hartmann) would offer abundant material for the specific importance of emotions in their lives. See Joachim Radkau, *Max Weber: A Biography*, trans. Patrick Camiller (Cambridge: Polity, 2009); and Peter Gay, *Freud: A Life for Our Time* (New York: Norton, 1987).
17. Lepenies, *Between Literature and Science*; Mark Frezzo, "The Ambivalent Role of Psychology and Psychoanalysis," in *Overcoming the Two Cultures: Science versus the Humanities in the Modern World-System*, ed. Richard E. Lee and Immanuel Wallerstein (Boulder: Paradigma, 2005), 73–86.
18. Jensen and Morat, eds., *Rationalisierung des Gefühls*.

19. Émile Durkheim, *The Elementary Forms of the Religious Life: A Study in Religious Sociology*, trans. Joseph Ward Swain (London: Allen & Unwin, 1926); Georg Simmel, *Sociology*, trans. Kurt H. Wolff (New York: Free Press, 1964), and *On Women, Sexuality, and Love*, trans. Guy Oakes (New Haven, CT: Yale University Press, 1984); William James, "What Is an Emotion?" *Mind* 9 (1884): 188–205; Aby M. Warburg, "Dürer and Italian Antiquity," in *The Renewal of Pagan Antiquity: Contributions to the Cultural History of the European Renaissance* (Los Angeles: Getty Research Institute for the History of Arts and the Humanities, 1999), 553–58; Lucien Febvre, "Sensibility and History: How to Reconstitute the Emotional Life of the Past," in *A New Kind of History: From the Writings of Febvre*, ed. Peter Burke (London: Routledge & Kegan Paul, 1973), 12–26.
20. Jürgen Gerhards, "Affektuelles Handeln: Der Stellenwert von Emotionen in der Soziologie Max Webers," in *Max Weber heute: Erträge und Probleme der Forschung*, ed. Johanna Weiss (Frankfurt: Suhrkamp, 1989); Helena Flam, *Soziologie der Emotionen: Eine Einführung* (Konstanz: UVK, 2002).
21. Ruth Leys, "How Did Fear Become a Scientific Object and What Kind of Object Is It?" *Representations* 110 (2010): 66–104; Ruth Leys and Marlene Goldman, "Navigating the Genealogies of Trauma, Guilt, and Affect: An Interview with Ruth Leys," *University of Toronto Quarterly* 79 (2010): 656–79.
22. Jack M. Barbalet, *Emotion, Social Theory, and Social Structure: A Macrosociological Approach* (Cambridge: Cambridge University Press, 2001); Anthony Kauders, "The Mind of a Rationalist: German Reactions to Psychoanalysis in the Weimar Republic and Beyond," *History of Psychology* 8 (2005): 255–70. For an attack on the Cartesian tradition and the presentation of Spinoza as an alternative, see Antonio R. Damasio, *Descartes' Error: Emotion, Reason and the Human Brain* (New York: Avon, 1994), and *Looking for Spinoza: Joy, Sorrow, and the Feeling Brain* (Orlando, FL: Harcourt, 2003).
23. Such concepts originated from notions in thermodynamic physics that were popularized in the late nineteenth century and became virtually hegemonic across a wide range of disciplines in the early twentieth century. On this, see Anson Rabinbach, *The Human Motor: Energy, Fatigue, and the Origin of Modernity* (Berkeley and Los Angeles: University of California Press, 1990).
24. Volker Heins, "Demokratie als Nervensache: Zum Verhältnis von Politik und Emotion bei Max Weber," in *Masse—Macht—Emotionen: Zu einer politischen Soziologie der Emotionen*, ed. Ansgar Klein and Frank Nullmeier (Opladen: Westdeutscher, 1999), 89–101.
25. The significance of energetic metaphors is debated in the historiography about psychoanalysis. See, e.g., Don R. Swanson, "A Critique of Psychic Energy as an Explanatory Concept," *Journal of the American Psychoanalytic Association* 25 (1977): 603–33; Yehuda Elkana, "Die Entlehnung des Energiebegriffs in der Freudschen Psychoanalyse," in *Anthropologie der*

Erkenntnis: Die Entwicklung des Wissens als episches Theater einer listigen Vernunft (Frankfurt: Suhrkamp, 1986), 376–97 (translated as Yehuda Elkana, "The Borrowing of the Concept of Energy in Freudian Psychoanalysis," in *Psicoanalisi e storia della scienze* [Florence: Leo S. Olschi, 1983], 55–80); and José Brunner, *Freud and the Politics of Psychoanalysis* (Oxford: Blackwell, 1995). On the more general problem of the deployment of metaphors in Freud, see Jonathan T. Edelson, "Freud's Use of Metaphor," *Psychoanalytic Study of the Child* 38 (1983): 17–59. For the importance of energetic notions in the history of emotions, see Robert C. Solomon, *True to Our Feelings: What Our Emotions Are Really Telling Us* (Oxford: Oxford University Press, 2007), 142–49.

26. Max Weber, *The Protestant Ethic and the Spirit of Capitalism*, trans. Talcott Parsons (London: Routledge, 1930).
27. Max Weber, "Wahlrecht und Demokratie in Deutschland," in *Zur Politik im Weltkrieg: Schriften und Reden, 1914–18*, vol. 15 of *Max Weber—Gesamtausgabe*, ed. Wolfgang J. Mommsen and Horst Baier (Tübingen: Mohr Siebeck, 1984), 345–96, 391. For an English translation, see Max Weber, "Suffrage and Democracy in Germany," in *Weber: Political Writings*, ed. Peter Lassmann and Ronald Speirs (Cambridge: Cambridge University Press, 1994), 80–129.
28. This reading places Weber in the tradition of the Kantian distinction between affect and passion. See Rainer Schützeichel, "Der Wert der politischen Leidenschaften—Über Max Webers 'Affektenlehre,'" in *Politische Leidenschaften: Zur Verknüpfung von Macht, Emotion und Vernunft in Deutschland* (Tel Aviver Jahrbuch für deutsche Geschichte 38), ed. José Brunner (Göttingen: Wallstein, 2010), 103–14.
29. *Max Weber: The Vocation Lectures*, ed. David S. Owen and Tracy B. Strong (Indianapolis: Hackett, 2004).
30. Schützeichel, "Wert der politischen Leidenschaften," 108.
31. *Max Weber: The Vocation Lectures*, 76.
32. To be sure, Freud's position on affect was complex and inconsistent: "Freud struggled with the problem of affect all his life." André Green, "Conceptions of Affect," *International Journal of Psycho-Analysis* 58 (1977): 129–56, 129. See also Uffa Jensen, "Freuds unheimliche Gefühle: Zur Rolle von Emotionen in der Freudschen Psychoanalyse," in Jensen and Morat, eds., *Rationalisierung des Gefühls*, 135–52.
33. Sigmund Freud, "On The Grounds for Detaching a Particular Syndrome from Neurasthenia under the Description 'Anxiety Neurosis'" (1894), in *The Standard Edition of the Complete Psychological Works of Sigmund Freud*, ed. James Strachey, 24 vols. (London: Hogarth, 1956–74), 3:85–115.
34. "What will weigh heavily on the future of the conception of affect is the subordination of the subjective quality of affect to its objective expression: the quantity whose measurement escapes our knowledge." Green, "Conceptions of Affect," 130.

35. Jensen, "Freuds unheimliche Gefühle," 143–46.
36. "There is no question that the problem of anxiety is a nodal point at which the most various and important questions converge, a riddle whose solution would be bound to throw a flood of light on our whole mental experience." Sigmund Freud, "Introductory Lectures on Psycho-Analysis," in Strachey, ed., *Complete Works*, 16:393. Freud admitted himself that his later concept of anxiety could not be formulated in quantitative terms. Sigmund Freud, "Inhibitions, Symptoms and Anxiety," in ibid., 20:140.
37. Freud, "Inhibitions, Symptoms and Anxiety."
38. This reinterpretation can be seen in Talcott Parsons, *The Structure of Social Action* (New York: McGraw-Hill, 1937).
39. Jere Cohen, Lawrence E. Hazelrigg, and Whitney Pope, "De-Parsonizing Weber: A Critique of Parsons' Interpretation of Weber's Sociology," *American Sociological Review* 40 (1975): 229–41. On Weber's reception in the United States, see Lawrence Scaff, "Max Weber's Reception in the United States, 1920–1960," in *Das Faszinosum Max Weber: Die Geschichte seiner Geltung*, ed. Karl-Ludwig Ay and Knut Borchardt (Konstanz: UVK, 2006), 55–89. In a different, more sympathetic reading, Uta Gerhardt interprets Parsons's commitment to Weber as a twofold rescue mission from distortions by (*a*) an unreceptive audience in the 1930s and (*b*) the protagonists of the Frankfurt school during the 1960s. Gerhardt's position, however, seems to be informed by a Parsonian preference for a rationalized sociology. See Uta Gerhardt, "The Transatlantic Origin of the Modern Reception of Max Weber's Work in the 1960s," in *Transatlantic Voyages and Sociology: The Migration and Development of Ideas*, ed. Cherry Schrecker (Aldershot: Ashgate, 2010), 21–37.
40. See Wilhelm Hennis, *Weber und Thukydides: Die "hellenische Geisteskultur" und die Ursprünge von Webers politischer Denkart* (Göttingen: Vandenhoeck & Ruprecht, 2003), 128. Interestingly, it was this American Weber who got reimported to West Germany and served as the basis for a new *historische Sozialwissenschaft*.
41. See Nathan G. Hale, *Rise and Crisis of Psychoanalysis in the United States: Freud and the Americans, 1917–1985* (New York: Oxford University Press, 1995); and Edith Kurzweil, *The Freudians: A Comparative Perspective* (New Haven, CT: Yale University Press, 1989).
42. Heinz Hartmann, *Essays on Ego Psychology: Selected Problems in Psychoanalytic Theory*, 2nd ed. (New York: International Universities Press, 1965), esp. chap. 3, "On Rational and Irrational Action" (1947).
43. David Riesman, Nathan Glazer, and Reuel Denney, *The Lonely Crowd*, 3rd ed. (New Haven, CT: Yale University Press, 1989), xxxii. A different, later example would be Robert N. Bellah, Richard Madsen, William M. Sullivan, Ann Swidler, and Steven M. Tipton, *Habits of the Heart: Individualism and Commitment in American Life* (Berkeley and Los Angeles: University of California, 1985).

44. Stephen Kalberg, "The Spirit of Capitalism Revisited: On the New Translation of Weber's Protestant Ethic," *Max Weber Studies* 2 (2001): 41–58; Darius G. Ornston, "Freud's Conception Is Different from Strachey's," *Journal of the American Psychoanalytic Association* 33 (1985): 379–412.
45. Fritz K. Ringer, *Max Weber's Methodology: The Unification of the Cultural and Social Sciences* (Cambridge, MA: Harvard University Press, 1997); Sarah Winter, *Freud and the Institution of Psychoanalytic Knowledge* (Stanford, CA: Stanford University Press, 1999).
46. Barbalet, *Emotion, Social Theory, and Social Structure*; Eli Zaretsky, *Secrets of the Soul: A Social and Cultural History of Psychoanalysis* (New York: Knopf, 2004).
47. This political dimension explicitly informed Parsons's and Hartmann's scholarship. See Talcott Parsons, "Max Weber and the Contemporary Political Crisis," in *Talcott Parsons on National Socialism*, ed. Uta Gerhardt (New York: De Gruyter, 1993), 159–87; and Heinz Hartmann, *Psychoanalysis and Moral Value*, trans. Marianne von Eckardt-Jaffé (New York: International Universities Press, 1960).
48. Daston und Galison, *Objectivity*, 372.

ELEVEN

Decolonizing Emotions: The Management of Feeling in the New World Order

JORDANNA BAILKIN

This chapter focuses on British efforts to manage emotions during the era of decolonization: a process that took place in multiple metropolitan and overseas locales. Colonial regimes had imposed a wide range of interventions in the emotional lives of those they ruled.[1] How, then, did Britons anticipate that independence might change these interventions, render them unnecessary or out of date, or warrant new forms of intrusion? Would decolonization conjure new feelings along with new states and new leaders? What exactly was the status of emotion in the context of imperial collapses and reconfigurations of imperial power? What role was emotion expected or hoped to play in the making of the new world order?

In posing these questions, I also want to ask which moments and sites have typically drawn the attention of scholars of emotion and why. With the possible exception of its violence, the history of decolonization has been strangely devoid of inner life. This omission is particularly striking to anyone acquainted with the scholarship on colonialism and the passions and drama that have come to infuse that scholarship. How might we explain the industry of attentiveness around colonial emotion as well as the relative absence of such scholarship for the postcolonial era? Was decolonization in fact an emotional process—

and, if so, then why has it been so effectively stripped of this content by its chroniclers? Part of what is at stake here is the question of what is meant by the term *decolonization*, where and when we believe it takes place, and how we evaluate its effects. Beginning to think about decolonization as a process that restructured the emotional lives of its participants requires that we reframe the end of empire beyond the diplomatic or "high" political realm and begin to explore its other, more social dimensions: namely, its impact on individual experience, daily life, and personal interactions.

This chapter is drawn from my book, *The Afterlife of Empire*,[2] and it amplifies some of the book's implicit themes. *The Afterlife of Empire* charts how decolonization transformed British society in the 1950s and 1960s, primarily by reshaping its structures of welfare. One aim of the book was to take two of the behemoths of post-1945 British history—the end of empire and the rise of the welfare state—and bring them into closer conversation with one another. Although these two phenomena have long been juxtaposed, they have rarely, if ever, been integrated. Perhaps one reason that these histories have been so disconnected has been precisely because they seem to have been conducted in such disparate, and even opposed, emotional registers: that is, the pessimism of the end of empire versus the optimism of the welfare state. Here, I wish to illustrate how these apparently discordant conditions were actually intertwined—simultaneous, rather than successive.

Numerous scholars have argued that the management of the contemporary self, shaped by the discourses of the therapeutic state, emerged as an increasingly coherent project in Britain after 1945, while the subjective capacities of British citizens became an object of public scrutiny.[3] The literature has focused overwhelmingly on the ways that profound shifts in the class structure of British society after 1945 altered individual expectations and experiences of the relations between states and selves. Yet this story was not exclusively about class, nor was it exclusively metropolitan. As this chapter suggests, British perceptions of the centrality of emotion in political life were reframed in response to global transformations as well as domestic ones and a growing awareness of how the two might be intertwined. Welfare and decolonization were both expected to bring about dramatic changes in material and political conditions but also to inaugurate new emotional experiences and new types of social bonds.

I begin to explore these issues by looking at the founding of Voluntary Service Overseas (VSO), which served as the British predecessor to the Peace Corps. I take VSO as a case study both of the centrality of emotion in the intersecting histories of welfare and decolonization and of the diversity of ways in which social scientists, their critics, and the decolonizing state understood the category of emotion itself. The founders of VSO were Alec Dickson, a journalist who had led a mobile propaganda unit to encourage African recruitment during the Second World War, and his wife, Mora Dickson, a Scottish-born painter and poet. The Dicksons conceived of the project of VSO as an outlet for the energies of elite British youths after National Service began its phase-out in 1957.[4] When National Service was implemented in 1949, every healthy male between the ages of seventeen and twenty-one was required to serve in the armed forces for eighteen months and to remain in the reserves for four years. This peacetime conscription absorbed some 2.5 million young men, a transformative social experience for this age group.[5] National servicemen were deployed in the military operations that signaled the empire's final days in Cyprus, Malaya, and Kenya. Before its demise, National Service had been criticized for breaking up family loyalties, unsettling the young, and raising the juvenile crime rate. But, in its absence, these problems did not disappear, and calls for a new form of overseas service set in quickly.[6]

VSO diverged from National Service in important ways. National servicemen frequently oversaw the violence of decolonization. By contrast, VSOs were supposed to preserve the spirit of imperial adventure while also forging new types of emotional experiences: namely, through the rubric of postimperial friendships and mentorships. Furthermore, one of the key elements of National Service was its universality. But VSO was never intended to encompass all young people, and its leadership constantly debated exactly which youths should be drawn into its orbit. The premise of VSO was that British teenagers lacked the chance to develop their characters fully because increased affluence and the provisions of the welfare state had sheltered them from opportunities to test themselves. At the same time, the Dicksons argued, young elites in the newly (or soon-to-be) independent nations of Africa and Asia had ample opportunities for service but little desire to devote themselves to it. The aim of VSO was not simply to improve young British elites by exposing them to the poverty of the Third World but also to bond them with other elite populations that happened to be located in Africa and Asia. This proposed symmetry of interest between First and Third World young elites proved difficult to maintain.[7]

The existence of groups that aimed to engage young people with international affairs was not unique to Britain. Between 1958 and 1965, nearly every industrialized nation started volunteer programs to spread the message of economic development and goodwill. These programs were linked by their efforts to inculcate the martial virtues of self-sacrifice while also promoting international amity during the Cold War.[8] To some extent, VSO reflected this broader culture of international aid. But it also reflected a distinctively British set of anxieties and aspirations. If the Peace Corps sought to reassure critics that America's power could be matched by its humanitarianism, then VSO championed a very different impulse: shaped by the loss of global supremacy rather than its acquisition.

Against this backdrop of international aid, VSOs were distinctive not only for the tangled colonial histories they faced abroad but also for the emotionally transformative constellation of welfare services they encountered at home. One question here is whether welfare and decolonization sought to revamp the emotional lives of their participants in ways that were compatible or divergent. Whereas the welfare state mediated relations between people, decolonization took the colonial state out of the equation, replacing it with as yet unknown forms of political power. For the Dicksons, both welfare and decolonization required new forms of social connectedness that must be enacted on a global stage. On this point, they were in line with many of their contemporaries who believed that welfare's true potency (along with its pitfall of dependency) was psychological and social rather than fiscal or material. Economic redistributions had been anticipated in earlier eras, but the promise of truly universal citizenship in a social democracy that institutionalized the principle of mutual care was a more compelling change.[9]

Under the regime of the welfare state, all Britons were transformed into both givers and receivers of aid. Many of the welfare state's architects viewed it as empowering people of all ranks to express and enact concern for each other, concern that would be manifested in their daily interactions.[10] The reduction of economic inequality was important, of course, but primarily to ensure the basis for a qualitative advance in human relations. According to Richard Titmuss, for example, one of welfare's chief theorists, the epitome of welfare was not a financial benefit but the blood donor service, in which one citizen voluntarily gave a lifesaving resource to another.[11] For the Dicksons, more specifically, the demands of decolonization provided an ideal opportunity for Britons to correct the lassitude that welfare had engendered in themselves and for Third Worlders to translate the lofty promises of political independence into socially useful action.

The first VSOs were widely celebrated in Britain as the new heroes of a postimperial age. Eager reporters tracked their journeys, and they were feted with interviews on the BBC, luxurious luncheons with the Colonial Secretary, and tea with the queen. They were lionized specifically for their youth, enthusiasm, and absence of expertise, qualities that would soon contrast sharply with the technocratic juggernaut of the American Peace Corps. The first eighteen volunteers who went out in 1958 to Ghana, Nigeria, and Sarawak were succeeded by sixty more the following year, aided by a £9,000 grant from the Colonial Development and Welfare Fund and free passages from the Royal Air Force.[12] Industrial firms, such as Esso and Shell, also supported the volunteers, who cost around £550 each.[13]

The largest group of volunteers was in Nigeria, but the sites of service quickly diversified even beyond the confines of the former empire, striving to ensure that Britain's role in the new world order would not be delimited by the imperial past. In the former colonies, VSOs confronted the dizzying array of tasks that accompanied decolonization, from midwifery to soil conservation, census taking, and police work.[14] In this sense, decolonization provided an outlet and an opportunity to counter the idleness of elite youths in a variety of locales. Most VSOs worked in education, but volunteers also built boats in Papua, arranged libraries on the Dewey Decimal system in Ghana, and supervised elections in Bechuanaland.[15]

Volunteers were supposed to be at least eighteen years old, although one member of the first group was actually seventeen. They served for twelve months and were provided with pocket money but no other pay. Typically, volunteers applied to VSO when they were in their final term at school and landed overseas in September. Alec Dickson preserved the spirit of adventure by keeping the assignments secret until the final moments of departure.[16] Training was minimal; one volunteer recalled being taught only how to build trench latrines.[17] Methods for teaching English or other subjects were not discussed, the assumption being that the volunteers' own educational background had prepared them to teach others.[18] In some sense, the sites of service were irrelevant; VSOs were not supposed to become experts in particular fields and thus did not require knowledge tailored to particular regions. For this reason, although the Dicksons were concerned generally with the new social relations that they believed were demanded by independence, their vision of decolonization was largely undifferentiated; the political specificities of how decolonization worked in different parts of the empire did not weigh heavily on them.

From 1958 to 1962, Eton boys represented the largest number of volunteers to VSO.[19] Most of the early VSOs were from public school backgrounds. More than half had been head prefects, and nearly all had places to take up at Oxford or Cambridge within a year.[20] At a training session, one public school volunteer amiably remarked that this was the first time he had ever met a grammar school boy. The motives for service were diverse; Michael Talibard, an eighteen-year-old from Jersey, looked forward to facing greater physical challenges overseas: a welcome change from the affluent society, "'where all you can die of is lung cancer or thrombosis.'"[21] As the explicit absence of preparation would suggest, VSOs were supposed to arrive at their sites of service not as trained technicians but as blank slates.

In 1959, VSO accepted its first female volunteers, who attracted considerable media attention. Althea Corden taught English, math, and first aid in Sarawak, and Bronwyn Quint taught domestic science in Kenya.[22] The experience of volunteering was intended to be transformative for girls as well, although these transformations were unpredictable. One VSO supervisor was shocked to see a teacher whom Alec Dickson had described as a "church mouse" turn up in Malaya in decidedly new spirits; she was now wearing green eye shadow and declaring that she intended to go on the stage.[23]

The Dicksons saw their persistent emphasis on personal relations as differentiating their ventures from the more formalized realm of economic development. In rejecting the mantra of technical aid, Alec wrote: "No amount of foreign capital or gigantic dams can create a sense of nationhood, unless or until the educated youth of the country feel themselves to be involved, physically and emotionally: yet the young elite is not going to be prised away from its Baghdad-Bloomsbury axis through exhortations by middle-aged, expatriate experts."[24] Because large-scale development schemes relied on adult experts, they overlooked the emotionally transformative power of youth and ignored the individual character of both the giver and the receiver of aid. Development, in Alec's view, either neglected the realm of emotion altogether or cultivated the wrong kind (i.e., pride in foreign investment and a desire to emulate the West rather than engagement with those lower on the social scale within one's own nation).

Alec urged that VSOs should be rigorously overworked rather than underemployed.[25] He believed that the true perils of service lay not in remote jungles but on university campuses, where VSOs were "exposed to the hypercritical questioning of a sophisticated intelligentsia."[26] Because of the frustrations Alec had experienced in trying to interest elite Africans in physical labor, he stressed the significance of VSOs engaging in this type of work. A report on one VSO, William Crawley, at a boys' school in Nigeria, stressed his "welcome preparedness to engage in taxing manual activities," which was said to be slowly overcoming the resistance to such labor among the indigenous youths. William's readiness to cut grass, repair bicycles, and perform carpentry, along with his "engaging boyishness" and "artless lack of any form of Superiority," was cited as inspiration for educated Africans.[27] These tasks were important less for the material benefits they offered to others than for the humility and spirit of service they might inspire in those who undertook them.

The Dicksons believed that *what* volunteers did was less important than *how* they did it.[28] Mora admitted that trained adults might have undertaken many of the VSO's tasks more effectively. Indeed, the receiving territories often derided the quality of VSO's material aid.[29] But teenagers, she suggested, possessed the priceless qualification of love that ran freely outside set channels, the springs of affection being still unblocked.[30] Those who embodied the authority (such as it was) of the decolonizing state echoed Mora's views. The high commissioner of Lagos, for example, argued that an enthusiastic young volunteer "is a more effective evangelist for his country than the average technical aid man who comes out here to show Nigerians how to spray for capsid beetles or how to file secret documents." Middle-aged civil servants held no sway over the hearts and minds of young nationalists coming to power.[31]

On the basis of their experiences with young men and women in Britain and overseas, the Dicksons believed that both decolonization and welfare had disrupted the natural progression of the life cycle. Because young people in the First and Third Worlds were growing up either too quickly or too slowly, emotional and national development had gotten out of sync. One aim of VSO was to encourage young people in both sites to restore each other to their appropriate phase of development through carefully scripted (though seemingly natural) interactions. If Third World adolescents had no real childhood as it was understood in Europe, then their British counterparts seemed to suffer the opposite problem: that of prolonged juvenility. Until they acquired professional training, they were set apart from the real business of living, except for their disproportionate consumer power.[32] Indeed, one reason that the Dicksons chose British teenagers as the first group of VSOs was precisely because this group seemed to be in such desperate need of redemption.[33] As Mora wrote of the first VSOs: "They were boys who had thought of themselves as young men before they left England:

now, when they were in truth young men, this new world brought out in them the humble recognition that they were still boys."[34]

It is important to emphasize the specificity of this promise. VSO was pledging not to turn boys into men but rather to capitalize on their boyishness. The volunteer's distinctive contribution was his precarious balance between immaturity and adulthood, which was thought to mimic the psyche of developing countries. The Dicksons believed that adolescents were peculiarly suited to bridging the gap between First and Third Worlds in that they were nearest to the "emotional turbulence, yearnings, and perplexities" that characterized the collective emotional state of the citizens of newly independent nations.[35] In this sense, VSOs were the ideal group to undertake a new form of connection with a population experiencing political independence for the first time. The volunteers were vital because of their presumed capacity for forging emotional bonds with people who were—regardless of their biological age—undergoing a profound transition from one phase of political life to another and were thus in a state of uncertainty that mirrored adolescence itself. British teenagers and independent Africans (of any age) could revel in and temper each other's enthusiasms.[36] Such interpretations both naturalized the political tensions of decolonization and repackaged them as a generation gap.

This valorization of the adolescent marked a divergence from the paternalistic tropes of imperial days in which indigenes were routinely infantilized. Adolescence was replacing childhood as the dominant rubric for conceptualizing political action in the former empire. Essential to the new world order was the notion that individuals and nations must move appropriately through developmental stages. In this sense, development was individual and emotional as well as economic and political. Youth, as the Dicksons saw it, offered a corrective to the imperial past, which had been grounded in unequal relations between colonial authorities and indigenes.

Prior to their founding of VSO, the Dicksons had established a leadership training program for young, educated Nigerians at Man O'War Bay, a derelict banana plantation located four miles from Victoria. In 1954, the Man O'War Bay center began a course for schoolboys. The curriculum included "Smuggler's Evenings," at which Alec disguised himself as a villain in black greasepaint and surprised the boys while they were dining. After suffering this "attack," the boys engaged in group sing-a-longs, with American slave songs as part of the entertainment. Students and teachers dressed up as pirates, a venture that Mora acknowledged most Britons would find incomprehensible and childish. She zealously defended the technique of "shared fun," which "makes men forget what colour or race they are and returns them to an innocence which takes no account of these things."[37] The Dicksons saw these racialized masquerades as distinguishing their own ventures from the more formalized realm of economic development and technical aid. In their view, postimperial relationships were best forged not only *by* the young but also by acting as children.[38]

In all his projects, Dickson was interested in psychological conceptions of youth as a period of preparation, of waiting, of enthusiasm, impatience, and idealism.[39] Yet his definition of *youth* was also site specific. For him, *youth* in Britain meant anyone from age fourteen to age twenty-five, but then again some of the African participants in his leadership programs were as old as thirty-nine—and difficult to describe as *young*, especially given life expectancies in Africa. Dickson adopted what he saw as a canonical sociological definition of *youth*: that is, having completed compulsory schooling but not having fully entered into adult responsibilities. As Dickson recognized, *youth* carried a considerable charge, both in terms of attracting resources and as a way to understand new forms of power in Africa. He did not frequently cite social science experts in his own writing, and, indeed, VSO was much less closely engaged with the apparatus of social science—such as psychological testing for applicants to the program—than was the Peace Corps. But he did read widely about individual development. In some sense, developmental psychology provided a language for him to intervene in the technocratic approach to economic development that he feared might dominate British endeavors abroad.

The trajectory of VSO raises questions about precisely who has the power to make or remake emotional regimes in an era of decolonization. As the present volume seeks to demonstrate, although the realm of emotion was often neglected within particular domains of postwar social science, this neglect was never without its critics. There were always figures who advocated for treating emotion as worthy of greater attention both within and outside the academy. At VSO, social scientists certainly played a vital role, and Dickson was not himself entirely independent of their influence. But his critique of these particular forms of social scientific expertise, which seemed to ignore emotion at their peril, found its own enthusiastic audience with the decolonizing state.

What kinds of new people would decolonization and welfare create? How would independent Africans and Asians relate to and interact with Britons who themselves were engaged in unprecedented levels of mutual dependency at home and abroad? Social scientists, especially mental health experts, contemplated how decolonization might prove beneficial or harmful to emotional states of well-being in the former colonies. Most scholars focused on the potentially destabilizing feelings of aggression or dislocation that independence might engender in formerly colonized individuals.[40] The Dicksons, who had no formal training in psychology or psychiatry, were less concerned with the negative emotions that might accompany or be engendered by independence than with how they might create a new spirit of communitarian feeling that colonialism itself had weakened. According to them, one of the most pernicious legacies of colonial rule was that it had encouraged the elites of the colonized territories to think of themselves as having more in common with Britons than with their compatriots lower on the social scale. What was the cure? How could the feelings of these elites be reoriented vertically (i.e., down their own social system) rather than horizontally (with people of their same social class around the world)? Within VSO, elite Africans or Asians and elite Britons would be bonded through their performance of tasks to serve the masses in order to encourage a feeling of attachment to non-elite Third Worlders that would not otherwise exist.

The crux of the VSO idea, then, was not technical aid, but international friendship. Whatever technical aid took place was secondary to emotional intimacy. Feelings of humility, followed by mutual amity, ultimately produced material results, but these concrete gains were largely epiphenomenal. Volunteers for the Dicksons asserted that the gains of political independence were incomplete because they left the social prejudices of the colonizers and the colonized untouched. Thus, they sought to conduct "a wholly original experiment in human relations," transcending the psychological limits of independence.[41] "'You incredible British,'" an Indian education officer said to Alec in Madras: "'You left as rulers—and return as friends.'"[42]

Friends, perhaps, but of what kind? And how was the anticipated conversion from ruler and ruled to friends to take place? Leela Gandhi has fruitfully parsed the politics of the often secret and unacknowledged friendships and collaborations that united anticolonial individuals in the metropole and colonial worlds in the late nineteenth century and the early twentieth.[43] In some sense, the Dicksons sought to bring the rubric of colonial friendship into the light and to promote it in new forms. In the colonial era, *friend* could serve as a metaphor for dissident cross-cultural collaboration.[44] But the trope of friendship could take on new meanings in an age of decolonization. The friendships undertaken through VSO sought to replace a paternalistic regime with an (ultimately failed) regime of shared adolescence. In earlier eras, anti-colonial activists had conceived of friendship as a powerful rejoinder to political structures of inequality. Now, amity and sympathy were reframed as responses to new and rapidly equalizing political conditions that might synchronize the psyches of former colonizers and former colonial subjects. Such friendships were not simply critiques of a bygone imperial state. Rather, they were themselves linked to the rapidly evolving aims of the decolonizing state, which was tremendously ambivalent toward the rubric of economic development.

The emphasis at VSO was on games and play that would cultivate intimate personal interactions while also cultivating particular forms of Anglicization. VSOs were counseled to take an interest in boxing, especially in West Africa, "the cradle of British Empire featherweight champions."[45] A musical instrument could be invaluable to "unlock" people in the West Indies. Mora suggested that the music need not be of the "Olde English" folk-tune sort: "Even the songs from 'My Fair Lady' can be a hit." And she fondly recalled hearing "Que Sera Sera" sung by the local boys as they cut grass in Sarawak.[46] Scottish country dancing was also highly recommended, though Mora remarked disapprovingly that young people abroad were often more interested in English ballroom dancing. There could be significant tensions between what African and Asian elites wanted to learn and what VSO was willing to teach. The friendships that were envisioned as part of this program were explicitly *not* meant to legitimate the social aspirations of Third World elites. Rather, the purpose was to engage them in activities that stressed their connectedness to other social groups in their own society. Once the affections of these elites had been engaged by VSOs, they could be redirected downward, at which point VSOs could retreat and withdraw.

The concern with revitalizing elite youths and bonding them to one another—rather than, for example, succoring the poor—continued to differentiate the Dicksons' projects from contemporary economic development schemes. Perhaps what is most startling about the Dicksons' projects is how they captured the attention and resources of the state at a moment more typically understood as invested in technocratic development. The success of VSO's explicit critique of large-scale economic development schemes in favor of the crafting of personal relationships highlights the flexibility of the decolonizing state's own understanding of development and the players that state called on in order to promote it. The fact that VSO stressed an emotional component as a necessary part of development schemes was not a detriment in the eyes of the state but rather a reason to invest in VSO more heavily.

VSO warned receiving territories that volunteers must not be used as status symbols. But some of its own projects belied these claims. In 1965, there was a public outcry in Britain over VSO's role in India's Sainik Schools, wealthy institutions created on the British pattern to produce an elite cadre of civil servants. VSO's participation in these schools seemed to counter the democratic tide that had accompanied independence. The VSO leadership claimed that these prestigious educational enterprises could actually make India more egalitarian by helping create a stable middle class. That same year, the writer and broadcaster Ludovic Kennedy unflatteringly reported for the *Daily Telegraph* on his visit with two VSO girls in Nigeria who kept ringing a bell for their servant (because, they said, it was too unladylike to yell for him) and enjoyed a four-course luncheon with soufflé for dessert.[47]

According to VSO field reports, some of the sharpest tensions that arose in aid work were between volunteers and their older white compatriots, who were reluctant to accept dramatic changes in race relations. Owing to housing shortages, volunteers in Zambia often had to lodge with these expatriates, who ridiculed their idealism and placed them under "severe psychological strain."[48] Several VSOs were urged by white settlers to sign lucrative contracts and "forget all that volunteer stuff."[49] British settlers objected that VSOs turned up at formal gatherings disheveled and unwashed, often in indigenous dress. In Sarawak, VSOs gravely offended the white settler community by dining out in shorts. As Mora saw it: "[They had] let the side down because they were not on anybody's side. . . . [T]hey were a transition kind of Briton, no longer a master, not quite a partner."[50]

For Alec, these concerns about racial prestige were an outdated distraction from more urgent Cold War demands. He dutifully cautioned volunteers to be careful about their dress and hygiene, but he mused: "I cannot help feeling that what clothes are worn at night is not much of an answer to Mao Tse Tung." He quoted one VSO in Northern Rhodesia, who said: "'It is not Communism that threatens Central Africa so much as pessimism.'"[51] Alec believed that VSO offered a corrective to pessimism, which would in turn deter communism. Within the emotional economy of the Cold War, in which British and American authorities perceived disenchantment with or the loss of hope about the future as part of what made individuals susceptible to communism, the maintenance of optimism within Western development schemes was vital.

In 1960, VSO instituted a program for British industrial apprentices to serve overseas. The industrial apprentice program generated new tensions at VSO as the drive to democratize overseas service and include working-class Britons was ultimately incompatible with Dickson's vision of young elites engaged in mutual revitalization around the world. The apprentice program, sponsored by firms such as Rolls Royce and Shell, promised to strengthen Britain's overseas trade connections and "to make it possible for the ordinary young man in industry to feel that he, personally, could have a stake in this kind of assistance to the under-developed countries." The hope was to ensure that faith in internationalism would not be limited to elites but would spread to anyone who worked with "our Stan in Sarawak" or "our Jim in Jamaica."[52]

Dickson was confident that working-class youths had voluntarist impulses and that existing stereotypes of this group required revaluation. "Between the technician and the teddy-boy," Alec exhorted, "there is a gigantic middle stratum, which must be more significant than either."[53] The question was how to turn individuals who were seen as fiercely local in their sympathies toward the broader world: to pledge young industrial workers to an internationalism that was untainted by radical politics. If the problem with educated Africans was that they were too apt to forget their villages, then the young British worker was seen as too apt to remember his. Again, VSO aimed to correct both tendencies at once.

Yet the inclusion of working-class British youths would require a rethinking of VSO's emotional regimes. The scheme of VSO was based on the idea that elite Britons could be deployed to encourage elite Africans and Asians to develop sympathy for (and service to) their non-elite compatriots. In contrast to earlier metropolitan schemes, elite Britons of VSO were not offering direct aid to the poor; rather, they were tutors in leadership for those who were quickly ascending to positions of power themselves.[54] In having elite British youths undertake tasks that were so clearly beneath them, it was the gap between their social position and the labor they were willing to perform that was supposed to be instructive. For working-class VSOs, if they could be found, this gap would be less visible, as would the powerful feelings of sympathy that were supposed to be aroused by it.

Here, we can see how the politics of international aid was constructed around particular (and often disunified) understandings of the emotional capabilities of various populations within and outside Britain. Dickson conceived of the industrial volunteer program as a rebirth of the *Wanderjahre*, the continental concept of apprenticeship enriched by experience in an unfamiliar setting.[55] Ideally, apprentices would return to Britain with "an added sense of responsibility, and an understanding of the fundamentals of life—poverty, hunger, sickness, etc. from which they are shielded to-day in our Welfare State."[56] The first industrial volunteers were deployed to Kenya, designing artificial limbs for leper colonies.[57] James Hill, a sheet metal worker, maintained the electrical equipment at a community development project in Ghana and helped breed disease-resistant poultry.[58] Colin Stevens served at the Nigerian Ministry of Works and Transport, where he rewired government buildings and aided in the fireworks displays for the independence celebrations.[59]

The Times reported that industrial apprentices proved "just as good in their own idiom" as public school volunteers.[60] But these apprentices did not always fare well in VSO's selection process, which weighted voluntary service (such as Scouting) over technical skills. The VSO leadership believed that overseas service was more difficult for industrial apprentices than for public school boys, especially the experience of reentry to British life. The elite school leaver came back to the new experience of university that helped him over the difficult transition home. But for the apprentice, Mora warned, life in Britain could suddenly feel routine or flat. He returned to the workbench, often having lost seniority, and felt that his peers did not value the service he had undertaken. Mora cited a high rate of "defections" among industrial volunteers, who turned to the church rather than continuing at their firms.[61] If VSO converted elite public school boys into virtuous leaders, then it seemed to ruin industrial apprentices for their work in Britain.

The idea that apprentices did not settle back into home life as easily as their public school counterparts was echoed at the Colonial Office and in the press.[62] Industrial volunteers were accustomed to a "firmly stratified" working environment, in which the young were seldom encouraged to exercise initiative and the rules of unions and management tended to frustrate any independent thinking. When such young men found themselves in positions of considerable authority in VSO, they were unlikely to take kindly to returning to "being a kid on the shop floor, working for set hours at a limited pace for modest material rewards."[63]

From its inception, the apprentice program struggled. Although the idea of the philanthropic teddy boy captivated the British press, the alliance of elite school leavers and industrial workers was always uneasy. By 1967, despite numerous publicity campaigns, the number of apprentices had dropped sharply, and VSO's relations with trade unions had deteriorated.[64] Focusing on the socially elite background of VSO boys and girls was one way to distinguish British aid organizations from the Peace Corps. On the virtues of elite youths, Alec wrote: "Available for immediate service now is a boy from Eton with four 'S' levels in science and an Open Scholarship at Cambridge College: a superb musician and plucky games player—the grandson of [a well-known postwar cabinet minister]. To suppose that he is not good enough to teach at some secondary school in East Africa for a year—whereas some youth who has taken a heavens-knows-what degree at some university in the States is regarded as superior—seems madness."[65] Clearly, the young industrial worker had much to offer in terms of technical skills. But Alec was wary of stressing this point, which did not mesh well with his vision of international amity between young elites. The goal was not to teach the recipients of aid how to perform specific tasks, as in traditional development schemes, but to create the impulse in them to serve others. Here, Alec feared that industrial apprentices were less than inspirational. After all, technical aid was designed for the masses. Aimed at creating an orderly and prosperous peasantry, this form of aid belonged to precisely the same culture of development that Alec sought to reject or revamp, in which the cultivation of emotion played little part.[66]

Elsewhere, the Dicksons had proved remarkably naive in trying to convince educated elites to act as laborers, a deeply resented aspect of the program. The industrial apprentice program proposed to reverse this equation. Although Dickson perceived manual labor as an important way to rehabilitate British and African elites, he was less sure about how to utilize those who had actual skills in this field. He was never confident that industrial apprentices had any kinship with the sophisticated young Africans and Asians they were supposed to reform. The centrality of sympathy in his scheme was supposed to be rooted in a sameness of privilege and status, even if *privilege* might mean different things in London and Lagos. Lacking this socioeconomic status in their communities of origin, industrial apprentices would be unable to deliver on VSO's promises of emotional parity.

Nor did apprentices fit easily into Dickson's conception of youth as a state of potential and malleability. Despite their biological age, they did not conform to his view of the young. Again, the problem was not with what the apprentices *did* but with how they *felt* or failed to feel. The element of amity and kinship between elites was absent, Dickson believed, and, in its absence, the apprentices' other, more material contributions were rendered less meaningful. Essential to the initial VSO scheme was the notion that the young people on both sides were natural companions because they were of equivalent social rank. In this sense, the apprentices were disruptive to the emotional ideals of the program.

One can see here the resurfacing of socioeconomic tensions in what was supposed to be the universalizing category of youth and how overseas aid could serve as a staging ground for recurrent forms of class conflict. Imagining the agents of development proved as fractious as determining its recipients. For Dickson, the bonds of youth traveled far more easily across cultures and oceans than class. Undertaken precisely at the moment of social democracy's ostensible triumph, his schemes point to the reassertion of class hierarchies within the postcolony and the contestedness of the social democratic moment.

In an age of decolonization, VSOs bore an ambiguous relationship to British officialdom. Alec described VSOs as wielding an indirect form of authority, "like being a prefect in a good school."[67] Some volunteers in Nigeria and Northern Rhodesia were deployed as assistant district officers and were closely associated with the interests of law and order.[68] But volunteers also celebrated independence with Africans, and many were directly involved with the transfer of power. When Somaliland held its first general election on the eve of independence in 1960, a VSO helped with the polling arrangements. He found it "amusing to be given a most friendly welcome by the villagers, who presented him with a live sheep—and then went on to record a 90% vote for the anti-British independence party."[69]

Another VSO, a Cambridge undergraduate named William, promised his African peers that he would mark Nigeria's independence by wearing full Yoruba dress when he was back at university. In order to honor his vow without embarrassment, William found himself bicycling furiously around Cambridge at dawn on Independence Day, dressed in his African robes.[70]

In 1963, volunteers were evacuated from Indonesia after anti-British disturbances in Jakarta, and individual cadets were removed from Cyprus and Cambodia in 1964. The single biggest emergency was prompted by the civil war in Nigeria in 1967, when 150 volunteers were withdrawn.[71] During the war in the Middle East in 1967, there were evacuations from Israel, Lebanon, Algeria, Tunisia, and Jordan.[72] Perhaps the most contentious issue was whether VSO should remain in southern Rhodesia after the unilateral declaration of independence. This discussion spoke directly to the question of whether VSOs were counted as official aid because, if they were, then they would legally have to be withdrawn from a rebel country. After much anguished debate, VSO elected to keep cadet volunteers at their posts teaching black

African students in mission schools.[73] For Mora, countries with virulent racial tensions such as Rhodesia imperiled volunteers by exposing them to moral corruption as well as physical danger.[74]

Despite the eclecticism of their schemes, the Dicksons were neither outliers nor outcasts. They were intimately connected to the centers of state power and highly effective in publicizing their vision of the transnational redemption of youth.[75] One Colonial Office representative spoke of Alec Dickson as embodying "a new sort of romantic evangelism" that would, he expected, establish a new way of being British in the world.[76] The Dicksons' organizations survived and thrived, with VSO sending tens of thousands volunteers abroad.[77] The zeal surrounding the Dicksons' schemes illustrates that it may be apt to read this era not as a withdrawal from empire but rather as a reinvestment in a new internationalism in which the former empire played a significant part.[78] During these decades, Britain did not retreat from the world; rather, it undertook new forms of engagement that were sometimes fearful, sometimes enthusiastic. Indeed, decolonization could bring Britons into closer—and more intimate—contact with their former colonial subjects than ever before. Yet the new relationships prompted by decolonization could be difficult even for their proponents to characterize, and to conceive of the ways that they both evoked and reworked the colonial. It is one contention of the present chapter that we need to develop a richer vocabulary—as well as a more nuanced conceptual apparatus—to make sense of the relationships that were galvanized or intensified by independence.

The inner transformations that VSO sought to achieve differed for each of its target populations. For elite British youths, VSO was a way to provide opportunities for adventurous service that the welfare state had rendered unnecessary or outmoded within Britain. At the same time, for elite young men and women in the former colonies, VSO could acquaint them with the more menial tasks of constructing new nations, thus inspiring a sense of humility and sympathy toward those of their compatriots whose education and manners differed from their own. For industrial volunteers, the issue was more complicated. Initially, the Dicksons imagined the construction of a newly internationalized British working class whose expanded sympathies would mesh with the needs of the newly global economy. But this population proved the most problematic from the Dicksons' point of view. When they worked in former colonial sites, their contributions seemed tainted by (or difficult to sever from) the specter of traditional development, overly grounded in the material. When they returned to Britain, they were perceived as unable to reintegrate into their appropriate social roles. Thus, VSO saw its participants in uneven terms, raising questions about how class could fracture the emotional economies of decolonization and welfare alike.

Perhaps Alec Dickson's most persistent idea, in all the locales in which he worked, was that the end of empire required new kinds of personal relationships and emotional experiences. This reconstruction of personal relationships was, in his view, crucial to the West's success in the Cold War as well as to Britain's peaceful management of decolonization and the new challenges to personal character brought by the rise of the welfare state.

Yet these battles were not to be waged by just anyone. Dickson's critique of development was driven by many factors, but key among them was his focus on reforming the inner lives of elites rather than the material conditions of the masses. It was this element of his vision that explained the early success of VSO at a moment when the plight of the teenager in the affluent society seemed to demand a global solution and the threat of the young African or Asian politician appeared to be accelerating. In this environment, the Dicksons' promise that they could create new Britons—along with new Africans and Asians—held tremendous appeal, as it has continued to do through the present day. Their remarkably durable work can, I hope, help us begin to isolate what is distinctive about emotion in the age of development and decolonization and to diversify the sites, moments, and registers in which we perceive the history of emotion to reside.

NOTES

1. This literature is vast, but see, e.g., Tony Ballantyne and Antoinette Burton, eds., *Moving Subjects: Gender, Mobility, and Intimacy in an Age of Global Empire* (Urbana: University of Illinois Press, 2009); Durba Ghosh, *Sex and the Family in Colonial India: The Making of Empire* (Cambridge: Cambridge University Press, 2006); Philippa Levine, *Prostitution, Race and Politics: Policing Venereal Disease in the British Empire* (New York: Routledge, 2003); Matt Matsuda, *Empire of Love: Histories of France and the Pacific* (Oxford: Oxford University Press, 2005); Vicente L. Rafael, *White Love and Other Events in Filipino History* (Durham, NC: Duke University Press, 2000); Emma Rothschild, *The Inner Life of Empires: An Eighteenth-Century History* (Princeton, NJ: Princeton University Press, 2011); Ann Laura Stoler, *Carnal Knowledge and Imperial Power: Race and the Intimate in Colonial Rule* (Berkeley and Los Angeles: University of California Press, 2002), and *Along the Archival Grain: Epistemic Anxieties and Colonial Common Sense* (Prince-

ton, NJ: Princeton University Press, 2009); and Ann Stoler, ed., *Haunted by Empire: Geographies of Intimacy in North American History* (Durham, NC: Duke University Press, 2006).

2. Jordanna Bailkin, *The Afterlife of Empire* (Berkeley and Los Angeles: University of California Press, 2012).
3. Becky Conekin, Frank Mort, and Chris Waters, introduction to *Moments of Modernity: Reconstructing Britain, 1945–64*, ed. Becky Conekin, Frank Mort, and Chris Waters (London: Rivers Oram, 1999), 1–21. See also Stephen Brooke, "Gender and Working-Class Identity in the 1950s," *Journal of Social History* 34.4 (2001): 773–95; Marcus Collins, *Married Love: An Intimate History of Men and Women in Twentieth-Century Britain* (London: Atlantic, 2003); Martin Francis, "A Flight from Commitment? Domesticity, Adventure, and the Masculine Imaginary in Britain After the Second World War," *Gender and History* 19.1 (April 2007): 163–85; Liz Heron, ed., *Truth, Dare or Promise: Girls Growing Up in the Fifties* (London: Virago, 1985); Patrick Joyce, "More Secondary Modern than Postmodern," *Rethinking History* 5.3 (December 2001): 367–82; Annette Kuhn, *Family Secrets: Acts of Memory and Imagination* (London: Verso, 1995); Frank Mort, "Social and Symbolic Fathers and Sons in Postwar Britain," *Journal of British Studies* 38.3 (July 1999): 353–54; Nikolas Rose, *Governing the Soul: The Shaping of the Private Self* (London: Routledge, 1990); and Carolyn Kay Steedman, *Landscape for a Good Woman: A Story of Two Lives* (New Brunswick, NJ: Rutgers University Press, 1986).
4. National Service operated on a reduced scale until 1963. David Lodge, *Ginger, You're Barmy* (London: Macgibbon & Kee, 1962); Martin S. Navias, "Terminating Conscription? The British National Service Controversy, 1955–56," *Journal of Contemporary History* 24.2 (1989): 195–208; L. V. Scott, *Conscription and the Attlee Governments: The Politics and Policy of National Service, 1945–1951* (Oxford: Clarendon, 1993); Leslie Thomas, *The Virgin Soldiers* (Boston: Little, Brown, 1966).
5. Brian Harrison, *Seeking a Role: The United Kingdom, 1951–1970* (Oxford: Oxford University Press, 2009), 91.
6. David Wainwright, *The Volunteers: The Story of Overseas Voluntary Service* (London: Macdonald, 1965), 27.
7. On the terminology of First and Third Worlds, see Carl E. Pletsch, "The Three Worlds; or, The Division of Social Scientific Labor, circa 1950–1975," *Comparative Studies in Society and History* 23.4 (1981): 565–90.
8. Elizabeth Cobbs Hoffman, *All You Need Is Love: The Peace Corps and the Spirit of the 1960s* (Cambridge, MA: Harvard University Press, 1998), 8, 15.
9. Harry Hopkins, *The New Look: A Social History of the Forties and Fifties in Britain* (London: Secker & Warburg, 1963); Rodney Lowe, *The Welfare State in Britain since 1945*, 2nd ed. (Basingstoke: Macmillan, 1999).
10. Lowe, *Welfare State*, 21; Richard Titmuss, *The Gift Relationship: From Human Blood to Social Policy* (New York: Pantheon, 1971).

11. S. M. Miller, "Introduction: The Legacy of Richard Titmuss," in *The Philosophy of Welfare: Selected Writings of Richard Titmuss*, ed. Brian Abel-Smith and Kay Titmuss (London: Alen & Unwin, 1987), 1–7.
12. Michael Adams, *Voluntary Service Overseas: The Story of the First Ten Years* (London: Faber, 1968).
13. Wainwright, *The Volunteers*, 33.
14. *The Times*, September 29, 1959.
15. Christopher Tipple to Alec Dickson, October 28, 1959 (?), DO 33/8199, National Archives, Kew.
16. Interview with Dick Bird, September 24, 2008.
17. "In Living Memory: VSO," BBC Radio 4, December 3, 2008.
18. Interview with Chris Tipple, July 12, 2009.
19. Robert Birley to Alec Dickson, November 13, 1962, PREM 11/5007, National Archives.
20. Dick Bird, *Never the Same Again: A History of the VSO* (Cambridge: Lutterworth, 1998), 25.
21. *The Times*, April 4, 1962.
22. "Volunteers Sent Abroad in 1959–1960," n.d., OD 10/3, National Archives.
23. J. R. Williams to A. J. Brown, December 4, 1961, DO 163/22, National Archives.
24. Quoted in Bird, *Never the Same Again*, 18.
25. Alec Dickson to P. F. Walker, May 23, 1961, DO 196/19, National Archives.
26. Alec Dickson, "A Great Voluntary Movement," *Guardian*, January 29, 1962.
27. Report of M. A. Marioghae on William Crawley, July 2, 1960, CO 859/1445, National Archives.
28. Mora Dickson, *A Season in Sarawak* (Chicago: Dobson, 1962), 216.
29. Basil Shone to C[harles] Y. Carstairs, October 8, 1960, CO 859/1445, National Archives.
30. Dickson, *A Season in Sarawak*, 192.
31. High Commissioner of Lagos to Duke of Devonshire, February 16, 1963, DO 163/22, National Archives.
32. Dickson, *A Season in Sarawak*, 154.
33. Mora Dickson, *A World Elsewhere: Voluntary Service Overseas* (Chicago: Rand McNally, 1964).
34. Dickson, *A Season in Sarawak*, 171.
35. *The Times*, May 26, 1962.
36. Margery Perham, *The Colonial Reckoning: The End of Imperial Rule in Africa in the Light of British Experience* (New York: Knopf, 1962), 201.
37. Nancy Rose Hunt has argued that this type of "topsy turvy" performance in colonial locales could conceal terror and sadism just beneath the affection it purported to promote. Nancy Rose Hunt, *A Colonial Lexicon: Of Birth Ritual, Medicalization, and Mobility in the Congo* (Durham, NC: Duke University Press, 1999).
38. Dickson, *A Season in Sarawak*, 161.

39. Alec Dickson, "Training of Youth Leaders for Work in Fundamental Education," *Fundamental and Adult Education* 10.2 (1958): 45–54.
40. J. C. Carothers, *The Psychology of the Mau Mau* (Nairobi: Government Printer, 1955); E. B. Forster, "The Theory and Practice of Psychiatry in Ghana," *American Journal of Psychotherapy* 16 (1962): 7–51; C. V. D. Hadley, "Personality Patterns, Social Class, and Aggression in the West Indies," *Human Relations* 2.4 (1949): 349–62; Sloan Mahone, "The Psychology of Rebellion: Colonial Medical Responses to Dissent in East Africa," *Journal of African History* 47.2 (July 2006): 241–58; Raymond Prince, "The Changing Picture of Depressive Syndromes in Africa: Is It Fact or Diagnostic Fashion?" *Canadian Journal of African Studies* 1 (1967): 177–92.
41. Adams, *Voluntary Service Overseas*, 56. See also Michael Courage with Dermot Wright, *New Guinea Venture* (London: Hale, 1967); and Colin Henfrey, *The Gentle People: A Journey among the Indian Tribes of Guiana* (London: Hutchinson, 1964).
42. Alec Dickson, "Southeast Asia Training Project in Work Camp Methods and Techniques," December 5, 1959, CO 859/1445, National Archives.
43. Leela Gandhi, *Affective Communities: Anticolonial Thought, Fin-de-Siècle Radicalism, and the Politics of Friendship* (Durham, NC: Duke University Press, 2006), 9.
44. Ibid., 10.
45. Dickson, *A World Elsewhere*, 22–23.
46. Ibid., 23.
47. Bird, *Never the Same Again*, 70.
48. A tiny minority of VSOs (about 1 percent) experienced breakdowns during their period of service. VSO was occasionally criticized for failing to put its volunteers through the same elaborate psychological screening as their Peace Corps counterparts. Adams, *Voluntary Service Overseas*.
49. J. D. G. Isherwood, "Report of a Field Visit to Zambia, Malawi and Tanzania," January 31–March 12, 1966, OD 10/79, National Archives.
50. Dickson, *A Season in Sarawak*, 173–74.
51. Alec Dickson, "Voluntary Service Overseas: Thomas Holland Memorial Lecture," *Overseas Quarterly* 2.2 (June 1960): 47–48.
52. Dickson, *A World Elsewhere*, 95.
53. Dickson, *A Chance to Serve*, 78.
54. See, e.g., Seth Koven, *Slumming: Social and Sexual Politics in Victorian London* (Princeton, NJ: Princeton University Press, 2004).
55. Dickson, "Voluntary Service Overseas."
56. C. N. F. Odgers, "Recruitment of Volunteers from Industry," June 30, 1967, OD 8/473, National Archives.
57. Wainwright, *The Volunteers*, 112.
58. "A Year Abroad for Young Employees," *The Guardian*, June 4, 1960.
59. "Volunteers Sent Abroad in 1960–1961," 1961–63 (?), OD 10/3, National Archives.

60. *The Times*, January 13, 1964.
61. Wainwright, *The Volunteers*, 134.
62. Patrick Denison to C[harles] Y. Carstairs, August 24, 1960, CO 859/1445, National Archives.
63. Wainwright, *The Volunteers*, 134.
64. C. N. F. Odgers, "Recruitment of Volunteers from Industry," June 30, 1967, OD 8/473, National Archives.
65. Bird, *Never the Same Again*, 38.
66. Monica Van Beusekom, *Negotiating Development: African Farmers and Colonial Experts at the Office du Niger, 1920–1960* (Portsmouth, NH: Heinemann, 2002).
67. Dickson, *A World Elsewhere*, 31.
68. Bird, *Never the Same Again*, 75.
69. Adams, *Voluntary Service Overseas*, 188–89.
70. Dickson, *A World Elsewhere.*
71. The returned volunteers protested at the VSO's London office. Adams, *Voluntary Service Overseas*, 186.
72. Bird, *Never the Same Again*, 57.
73. W. J. Smith, "VSO and U.D.I.," October 6, 1965, OD 8/473, National Archives.
74. Dickson, *A World Elsewhere*, 66.
75. The Albemarle Report (1960), which urged a major expansion in spending on youth organizations, praised VSO as a useful form of "social training."
76. R. Terrell to Smith, January 21, 1960, CO 859/1445, National Archives.
77. "In Living Memory: VSO."
78. Martin Lynn, ed., *The British Empire in the 1950s: Retreat or Revival?* (Basingstoke: Palgrave Macmillan, 2006).

TWELVE

Passions, Preferences, and Animal Spirits: How Does *Homo Oeconomicus* Cope with Emotions?

UTE FREVERT

During the second half of the twentieth century, economic science saw an enormous surge in public interest and approval. In more recent times, however, it has become the object of sharp criticism for its inaccuracy and outright failure.[1] Both stories of success and disenchantment are deeply connected with the rise and fall of (neo) classical economists' favorite concept: *homo oeconomicus*. This concept, I will argue, has been enormously influential in underlying economic modeling and theorizing. At the same time, however, it is riddled with serious flaws and shortcomings, mainly as far as emotions and "animal spirits" are concerned. Those flaws and vague points seem to bear a negative impact on economists' ability to accurately analyze and predict human behavior in the economic sphere.

This chapter explores how and why mainstream economists tended to build their theories on the assumption that human actors rationally follow the motives of self-interest. It reconstructs the reasons why competing assumptions were neglected and pushed to the margins. Moreover it investigates under which conditions and to what extent the concept of *homo oeconomicus* was ultimately challenged and modified by a new generation of economists, psychologists, and neuroscientists.

Economics on the Transnational Road to Triumph

The history of the Nobel Prize is an excellent case with which to illustrate the fame and glory of economics in the post-1945 era. When Alfred Nobel endowed the prize in the late nineteenth century, it had not occurred to him that one day the recipients might include economists. Physicists, chemical engineers, and physiologists were the scientists that he deemed worthy of an award bestowing reputation and money. They worked in fields that, in his opinion, "have conferred the greatest benefit to mankind." Economic theory was definitely not among them since, from his own experience, Nobel did not believe it to be helpful.[2] As a highly successful inventor and entrepreneur, he deliberately kept his distance from academics who claimed that they could produce authoritative knowledge on how market operations functioned.

The Swedish Riksbank obviously thought differently when they instituted a new prize for economics in 1968, in memory of Nobel. Between then and 2011, one female and sixty-eight male economists have been honored, mostly for their work on modeling. The great majority come from the United States and, albeit to a much lesser degree, from Great Britain, a fact that points to the overriding importance of Anglo-American economic theory after World War II.

The reason why this theory became so strong and influential can be attributed partly to the general leading position of American science from the 1940s and 1950s onward. Whereas European universities and research institutes were all in some way or other plagued by the accumulated effects of two major wars within one generation, their American counterparts managed to reconfigure themselves in a profound and dynamic manner. Benefiting from immigrant scholars fleeing from Nazi persecution and postwar material hardships, they used government incentives and funding to extend the reach of higher education and invested in groundbreaking research, basic as well as applied. The efficiency of this strategy was soon demonstrated by the firmly established American dominance among Nobel Prize winners in the natural and life sciences.

Regarding the field of economics, however, additional factors were involved in America's prominence and influence. On the one hand, the challenge posed by the German-led school of historical economics on (neo)classical theory had already weakened and faded away after the 1920s. In the German-speaking world, the Younger Historical School's intellectual hegemony, which had reached its peak in the 1890s, did not outlive World War I. The manifold attempts to develop new and convincing paradigms of economic theory during the interwar period resulted in a wide array of rivaling approaches and "last principles." Apart from the neoclassical Austrian school, there were numerous others seeking to reinvent organic concepts or favoring mildly socialist views. Joseph Schumpeter, who left his professorship at Bonn University in 1932 to join Harvard's faculty, deplored the radical sense of enmity and noncommunication among his German colleagues, who, like philosophers, competed for the absolute truth instead of progressing in the field of positive knowledge and "pure" economic theory.[3]

Pure theory and analysis had, on the other hand, found an early and comfortable home in North America. Since Adam Smith's times, the classical approach had supported liberal market society. Its underlying assumptions—mainly the notion of free individuals pursuing happiness (defined as the fulfillment of strong material desires) unhampered by political constraints and state interventions—seemed to reflect both the history of the United States and its continuous promise to people all over the world. It is true that there had been occasions when those promises had been jeopardized by economic stagnation or depression and when the state struck a "new deal" to ensure economic recovery. Still, the United States continued to attract new immigrants who pursued the American dream, "from rags to riches," from dishwasher to billionaire. For Europeans suffering the consequences in the aftermath of two consecutive wars, North America offered a paradise of unlimited wealth and freedom. Economic theories that closely related wealth to unrestricted mobility of labor and capital in a free market were thus highly respected. American-led programs of economic recovery in Europe further strengthened this tendency, as did fellowships that acquainted European students and academics with US universities and research institutions. The combination of these factors and others was the condition for the possibility of neoclassical theory's triumphant success in postwar non-Communist Europe.[4]

Abstraction of Passion

What got largely lost on the transatlantic journey was a sense of economy's historical and political roots. This aspect had been strongly accentuated by the Younger Historical School in Germany around 1900 and had been extensively applied in American institutionalism. Initiated by Thorstein Veblen, the latter exerted considerable influence on US economists in the 1920s but later lost its importance.[5] Gustav Schmoller, the head of the German school and an influence of sorts on Veblen, had, time and again, stressed the degree to which the economy remained part and parcel of social life in general. Economic theory therefore had to ponder the relationship between genuinely economic matters—like the division and organization of labor, transactions and mobility, income distribution, and the formation of prices—and matters of state, law, customs, and morality. According to Schmoller, every economic phenomenon, such as the increase in grain prices or wages, consisted of people's "feelings, motives and actions" and was shaped by "morals and institutions" with widely diverging causes. Historical knowledge was thus necessary in order to account for those institutions, worldviews, and attitudes. The assumption that a person was an individual exclusively governed by egotistical interests and selfish preferences was synonymous with misreading the complex frames of reference in which human motives and drives were formed and acted on. Even when it came to acquisitive impulses (*Erwerbstrieb*) and the desire to get rich and richer, substantial differences can be identified within and among various cultures, social milieus, and nations.[6]

Schmoller here directly addressed and attacked the underlying propositions of (neo)classical theory, mainly that of *homo oeconomicus*. "Economic man" was driven by what Adam Smith had called *self-love* and the ensuing attempt to maximize his gains while minimizing the amount of labor necessary. Such behavioral patterns were deemed natural and universal; they constituted the building blocks of economic reasoning from Smith on.

Both Smith and his followers were certainly aware of the fact that human behavior was far more complex and informed by many other motives and feelings. As much as Smith, in an earlier treatise on "Moral Sentiments," knew about the power of sympathy in fostering social cooperation and communication, John Stuart Mill cited "those laws of human nature" that called forth "the *affections*, the *conscience*, or feeling of duty, and the love of *approbation*" among human beings. No political economist, Mill claimed, "was ever so absurd as to suppose that mankind are really thus constituted"—that is, "determined, by the necessity of his nature, to prefer a greater portion of wealth to a smaller in all cases." Economic science, however, had to "necessarily proceed" in this mode of narrow determinism in order to reach proper conclusions in its own field. Needless to say, in other fields other laws pertained.[7]

What was to be gained by this exclusive concern with man "as a being who desires to possess wealth, and who is capable of judging of the comparative efficacy of means for obtaining that end" was obvious.[8] Only by neglecting other desires and ways of conduct could political economy establish itself as an independent and autonomous field of scientific research. The decision to make "entire abstraction of every other human passion or motive"[9] was rewarded by a considerable increase in disciplinary acumen and methodological sophistication.

This was not lost on Schmoller and his followers. At the same time, however, they were reluctant to discard the notion of economics as a profoundly social and cultural science. They cast a critical eye on tendencies of formalization and mathematization that were concomitant with isolating economics from the wider web of human interaction. Schmoller even went as far as to suggest that economic phenomena had to be perceived as deeply psychological. Economists thus had to account for the "transformation of psychological causes and how they connected to ethnological and class differences"; as a second step, they should then study how those transformations shaped the concrete ways in which people acted in economic relations.[10]

The price to be paid for such a complex perspective was high. Perceiving economics as a deeply historical science meant that it lacked clearly discernible paradigms and ran short of deductive theorizing. By reaching out to other disciplines in the humanities and the social sciences, it acquired an interdisciplinary character that rendered identifying its proper disciplinary profile difficult. Furthermore, it proved unwilling and unable to address the manifest problems haunting European economies after World War I.[11]

In contrast, it was much easier for classical and neoclassical economics to develop a distinct profile and the accompanying academic recognition by narrowing the field of scientific inquiry and disregarding human behavior. Even if many German economists continued to criticize their Anglo-American colleagues for putting too much effort into theoretical models that ignored reality, those models proved relatively successful in shedding light on complicated matters like business cycles, prices, and economic growth.

Thanks to its sophisticated quantitative methods, neoclassical theorizing even attracted scholars who did not share its underlying belief in the gospel of the free market. The famous British economist John Maynard Keynes, for example, established his ideas concerning the relationship between employment, interest, and money in radical confrontation with the neoclassical belief in economic self-regulation and full employment equilibrium. Still, his concepts could be—and were—easily translated into formal mathematical models and accessed by mainstream methodology. As early as 1938, economists like John R. Hicks integrated Keynes's assumptions into a general equilibrium model that has guided the analysis of capital and money markets ever since. Combining Keynes's theory of employment and income with classical theory eventually produced the "neoclassical synthesis" (Paul Samuelson) that dominated international economic research in the 1950s and 1960s.[12]

Economic Man's Rational Choices

It was during those early postwar decades that economic research became highly influential in politics and public life both in the United States and in Western European countries. In 1946, the US government set up the Council of Economic Advisors, whose aim was to counsel the president on economic policy and supply him with objective data and material on which decisions had to be based. The West German government emulated this model in 1963. Even prior to that, there was high demand for economic expertise, and the social reputation of experts was at its peak. On the one hand, this was due to the growing importance and complexity of economic policy in postwar capitalist societies. On the other hand, it reflected the self-confidence of economists to deliver theory-based knowledge on how the economy functioned and what the state could do to improve its function in Keynesian and ordoliberal terms. Such self-confidence was directly related to the increasing uniformity of theoretical paradigms on both sides of the Atlantic. The triumph of the neoclassical synthesis and the omnipresence of econometrics and mathematical modeling helped boost the self-perception of economists as scientists independent of other disciplines.

As economics increasingly gravitated toward natural science methods and severed its old ties to the humanities and social sciences, the notion of *homo oeconomicus* also grew in importance. Perceiving human beings as rational actors who were guided by self-interest became an ever more powerful assumption of economic modeling. The surge of microeconomics since the 1960s (reflected in the types of economists awarded Nobel Prizes) helped strengthen it even further, up to the point where it reached out to noneconomic spheres as well. Having children, raising a family, caring for older family members, voting for a party, or joining an association were now all framed according to the model of human beings making decisions in a rational, autonomous, and self-regarding way. *Rationality* was defined as an algorithmic relation between input and output, between means and ends. The more people were able to maximize their benefits while, at the same time, minimizing individual effort to obtain these benefits, the higher the degree of rationality involved in this operation. Benefits or gains were dictated by self-interest.

Linking the concept of the economic man with rational choice theory turned out to be an extremely successful move. It helped reinvent economics as a discipline that aspired to a "general unified theory," pretending that this theory was universally applicable.[13] As much as the economic man was thought to be a given and present at all times and in all places, his actions and nonactions could be modeled and predicted on a universal scale. Anyone questioning the assumption of rationality or pointing toward other factors apart from self-interest that might inform people's actions clearly curtailed the explanatory power of economics and threatened economists' newly acquired public status as makers of, or advisers on, economic policy.

The notion of economic man was rendered so powerful not only through economic theory's predilection for abstractions and generalizations. The concept also reflected the self-description of nineteenth- and twentieth-century capitalist societies. Since its invention as part of eighteenth-century positive anthropology, it developed into a popular self-representation of modernity. Instead of being haunted by unruly passions and the lust for tyranny, modern human beings learned to transform passions into interests and pursue the latter in a rational way. The capitalist marketplace taught them what *rationality* meant: to use the most efficient strategy to achieve one's ends. Since this involved other players as well, the strategy had to take into account and accommodate their interests. Hence, economic communications and transactions were supposed to function both as self-regulatory processes and as civil—and civilizing—building blocks of modern society.[14]

Even though this optimistic view was by no means universally shared, it increasingly managed to capture people's imagination and self-perception. Acting rationally, taking the right decisions, and pursuing one's goals in a calculating and calculable fashion became the leitmotif of modern and modernizing societies—a leitmotif that was praised as much as it was lamented. Instead of being an anthropological given, it became enshrined in modern bureaucratic institutions and turned into a normative concept that men—more than women—had to learn early on. Escaping it was possible, if at all, only in the spheres of religious mysticism, art, and erotic experience, as Max Weber argued.[15] In economic matters, the norm ruled ubiquitously and compellingly. All those entering the market, whether they were producers or consumers, entrepreneurs or workers, were considered rational agents. As Keynes put it in 1937, they desperately wanted to "behave in a manner which saves our faces as rational, economic men."[16] Instead of being a mere methodological tool and construct, as Schumpeter suggested, *homo oeconomicus* had become a social concept that actually informed people's self-perception and framed their preferences.[17]

Animal Spirits

As long as the economy developed in a relatively steady and positive manner, the norm was easy to follow. The experience of the 1920s, however, shattered the widely held belief in stable, self-regulating markets and business cycles. Risk was, as Keynes pointed out, a fundamental feature of the economic process. It was the rule, not the exception, as classical theory had believed. Decisionmaking thus happened under the assumption that the future was uncertain, unknown, and hardly calculable. How could people deal with this dilemma? How could they manage to cope with uncertainty and still pretend that they were acting rationally?

For this purpose, Keynes saw economic agents devise "a variety of techniques." Among them was projective and habitual behavior that could not, however, be reduced to "cold calculation." Instead, it followed the psychology of "human nature," motives of "satisfaction" and "temptation." Entrepreneurs, for instance, were often "not really relying on a precise calculation of prospective profit" but "embark[ing] on business as a way of life" and playing "a mixed game of skill and chance." As human beings and market players, they also responded to what Keynes called *animal spirits*. Those were of a "spontaneous" nature and urged one "to action rather than inaction," thus helping bypass a situation in which too much reasoning and calculating might lead to inaction. Instead of relying on algorithms and "mathematical expectation," animal spirits induced a "spontaneous optimism" that allowed people to act under uncertainty, thus enabling market transactions to continue.[18]

Keynes understood animal spirits as affects, short-lived but intense feelings not given to thorough reflection and self-control ("whim or sentiment"). They seemed to appear spontaneously, as a kind of anarchic and vital force. This force compelled people to act even if they lacked sufficient information to make a "rational," well-prepared decision. Consequently, Keynes attributed positive qualities to animal spirits since without them "enterprise will fade and die." Without those vital energies and instincts, any economic activity, and particularly in times of severe crisis, would cease, with dramatic and unforeseeable consequences.[19]

It is not entirely clear from where Keynes borrowed his concept of animal spirits. It might have been René Descartes, who, in his 1649 *Les passions de l'âme*, had described them as highly delicate and mobile blood particles that flow into the brain, then get passed on to nerves and muscles, and consequently set the body in motion. Sentiment, memory, imagination, desires, and passions all originate, according to Descartes, from those particles. Isaac Newton, who shared this interest in animal spirits that pervaded brain, nerves, and muscles, was even better known to Keynes, who was fascinated by Newton's manuscripts and, in the early 1940s, prepared a lecture celebrating the tercentenary of his birth.[20]

The notion that sentiments, passions, and affects were closely related to what might be translated as *the energy of mind and soul* (*spiritus animalis*) had been a staple in early modern philosophy. The manner in which this relation was perceived and judged, however, remained controversial. While some writers and commentators praised strong passions as dynamic and constructive forces, the majority stressed instead their negative impact. For them, the problem was that mankind had not too few but too many passions and showed too much reluctance to curb their overwhelming power. Emotions of any kind, whether short-lived or long-lasting, had to be controlled and channeled so as not to interfere with rational planning and decisionmaking. Even if they could not, and should not, be dismissed and suppressed altogether, they should be transformed into something mild and benevolent. Similar transformation strategies were suggested by that ultramodern strand of contemporary psychology that Keynes was well acquainted with: Sigmund Freud's theory of sublimation, which turned sex drives into cultural work.[21]

As a general rule, emotions (much more than drives) were held to be utterly unpredictable, their unpredictability rendering them hard to process for economic theory. As strong motivators of people's actions and nonactions, they definitely played a major role in all fields. Closely linked to value systems and notions of morality, they evidently informed human behavior, both in the private and in the public sphere. But how could they be integrated into theoretical models of economic processes? According to Schmoller and, to a lesser degree, Werner Sombart, psychology proved indispensable in order to make sense of men's economic actions.[22] Others, like Schumpeter, discarded the question of motives and actions altogether. Pure economics should not be concerned with human beings and what they wanted or did and for what reason. What mattered was the amount of goods that those beings possessed and traded: "We want to describe the changes, or better, a certain type of changes, as if they would happen automatically, without looking at the people who are responsible for those changes."[23]

Interestingly, in his own work Schumpeter did not practice what he preached. As someone fascinated by the innovative entrepreneur figure (he himself had been brought up in an entrepreneurial family), he was keen to explore the conditions under which men recombined the means of production and empowered economic development. In order to explain why entrepreneurs did not follow the traditional track of making money, he used psychological concepts and referred to emotional driving forces. Instead of calculating utility or satisfying his needs and wishes, the dynamic entrepreneur was obsessed by "the dream and the will to found a private kingdom"; he indulged in the "joy of creating, of getting things done, or simply of exercising one's energy and ingenuity," and he followed "the will to conquer: the impulse to fight, to prove oneself superior to others, to succeed for the sake, not of the fruits of success, but of success itself." Such an entrepreneurial spirit surpassed the inbuilt logic of the marketplace and could by no means be subsumed under pure economics.[24]

Consequently, those economists who adhered to the pure ideal and worked hard to turn economics into a precise science took their distance when it came to spirits, entrepreneurial or other. Keynes's cautious remark about animal spirits helping the economic process continue was not received well. Instead, postwar economists rapidly expanded their efforts to model individual behavior with the help of mathematical algorithms. The rise of microeconomics based on the notion of utility-maximizing individuals occurred within the framework of a general equilibrium theory that sought to derive macroeconomic regularities from the very same assumptions, that is, from individualistic ontologies.

Bounded Rationality

From the 1970s, however, it became apparent that those ontologies were heavily flawed. First, the hypothesis of individual rationality was increasingly questioned, and so were people's endless calculating capability and their skills to make superbly rational choices. Casting doubt on the cherished premises of both microeconomics and macroeconomics was concomitant with developments within the discipline of psychology. Cognitive psychology, which paid attention to processes of mental planning and prefiguration, became increasingly popular as orthodox behavioralism lost traction. This resulted in a renewed interest in how people interpreted and framed what they saw and encountered. In the wake of this epistemological shift, the concepts of habit and instinct—perfectly compatible with Keynes's animal spirits—were reinstated.[25]

As early as 1947, Herbert Simon (who was awarded the Nobel Prize in 1975) made use of cognitive psychology's insights. Questioning the assumption that individuals could ever possess complete information about how to best achieve certain goals and choose among alternative means, he underlined the limits to rational decisionmaking. Still "striving for rationality," individuals developed "working procedures that partially overcome these difficulties." They acted, therefore, under conditions of "bounded rationality" that prompted them, as Simon and James March stated in 1958, to accommodate goals to means. As a rule, actors who generally commanded a limited cognitive capacity resorted to a strategy of "satisficing" instead of maximizing. *Satisficing* was a neologism that combined *to satisfy* and *to suffice*. It described, according to Simon, an economic behavior that tries to achieve a less than optimal level of a particular variable but one that still suffices to satisfy the inbuilt logic of that behavior.[26]

The concept of bounded rationality was strengthened when Daniel Kahneman was awarded the Nobel Prize in 2002. Together with Amos Tversky, he had designed psychological experiments that showed how people tried to form judgments under uncertainty and take decisions under risk. Again, the experiments were not about making optimal decisions in perfect settings but modeling real-life choices. Their outcomes stressed the importance of intuition and instinct; furthermore, they questioned the assumption that actors are concerned only about the value that they themselves receive from their decisions. Instead, actors also seem to consider the value received by others.[27]

Kahneman's work forms part of the burgeoning field of behavioral economics that has been described as a "revolution in economics."[28] Instead of exclusively relying on the rational optimizing model, this approach stresses the frames of reference that inform people's actions beyond sheer utility. It also criticizes, and increasingly so, the standard model of self-regarding preferences. Through laboratory and field experiments, behavioral economists sought to prove that agents care not only about the outcome of economic interactions but also about the process. They value fairness and punish unfair players even when this reduces their own gains. Experimental evidence consequently gave rise to the concept of strong reciprocity, defined as "a predisposition to cooperate with others, and to punish (at personal cost, if necessary) those who violate the norms of cooperation, even when it is implausible to expect that these costs will be recovered at a later date."[29]

This concept does not, as its authors point out, "contradict the fundamental ideas of rationality"[30] but rather expands on what rationality might mean.[31] Even if they do not follow computational rules of rational behavior, people have reasons to act as they do. These reasons might exceed some narrowly defined interest in maximizing economic utility to include moral values, aesthetic tastes, and/or emotional needs. Desires, "fears and hopes" (Keynes), joyful impulses, as well as moral concerns about fairness and justice inform human conduct to a degree that economists no longer feel inclined—or compelled—to overlook.

This clearly reflects the current attention to emotions as ubiquitous and positively appreciated facets of social relations and communication. During the last two decades, most human and social sciences have taken an "emotional turn." This turn has also strongly influenced theories and practices of management. *Emotional intelligence* has become the catchword for those who aim for more efficient ways of handling business operations and ensuring job satisfaction.[32] All in all, emotions have been reconsidered as prime motivators of economic (among other) behavior. At a time when this behavior was increasingly and rapidly located in the service and finance sectors of modern economies, economic man was encouraged to remodel himself as a true "entrepreneur of his own self." As such, he (and, increasingly, she) has to pay close attention to emotional management and self-fashioning. Thus, emotions have acquired the status of a major individual and socioeconomic resource and have entered the purview of scholars as worthy objects of analysis and measurement.[33]

The recent surge of emotion studies also established the close ties of emotions to institutionalized modes of conduct and normative frameworks. Instead of being regarded as spontaneous, bodily instincts or drives, emotions are increasingly conceptualized as intimately linked to mental processes of appraisal and interpretation. Consequently, emotionality is no longer considered to be the opposite of rationality. Body and mind seem to work together instead of being radically separated. Both are informed by cultural patterns that strongly influence how emotions are felt, experienced, and expressed.

Such insights are easily compatible with the second shift that has happened in economics since the crisis of general equilibrium theory and its claim to explain macroeconomics through the rationality of individual agents. This shift has given rise to a renewed interest in institutional arrangements that, as the Nobel laureate Douglass North has argued, provide "both formal rules and informal norms of behavior." Institutions mediate between mental models and belief structures, on the one hand, and societal and economic structures, on the other. They shape and frame cognitive systems of interpretation that include both emotional and moral categories. Through institutions, individuals acquire a more homogenous perception of their environment; they learn to share certain patterns of thought and emotion, and they also become acquainted with different modes of rationality situated in specific contexts.[34]

New institutional economics—which builds on older traditions developed by Veblen and the Younger Historical School—thus offers a far more complex and holistic perspective on how economic interactions work and how economic performance changes over time.[35] By challenging neoclassical theory's narrow rationality assumption, it pays attention to the role of "ideas, ideologies, myths, dogmas, and prejudices" in informing people's choices and decisions.[36] This might—and should—include the role of emotions in spurring those decisions and preferences.[37]

Among the most popular attempts to conceptualize emotions in economics is George Akerlof and Robert Shiller's best-selling book on "how human psychology drives the economy." Loosely referring to Keynes's notion of animal spirits, the authors describe them as "thought patterns that animate people's ideas and feelings." As such, they refer to "a restless and inconsistent element in the economy" that is hard to model but nevertheless necessary to incorporate into macroeconomic theory. Although this wide-reaching claim is noteworthy and promising, the initial assumptions are not. First, animal spirits are introduced as a *catégorie poubelle* in which many different phenomena are grouped together: ideas and feelings, patterns of behavior, memories and narratives. The way in which these diverse phenomena are linked to each other is, however, left unexplored. Second, they are classified not only as *noneconomic motivations* but also as *irrational*. Such labels reproduce the narrow understanding of rationality that has already been criticized as a grand illusion of (neo)classical theory.[38]

Those labels also defy what has been stressed by psychologists and neuroscientists: that emotions are deeply and consistently involved in cognitive processes like forming preferences and judgments and attributing value and meaning. It is high time, then, that *homo oeconomicus* be thoroughly reconfigured as a far more complex entity whose rationality includes passions of *all* kinds that reach far beyond self-love and self-interest.[39]

NOTES

My thanks go to Benno Gammerl and Jan Plamper for helpful criticism, to Kate Davison and Kerstin Singer for bibliographic support, and to Christina Becher for polishing my English. I also benefited greatly from Martin Hellwig's critical comments, even though my main argument clearly diverges from the way he sees the development of economic theory and practice.

1. As a reaction to the inability of most economists to foresee the recent financial crisis, see the March 2012 German memorandum "for a reformation of economics" (www.mem-wirtschaftsethik.de). A similar initiative was launched in the United States in 2009 when economists like George Akerlof and Joseph Stiglitz, among many others, founded the New York–based Institute for New Economic Thinking (www.ineteconomics.org). For a wider historical account, see Hartmut Berghoff, "Rationalität und Irrationalität auf Finanzmärkten," in *Kapitalismus: Historische Annäherungen*, ed. Gunilla Budde (Göttingen: Vandenhoeck & Ruprecht, 2011), 73–96.
2. "Full Text of Alfred Nobel's Will," http://www.nobelprize.org/alfred_nobel/will/will-full.html; "Falskt pris i Nobels namn (open letter)," *Svenska Dagsbladet*, November 2, 2001, 22; "Cloud Hangs over Nobel," *Financial Times*, November 24, 2001.
3. Josef A. Schumpeter, "Das Woher und Wohin unserer Wissenschaft," in *Aufsätze zur ökonomischen Theorie* (Tübingen: J. C. B. Mohr, 1952), 598–608. As to Schumpeter's concept of pure economic theory—pure because it was interested neither in the essence of economic behavior nor in its motives but only in its empirical logic—see Josef A. Schumpeter, *The Nature and Essence of Economic Theory* (New Brunswick, NJ: Transaction, 2010), 49–55. Regarding the state of economics during the Weimar period, see Roman Köster, *Die Wissenschaft der Außenseiter: Die Krise der Nationalökonomie in der Weimarer Republik* (Göttingen: Vandenhoeck & Ruprecht, 2011).

4. For the West German transformation, with special reference to indigenous schools like ordoliberalism, see Jan-Otmar Hesse, *Wirtschaft als Wissenschaft: Die Volkswirtschaftslehre in der frühen Bundesrepublik* (Frankfurt a.M.: Campus, 2010); and Alexander Nützenadel, *Stunde der Ökonomen: Wissenschaft, Politik und Expertenkultur in der Bundesrepublik, 1949–1974* (Göttingen: Vandenhoeck & Ruprecht, 2005), chap. 1.
5. See Geoffrey M. Hodgson, *How Economics Forgot History: The Problem of Historical Specificity in Social Science* (London: Routledge, 2001), and *The Evolution of Institutional Economics: Agency, Structure and Darwinism in American Institutionalism* (London: Routledge, 2004), pt. 3.
6. Gustav von Schmoller, *Die Volkswirtschaft, die Volkswirtschaftslehre und ihre Methode, 1893* (Frankfurt a.M.: V. Klostermann, 1949), 12, 15–16, 31, 44, 56. See also Erik Grimmer-Solem, *The Rise of Historical Economics and Social Reform in Germany, 1864–1894* (Oxford: Clarendon, 2003), esp. chap. 4.
7. Adam Smith, *The Theory of Moral Sentiments* (1759; reprint, Amherst, MA: Prometheus, 2000); John Stuart Mill, "On the Definition of Political Economy; and on the Method of Investigation Proper to It," in *Essays on Some Unsettled Questions of Political Economy* (London: J. W. Parker, 1844), 134, 137–39.
8. Mill, "Definition of Political Economy," 137. For Mill, see Joseph Persky, "The Ethology of Homo Economicus," *Journal of Economic Perspectives* 9 (1995): 221–31.
9. Mill, "Definition of Political Economy," 137.
10. Schmoller, *Volkswirtschaft*, 70 (my translation). Schmoller was well acquainted with the work of Wilhelm Wundt and quoted him extensively. In a later edition—Gustav von Schmoller, *Die Volkswirtschaft, die Volkswirtschaftslehre und Ihre Methode* (Jena: G. Fisher, 1911)—he wrote even more approvingly about the need to make use of psychological insights for economic explanations (91ff.). For the concept of economics as social and cultural science and the long-lasting criticism of mathematical modeling, see Nützenadel, *Stunde der Ökonomen*, chap. 1.
11. Köster, *Die Wissenschaft der Außenseiter*, 31–59, 316–18.
12. John R. Hicks, *Value and Capital: An Inquiry into Some Fundamental Principles of Economic Theory*, 2nd ed. (Oxford: Clarendon, 1953); Nützenadel, *Stunde der Ökonomen*, 60.
13. Richard W. Kopcke, Jane Sneddon Little, and Geoffrey M. B. Tootell, "How Humans Behave: Implications for Economics and Economic Policy," *New England Economic Review* (First Quarter 2004): 3–35, 4.
14. Albert O. Hirschman, *The Passions and the Interests: Political Arguments for Capitalism Before Its Triumph* (Princeton, NJ: Princeton University Press, 1977); Emma Rothschild, *Economic Sentiments: Adam Smith, Condorcet, and the Enlightenment* (Cambridge, MA: Harvard University Press, 2001). These arguments were shared by such twentieth-century economists as John Maynard Keynes, who stated: "Moreover, dangerous human proclivities

can be canalised into comparatively harmless channels by the existence of opportunities for money-making and private wealth, which, if they cannot be satisfied in this way, may find their outlet in cruelty, the reckless pursuit of personal power and authority, and other forms of self-aggrandisement." John Maynard Keynes, *The General Theory of Employment, Interest and Money* (1936; reprint, London: Macmillan, 1960), 374. On the birth of *homo oeconomicus* as an anthropological construct, see Werner Plumpe, "Die Geburt des 'Homo oeconomicus': Historische Überlegungen zur Entstehung und Bedeutung des Handlungsmodells der modernen Wirtschaft," in *Menschen und Märkte: Studien zur historischen Wirtschaftsanthropologie*, ed. Wolfgang Reinhard and Justin Stagl (Vienna: Böhlau, 2007), 319–52.

15. Max Weber, "Richtungen und Stufen religiöser Weltablehnung," in *Soziologie—Weltgeschichtliche Analysen—Politik*, ed. Johannes Winckelmann (Stuttgart: Kröner, 1968), 441–83.
16. John Maynard Keynes, "The General Theory of Employment," *Quarterly Journal of Economics* 51 (1937): 209–23, 214.
17. Schumpeter, *Nature and Essence*, 55.
18. Keynes, "General Theory of Employment," (see n. 16) 214ff. (see n. 14) 150, 161–63. See also Berghoff, "Rationalität und Irrationalität," 80–83; and Jens Beckert, "Was tun? Die emotionale Konstruktion von Zuversicht bei Entscheidungen unter Ungewissheit," in *Kluges Entscheiden*, ed. Arno Scherzberg (Tübingen: Mohr Siebeck, 2006), 123–41, esp. 133–35.
19. Keynes, "General Theory of Employment," 162–63.
20. For Descartes, see Catherine Newmark, *Passion—Affekt—Gefühl: Philosophische Theorien der Emotionen zwischen Aristoteles und Kant* (Hamburg: Felix Meiner, 2008), chap. 5. For Newton, see Rob Iliffe, "'That Puzleing Problem': Isaac Newton and the Political Physiology of Self," *Medical History* 39 (1995): 433–58. As for Keynes's lecture on Newton, "Newton the Man," see http://www-history.mcs.st-and.ac.uk/Extras/Keynes_Newton.html. See also R. C. O. Matthews, "Animal Spirits," *Proceedings of the British Academy* 70 (1984): 209–29; Roger Koppl, "Animal Spirits," *Journal of Economic Perspectives* 5 (1991): 203–10; and Donald Moggridge, "The Source of Animal Spirits," *Journal of Economic Perspectives* 6 (1992): 207–9. On David Hume, another influence on Keynes, see David Hume, *A Treatise on Human Nature* (1739/40), ed. Ernest C. Mossner (Harmondsworth: Penguin, 1985).
21. Newmark, *Passion—Affekt—Gefühl*; Sidney Ochs, *A History of Nerve Functions: From Animal Spirits to Molecular Mechanisms* (Cambridge: Cambridge University Press, 2004); Ute Frevert et al., *Gefühlswissen: Eine lexikalische Spurensuche in der Moderne* (Frankfurt a.M.: Campus, 2011), esp. 26–31. As regards Freud, see his "Die 'kulturelle' Sexualmoral und die moderne Nervosität," in *Drei Abhandlungen zur Sexualtheorie* (Frankfurt: Fischer, 1961), 120–39, esp. 125.

22. Schmoller, *Volkswirtschaft*. As for Werner Sombart, see his *Der Bourgeois: Zur Geistesgeschichte des modernen Wirtschaftsmenschen* (1913; Reinbek: Rowohlt, 1988), on the development and sources of the capitalist spirit, as well as his *Der moderne Kapitalismus*, 3 vols. (Berlin: Duncker & Humblot, 1928), 3:23–41, on the modern economic leader's drives and energies.
23. Schumpeter, *Nature and Essence*, 54.
24. Josef A. Schumpeter, *The Theory of Economic Development*, 8th ed. (Cambridge, MA: Harvard University Press, 1968), 93.
25. Hodgson, *Evolution of Institutional Economics*, 401–3.
26. Herbert A. Simon, *Administrative Behavior*, 3rd ed. (New York: Free Press, 1976), 82, "A Behavioral Model of Rational Choice," *Quarterly Journal of Economics* 69 (1955): 99–118, "Reply: Surrogates for Uncertain Decision Problems," in *Models of Bounded Rationality*, 2 vols. (Cambridge, MA: MIT Press, 1982), 1:235–44, and "Rational Choice and the Structure of the Environment," *Psychological Review* 63 (1956): 129–38.
27. Daniel Kahneman and Amos Tversky, "Prospect Theory: An Analysis of Decision under Risk," *Econometrica* 47 (1979): 263–91; Daniel Kahneman, Paul Slovic, and Amos Tversky, eds., *Judgment under Uncertainty: Heuristics and Biases* (New York: Cambridge University Press, 1982). See also Gerd Gigerenzer, *Gut Feelings: The Intelligence of the Unconscious* (London: Viking, 2007). Gigerenzer goes beyond Kahneman in considering intuition a superbly efficient device to arrive at smart and fast decisions.
28. Robert J. Shiller, "Behavioral Economics and Institutional Innovation," *Southern Economic Journal* 72 (2005): 269–83. See also Luigino Bruno and Robert Sugden, "The Road Not Taken: How Psychology was Removed from Economics, and How It Might Be Brought Back," *Economic Journal* 117 (2007): 146–73.
29. Herbert Gintis, Samuel Bowles, Robert Boyd, and Ernst Fehr, eds., *Moral Sentiments and Material Interests: The Foundations of Cooperation in Economic Life* (Cambridge, MA: MIT Press, 2005), 8; Joseph Henrich et al., eds., *Foundations of Human Sociality* (Oxford: Oxford University Press, 2004).
30. Gintis, Bowles, Boyd, and Fehr, eds., *Moral Sentiments and Material Interests*, 5.
31. For the reverse, i.e., the reduction of reason to the mathematics of rationality, see Lorraine Daston, "The Rule of Rules; or, How Reason Became Rationality," in *How Reason Almost Lost Its Mind: The Strange Career of Cold War Rationality*, by Paul Erickson et al. (Chicago: University of Chicago Press, 2013).
32. Actually, this interest in emotions as enhancing (or reducing) productivity goes back to the early twentieth century, when applied psychology invented *industrielle Psychotechnik*. See, e.g., Hugo Münsterberg, *Psychologie und Wirtschaftsleben* (Leipzig: Barth, 1912); and, as an overview, Laura L. Koppes, ed., *Historical Perspectives in Industrial and Organizational Psychology* (Mahwah, NJ: Erlbaum, 2007).

33. This has been stressed by sociologists more than by economists. See, e.g., Mabel Berezin, "Emotions and the Economy," in *The Handbook of Economic Sociology* (2nd ed.), ed. Neil J. Smelser and Richard Swedberg (Princeton, NJ: Princeton University Press, 2005), 109–27, and "Exploring Emotions and the Economy: New Contributions from Sociological Theory," *Theory and Society* 38 (2009): 335–46; and Viviana A. Zelizer, *Economic Lives: How Culture Shapes the Economy* (Princeton, NJ: Princeton University Press, 2011). See also Jocelyn Pixley, *Emotions in Finance: Distrust and Uncertainty in Global Markets* (Cambridge: Cambridge University Press, 2004). On the entrepreneurial self, see Ulrich Bröckling, *Das unternehmerische Selbst: Soziologie einer Subjektivierungsform* (Frankfurt: Suhrkamp, 2007).
34. Douglass C. North, "Economic Performance through Time," *American Economic Review* 84 (1994): 359–68; Hodgson, *Evolution of Institutional Economics*, chap. 20.
35. Werner Plumpe, "Gustav von Schmoller und der Institutionalismus: Zur Bedeutung der historischen Schule der Nationalökonomie für die moderne Wirtschaftsgeschichtsschreibung," *Geschichte und Gesellschaft* 25 (1999): 252–75.
36. North, "Economic Performance," 362.
37. As to Veblen's—changing—concept of instincts and how they are related to institutions, see Harold Wolozin, "Thorstein Veblen and Human Emotions: An Unfulfilled Prescience," *Journal of Economic Issues* 39 (2005): 727–40.
38. George A. Akerlof and Robert J. Shiller, *Animal Spirits: How Human Psychology Drives the Economy, and Why It Matters for Global Capitalism* (Princeton, NJ: Princeton University Press, 2009), 1, 4.
39. For a similar conclusion, see Jocelyn Pixley, "Emotions and Economics," in *Emotions and Sociology*, ed. Jack Barbalet (Oxford: Blackwell, 2002), 69–89, 83, and *Emotions in Finance*, esp. 185–89.

THIRTEEN

The Transatlantic Element in the Sociology of Emotions

HELENA FLAM

After World War II, which cost more than sixty million human lives and ended with the revelations about the horrors of the Holocaust, people were emotionally petrified. In the United States, the first social scientific diagnoses spoke of exaggerated conformity with social norms and political apathy.[1] Emotions were kept under the lid. They were synonymous with the irrationality associated with the extremes of fascism and communism and the atrocities they had produced. Postwar societies put their trust in rationality, in the effort to rebuild the social order ravaged by the war.

Following World War II, American sociology, which posed as central the question of social order and stressed value rationality as a premise of human action, profoundly resonated with the sentiments of the people petrified by and seeking escape from the war.[2] It greatly influenced European sociology, in particular postfascist and ruptured Italian and German sociology as well as Swedish, Danish, and British sociology. Talcott Parsons at Harvard and Robert Merton at Columbia had earlier drawn from the European classics of sociology. Now the roles were reversed, and European sociologies built on Parsons and Merton.

After World War II, Parsons's research agenda focused inter alia on developing system theory and the "unit act."

The unit act posited a single individual frozen at the moment of choice about how to act. At first, Parsons pointed out that an individual is likely to have diffuse interests and act on values and emotions. Initially his individual could reject prescribed action options. But his final model instead envisioned set choices. He proposed pairs of values—the so-called pattern variables—on which the individual acts, choosing only one of two possibilities: universalistic or particularistic, specific or diffused, self- or collectively oriented, ascriptive or achievement oriented, affective or neutral, etc. As his critics charged, his actor became a social puppet,[3] his emotions a matter of choice.

Merton also searched for an elegant, parsimonious way of spelling out the parameters of action. But he was more sensitive to the socioeconomic and cultural conditions under which actors act. Unlike Parsons's single actor, whose action was voluntaristic, Merton's actor was more constrained. His choices were structured by whether he internalized societal values and whether he had the means to pursue these values according to social norms. His model encompassed the possibilities of value validating but also of routinized, deviant, and rebellious action. His actor was emotionless.

A single (American) culture loomed large on Parsons's and Merton's theoretical horizon. Their society knew no gender, "race," or class and no relations of domination. Their social order was a value-rational construct. Although their explicit theorizing did not make room for emotions, emotions did surface in surprising ways and publications. More importantly, Merton supported his colleagues who worked on emotions, while some of Parsons's and Merton's former students took up emotions when their careers became secured. Let me offer a few examples.

At Columbia University, C. W. Mills (along with H. H. Gerth, Mills's former Wisconsin professor and, at that point in time, coauthor) thanked Merton profusely for supporting the famous *Character and Social Structure*, which occasionally paid much attention to emotions,[4] much like Mills's *White Collar*. In *White Collar*, Mills underscored that the mushrooming department stores dazzled customers at a considerable cost to their employees, from whom they demanded strong control of negative emotions so as to display kindness, friendliness, and courtesy at all times.[5] Peter Blau's classic in the sociology of organizations, *Exchange and Power in Social Life*, recognized the importance of expressing gratitude for sustaining exchange relationships, up to a point of decreasing returns.[6] His book also included a surprising, "unfitting" chapter on love. A former student, coauthor, and protégé of

Parsons's, Neil Smelser underscored the role of social strain and anxiety in his *Theory of Collective Behavior*, a book accounting for the emergence of protest that he wrote after his appointment as professor at the University of California, Berkeley, in 1958.[7] Similarly, in his famous article on breaching norms, Parsons's former student Harold Garfinkel, who joined the faculty of sociology at the University of California, Los Angeles, in 1954, argued that norm breaches rarely generate joy but instead lead to great emotional discomfort, even suspicion, hatred, and breakups—a view that echoed the classical sociologist Émile Durkheim.[8] Merton's own text on ambivalence was inundated with emotion: he argued that anxiety, dependency, admiration, hostility, and an oscillation between trust and mistrust essentially shape the relationship of a patient to a doctor.[9] Still, he called his book not *Emotional Ambivalence* but instead *Sociological Ambivalence*.

At this point, American sociology had no separate vocabulary for emotions. If it dealt explicitly with expressive conduct, it associated emotion with a historical past or the irrational and deviant—best to be relegated to and studied by the sociology of deviance or of collective action.

Beyond the Mainstream

According to the textbook history of American sociology, Harvard and Columbia formed an alliance rejecting a number of competing approaches, each of which offered its own answer to the question of how social order is possible and how it reproduces itself. They indeed dismissed all alternatives, creating a very real difference between the center (the Northeast) and the periphery (California), as I will soon show. Political developments of great importance undermined their hegemonic position.

The end of the social and political conformism characteristic of the immediate post–World War II period became apparent when millions of American citizens became drawn into highly charged national events. It seems highly plausible that these events played a role in the first expansion of the sociology of emotions. Black Americans were first marching and speaking, then demonstrating and militantly organizing for equal rights. Soon a passionate nationwide civil rights movement developed. On university campuses, outraged students condemned the political-military complex, colonial ventures of the past, and US military engagements of their time. They hotly debated American capitalism, class- and "race"-based inequalities, the social and political conformity of their parents' generation, and the two-party political system. Police entered campuses, shot at students, killing several, and generated even more outrage. The anti-Vietnam protests multiplied, mobilizing first the young, then the American public across the country. A new president brought hope; his assassination caused emotional devastation. These were intensely emotional times, times of social change for which sociology had no explicit vocabulary.

In this highly emotional, conflict-ridden, unsettling context of social change, Parsonian and Mertonian sociology was pushed to the margins of the academic agenda. Marxism addressed class inequalities, class conflict, colonialism, and imperialism head-on. Organizational sociology, harking back to Max Weber, made possible paying even more attention to exchange and power as key elements structuring organizations as well as interactions within and between organizations. Symbolic interactionism stressed the importance of meaningful human action—predicated on the human capacity to understand and engage symbols, such as language or gestures—for the ongoing orderly process of human communication and interaction. In some versions, it captured (and ridiculed) the pressures to conformity inherent in norms and institutions dictating the rituals of daily interactions. From a reversed perspective, that of interests, rational choice proposed that individuals are to be understood as rational egoists, moved to act only when the benefits of doing so outweigh the costs. None of these versions of sociology addressed emotions in an explicit manner. On the contrary, some, such as Marxism and rational choice, expressed their disdain for anything but the hard language of interests.

When in 1961 Erving Goffman, a creator of the dramaturgical version of symbolic interactionism, received the MacIver Award from the American Sociological Association (ASA) for his first and most famous book, *The Presentation of Self in Everyday Life*,[10] and in 1975 Lewis Coser, a conflict theorist, was elected president of the ASA, it became clear that, to quote a famous Bob Dylan song, "the times, they [were] changing." When Goffman was elected president of the ASA in 1981, his supporters both expressed and put the seal of approval on the turnabout taking place within American sociology. Sociologists were no longer willing to deny the importance of social inequality and conflict in American society. The also voted for the importance of theoretical and methodological pluralism in sociological research.

Since Goffman played a key role in the emergence of the California version of the sociology of emotions that exercised transatlantic influence, a few words are in order about his answer to the question of order, paying special attention to emotions. Although he never focused on them, he had an excellent ear for emotions. He cast everyday life as theater and individuals as actors, engaged in interactions loosely following scripts. His individuals are interested in competent playacting and fear the excruciating pain of the shame and embarrassment associated with incompetent performance.

Guilt, fear, and shame prevent deceptive playacting and counteract the very many reasons that individuals may have to deceive others.[11] Most individuals feel guilty when they act dishonestly, fear disclosure, and are ashamed when they are caught. These three emotions secure cooperative, mutually supporting playacting, countering dishonesty, suspicion, skepticism, and doubt. However, daily interaction rituals threaten to collapse, not only when cues are false, props are missing, or individuals do not know their roles, but also when an actor is emotionally off: cynical, disrespectful or hostile, displaying too much or too little pride or too much or too little self-assurance. When the interaction cannot be saved by any of the well-known strategies, such as changing the focus, ignoring the mishaps, or joking, and comes to a visible halt, confusion, anxiety, deep paralyzing embarrassment, shock, and even hostility can surface—all very discomforting and upsetting emotions.[12] For fear of "losing face" and of experiencing these upsetting emotions, individuals continue to play. By the same token, they affirm and (re-)create the social order. For Goffman, then, many painful emotions combine to motivate individuals to come together in sustaining the interactive and institutional order of society.[13]

Goffman was appointed professor of sociology at the University of California, Berkeley, in 1958. His emphasis on microlevel rules and rituals shaping everyday social interactions left a deep impression on three pioneers of the sociology of emotions—Arlie Hochschild, Randall Collins, and Thomas Scheff—who knew his work, not only as his readers, but also as his students or research assistants.

Pioneers in the Sociology of Emotions

Just as did other critics of structural functionalism, the pioneers in the sociology of emotions found its lack of attention to power and conflict in society unsatisfactory. The eventful national and the vibrant on-campus life spurred them on. With their research agendas, they contributed to the emerging sociological pluralism. While structural

Marxism and rational choice banned emotions from the purview of sociology, the pioneers of what became the sociology of emotions made emotions (more or less) central.

The sociology of emotions surfaced for the first time in 1975 when Arlie Hochschild published "The Sociology of Feelings and Emotions" and Thomas Scheff organized the very first session on emotions at the annual ASA congress.[14] Before I briefly present the first, constitutive debate, let me just note that, during the next six years, several important books and articles on emotions—articles published in key sociological journals—heralded the advent of the sociology of emotions as a sociological undertaking.[15] In 1983 Arlie Hochschild's *The Managed Heart* appeared and a year later Norman Denzin's *On Understanding Emotion*.[16]

In a debate that in many ways was constitutive of the sociology of emotions, Susan Shott and Hochschild argued for a *constructivist* perspective on emotions.[17] Shott referred to laboratory experiments to point out that individuals are often unable to say what they feel. Hence, the conclusion suggested itself that culture helps construct emotions. Hochschild stressed that culture supplies special types of norms dealing with emotions—the feeling rules. Both pointed out that early emotional socialization processes, more emphasized in middle- than working-class families, teach children the rules pertaining to emotions. Hochschild highlighted that they grow up learning the emotional alphabet: which emotions are appropriate for which social situation and with what intensity and for how long these are to be felt and expressed. Her research showed that most individuals are aware of when a mismatch develops between what they actually feel and what feeling rules suggest their emotions should be. And most are also willing to talk to themselves or to people they know when this happens and in the end rely on some emotion management strategies to generate the expected emotions. Following feeling rules by generating the expected emotions—emotional conformism—smoothes social interactions and helps people deal with each other in everyday life.

Theodore Kemper developed what he called a *positivist* perspective, rooting emotions firmly in our organic yet social selves.[18] He pointed to the same laboratory experiments as Shott did to say that it is only when the stimuli are weak that people do not know what they feel. Arguing against the constructivist perspective adopted by Hochschild and Shott, he proposed that individuals experience "real emotions" and that these can be explained in terms of outcomes of interactions between individuals: positive emotions emerge when an interaction is experienced as one between equals in power, autonomy, and social recognition; otherwise, negative, stressful emotions result. His table showed that emotional outcomes of interactions depend also on whether the individual attributes the responsibility for the outcomes of an interaction to himself or the other. When granted social recognition and power, individuals will feel happy—when they think they earned it. If they are denied power and status and think that denial was their own fault, they will feel frustration and shame. If they blame the other, they will feel anger. If they grant the other too little power and social recognition, they will feel guilt. When they are granted too much power and social recognition, they will feel shame and guilt. Kemper's real emotions emerge when individuals not only experience but also anticipate their future or relive their past experiences and their outcomes. Kemper later conceded that, although power and status differences determine our real emotions, culture influences whether and how we show these emotions.[19] If a person is denied power and social recognition and thinks it is the fault of her interaction partner, she will feel angry. But she will not express that anger if her interaction partner is socially superior since cultural rules (and fear of sanctions) proscribe such expressions. For Kemper, feeling rules determine not what one actually feels but only the emotions one puts on display.

Apart from debating and publishing on emotions, these and other pioneers established a section of the ASA devoted to the sociology of emotions in 1986. Kemper conceptualized and Steven Gordon organized the two sessions that led to a collection introducing the sociology of emotions. The volume, edited by Kemper, presented the contributors' very diverse research agendas. It also included some so-far-unnamed authors, such as Peggy Thoits, on the history of the sociology of emotions and the point at which emotions become defined as deviant, or Candace Clark, on emotions as constitutive of the microsociology of power.[20] It became a standard point of reference for all those in the United States and beyond curious about the burgeoning sociology of emotions.

Let me now narrow my focus to those few American sociologists who contributed to the emergence of the several European sociologies of emotion. This requires going back in time to locate the pioneers in the discourses on inequality and gender.

The Inequality Discourse

The handful of pioneers in the sociology of emotions who later came to inspire some Europeans shared in common a keen interest in the key post-1960s inequality discourse. They rejected the pre-1960s US-glorifying insistence of American sociology, its attempts to prove that the American occupational structure is exceptionally open to upward mobility and superior to all other occupational systems.[21] The pioneers—each in their own way—dismissed the idea that the United States amounts to an almost classless yet finely stratified society. They did not accept the idea that the unequal income distribution was functional, merely reflecting differences in demand for specific tasks/occupations.[22] Yet they did not join any of the groups of Marxist scholars that became widespread after the student revolution of 1968. Instead, even if they built on Marx, they sought to shape their own, quite distinct agendas when addressing the issue of inequality—they raised the question of how inequality affects emotions and how these emotions, in turn, influence inequality.

Scheff and Kemper were concerned to show that inequalities—whether real or imagined—account for the emergence of specific emotions. Power and status asymmetries generate an entire range of negative emotions, such as shame, frustration, and (unexpressed) hatred or anger. Kemper's approach was sketched out above, so let me turn to Scheff, who found shame particularly interesting. As he often mentions, leaning on his wife's consulting experience, he distinguished between different types of shame. He proposed that, when an individual in an interactive situation feels that social confirmation/validation is not forthcoming, he might deny the feelings the refusal to offer confirmation generates.[23] This will produce bypassed or overt, undifferentiated shame. Subordinates, as a rule, feel shame in face-to-face interactions with their social, cultural, or work superiors. Shame immobilizes and restrains them. In this manner, this particular emotion helps reproduce the established status and power structures that produce the feeling of shame in the first place. Scheff also explored vicious, mutually reinforcing spirals of shame, frustration, and anger developing between two interaction partners. He applied his theory to two nations, Germany and France, to explain why they engaged in war.

In his *Conflict Sociology*, Collins proposed his own theory of stratification. In contrast to his colleagues, he investigated how emotions are harnessed to class struggles. Taking from Marx the idea of property-based class conflicts but not the idea of only two antagonistic classes in each historical phase, he combined Weber, Durkheim, and Goffman to create a more complex explanatory model. He argued that many different forms of property generate multiple class cleavages and conflicts of interests in society. The relative power of various classes and their chances of gaining dominance over others depend on their ability to retain or get hold of not only property but also organizational, communicative, symbolic, and emotional resources. With these resources, they can transform themselves from classes into status groups that not only develop internal solidarity but also manage to project themselves as superior to and worthy of emulation by the other groups. One key precondition for the transformation of a class into a tightly knit status group is devising and orchestrating shared rituals "designed to arouse emotions . . . [that] create strongly held beliefs and a sense of solidarity within the community constituted by participation in these rituals." For Collins, rituals are the key means of emotional production. In his view, the struggle for domination entails the struggle over the means of economic production, mental production, emotional production, and social communication. Rituals create solidarity, and, as Collins stressed in a radical departure from the sociological standard, solidarity does not supplant conflict but is one of the main weapons used in it. Emotional rituals can be used for domination within a group or an organization; they are a vehicle by which alliances are formed in the struggle against other groups; and they can be used to impose a hierarchy of prestige in which some groups dominate others by providing an ideal to emulate under inferior conditions."[24]

The other key precondition for gaining dominance in a society is wielding control over coercive power, in particular that of the state, since it can be used to secure economic goods and emotional gratifications for one's own group while denying them to others. To sum up: in order to ensure dominance, force must be flanked by shared emotions and ideas: "Raw coercion and material wealth must operate within a substrate of emotional ties and common constructions of subjective reality. This is what the 'collective conscience' meant for Durkheim and 'legitimacy' for Weber."[25]

These were all very exciting ideas, and they have stood the test of time. But Collins devoted no more than just a couple of pages to linking the determinants of stratification and social conflict to emotions. Emotions receive an occasional mention elsewhere in *Conflict Sociology* but are nowhere a focal point.[26] He recognized the importance of, yet did not focus on, emotions. This perhaps accounts for why his ideas on emotions in *Conflict Sociology* did not generate much interest, although the book certainly did.

The Gender Discourse

While the discourse on inequality in its various forms became dominant much before World War II and has remained with us—although in new guises—the discourse on gender, until recently, had to be forced on the academic, media, and political agenda. Its first true public successes began in the 1980s.[27] Even then, it was the demand for gender equality in access to work and in the workplace itself that was the most readily accepted because it resonated with the dominant themes played out within the discourse on inequality. The feminist agenda was much richer, however, posing such issues as, for example, (i) the contrasting images of men and women, wherein men were seen as capable, intelligent, and rational while women were seen as incapable, dependent, corporeal, and emotional, and (ii) the separation of the private sphere from the public sphere, wherein women were pressed into invisible domestic labor and a confining, voice-depriving, and dumbing domesticity while men were free to engage in occupational pursuits and take on the role of a citizen and decisionmaker who comes home to enjoy his family and to rest.[28]

Along with Elisabeth Moss Kanter, in whose collection her very first full-fledged article on emotions appeared,[29] Hochschild was an extremely clear and well-articulated feminist at a time when the women's movement was still seeking for words in which best to put its concerns.[30] On gender issues, she could draw on the academic feminism that took off in the United States during the early 1970s. Several outspoken academic women were forging—alone and together—a gender-focused worldview and a corresponding research agenda. In 1970, after a confrontation with their colleagues at a national congress of sociology, they formed Sociologists for Women in Society (SWS).[31] In 1972, echoing the civil rights and the burgeoning women's movements, SWS made itself highly visible at the ASA meeting in New Orleans. Supported by Goffman, its members conducted a sit-in in the Monteleone Hotel's all-male lunchroom to underscore the segregation and marginalization of women within the ASA. They fought to include gender issues in teaching curricula, journals of sociology, and the congresses of the ASA. They confronted their male colleagues about too few female graduate students and the absence of women on sociology faculties, on the executive boards of the ASA, and in the pages and on the editorial boards of the journals of sociology. And they went on to elect a female ASA president, edit special issues on gender for well-established journals, and establish the journal *Gender and Society*. As early as 1973, the Woodrow Wilson Foundation started offering doctoral fellowships in women's studies. Perhaps this contextualization makes it less surprising that, in contrast to her male colleagues, Hochschild took on not only class and power asymmetries but also gender issues when addressing emotions.

Like other feminists, Hochschild deconstructed the quasi-natural binary that juxtaposes rational men and emotional women: in *The Managed Heart*, for example, she pointed out that (middle-class) women were socialized to be more sensitive to and willing to take on emotions. They were taught how to work on emotions, both their own and those of others, their husbands and children included. Their role was to make others feel good. Where the French sociologist Pierre Bourdieu supplanted economic capital with social, cultural, and symbolic capital, Hochschild spoke of emotions in a similar manner—as a habitus and resource as well. The rapidly expanding service sector of the economy provided completely new opportunities for women to exchange their specific resource—their ability to manage their own and others' emotions, to engage in emotional "deep acting" and to follow the special norms for emotions, the so-called feeling rules—for employment, income, and a modicum of independence. In this sector, the employer demanded that employees not only display but also actually feel the displayed emotions. This called for a new type of work—on one's own emotions. Emotional labor was a burden, but it could be turned into an object of exchange. Women could enter service firms, provided they agreed to follow the management-dictated feeling rules and, just like their male colleagues, paid the price of emotional alienation. In *The Second Shift*, Hochschild turned her attention to couples with children to investigate how they handle the burden of—severely undervalued—domestic labor and child care in a context in which (i) capitalist firms open labor markets to women but demand long working hours, (ii) traditional gender role models still prevail but feminism calls for more equality between men and women not just on the job but in child care and household tasks, and (iii) neither the capitalist enterprise nor the state take responsibility for this state of affairs.[32] Her conclusion was that love suffers when partners' gendered role expectations come into conflict with each other and when even well-meaning male partners fail to offer a helping hand. If partners do not develop a shared family myth or engage in strategic subterfuge, both come to feel frustration and anger instead of the mutual respect and gratitude that reinforce love. Conflicts, separations, and divorce become more frequent as a result.[33]

The Transatlantic Influence

At least a dozen American sociologists, both female and male, pioneered the sociology of emotions.[34] In the late 1980s and early 1990s, just a handful became well-known in Europe: Hochschild, Scheff, Collins and Kemper. The first three studied and/or taught in California and had been influenced in key ways by the microsociology of Goffman. I will focus on their influence on European sociologists of emotions before turning to Kemper. This is not to say that European sociologists were unfamiliar with or did not meet other authors. It is to say, however, that these four were on the Europeans' "must talk to" lists.

In the United States, Hochschild's books attracted immediate attention. Her first book inspired virtually hundreds of studies. It proposed a clear feminist agenda and offered a special take on the rapidly developing service sector. The research questions concerning management-imposed on-the-job feeling rules and emotions management were original, the research niche unexplored, and the qualitative research design relatively easy to replicate, thus lending itself to a low-budget doctoral project. The fact that women entered American sociology in ever greater numbers probably helped too. Already in the 1960s, women outnumbered men among sociologists, but not among those holding master's degrees or doctorates. However: "As the feminist protests [within the ASA and at some universities] began, this proportion began to rise significantly and . . . the shift at the PhD level accelerated substantially, going up to 20% in 1971 and jumping to 40% ten years later. By 1980, sociology granted a majority of its MA degrees to women, and by 1988, gave a majority of its PhD degrees to women as well."[35]

Hochschild's first book had great resonance in Europe as well. Among the early critics of her "managed-heart" thesis we find Jürgen Gerhards in Germany and Artur Bogner and Cas Wouters in the Netherlands.[36] Educated in Sweden and the United States but a visiting scholar in Germany at that time, I pinned down the main differences between the rational, normative, and emotional models of human action in "The Emotional 'Man,'" greatly assisted by Hochschild's as well as Kemper's and Scheff's work.[37]

As it turned out, it was Hochschild's feminist agenda—the issue of women's double work burden, or what she called the "second shift"—that at first generated considerable interest in Germany, not her focus on emotions. In Germany, neither gender studies, nor social movement studies, nor the sociology of emotions, all of which developed into grand sociological enterprises in the United States, could take off. German sociology has remained dominated by conservative men interested in advancing research frameworks stressing rationality and quantitative methods for much longer than the United States has. Even those working from other perspectives were concerned that the legitimacy of sociology would be undermined if it turned to such suspect and "frivolous" items as gender, social movements, and emotions.

In Germany today, Hochschild is becoming well-known as a sociologist of emotions, not just as a sociologist taking up labor and gender issues,[38] as a result of a great surge of interest in emotions in virtually all other related disciplines—neurology, management, organizational studies, etc. Perhaps the increasing number of women in sociology also accounts for greater interest in her sociology of emotions. In the 1970s and 1980s—later than in the United States—women in Germany began entering universities in numbers greater than ever before. But they have not fought or won as many battles as their US counterparts. In 1992, women occupied only 6.5 percent of all professorships; by 2007, that figure had grown only to 18 percent. In 1992, soon after Hochschild's *The Managed Heart* appeared in German, only 11.5 percent of sociology professors were women; by 2008, that figure had grown to 29 percent, making sociology the discipline most open to women.[39]

Hochschild and Scheff stayed on in the University of California system until their retirement. Their European admirers turned California into their Mecca. In the early 1990s, Hochschild's alma mater, Berkeley, became the destination of a number of European sociologists. The Briton Steven Fineman, who taught at the School of Management of the University of Bath, from then on made emotions in organizations his research agenda.[40] At the University of California, Santa Barbara, Scheff (and his wife, Suzanne Retzinger, a psychologist) became friends with the Swedish sociologist Bengt Starrin, who arrived there in 1979–80, with the British-Australian sociologist Jack Barbalet, who came from Australia in 1990, and with the Danish sociologist Charlotte Bloch, who came in 1994. In various ways, Starrin has investigated shaming, feeling shame, and discourses on shame in everyday life. He helped jump-start the sociology of emotions in Sweden in the early years of the twentieth century, emphasizing its role in everyday life. Bloch focused first on the experience of emotional flow and stress in everyday situations before conducting a large study of emotions in the alleged fortress of rationality—academe—while employing a very broad range of methods to unearth them.[41] Barbalet wrote initially on class resentment.[42] A much cited classic, *Emotion, Social Theory and Social Structure,* followed in 1998.[43] It dealt with the discourses that banned emotions from the social sciences in the past few centuries and discussed emotions in various historical periods and settings, reflecting Barbalet's historicized-contextualized take on emotions.[44] Collins's work on emotions needed longer to be noticed. Collins has been well-known for his *Conflict Sociology* as well as for his excellent research first on Max Weber and later on the philosophical schools of thought but not for what he had to say on emotions.[45] Even today, sociologists of emotion in Europe are only beginning to incorporate his concept of energizing solidaristic interaction ritual chains into their work—its strongest influence by far has been in Sweden and Denmark.[46] To larger sociological audiences his opus stands apart from his sociology of emotions.

The fourth important pioneer in the sociology of emotions, Kemper, obtained his doctorate from New York University. In the 1980s and 1990s, his insight that interactions generate real emotions became of interest to some German and UK-based sociologists.[47] In 1984, Kemper was visited by a German sociologist, Jürgen Gerhards, who, once in California, also met with Collins, Hochschild, and ("just for coffee") with Scheff.[48]

In 1990, Barbalet read Kemper's *Social Interactional Theory of Emotions.* A phone call resulted in an almost one-hour-long discussion of emotions between Westwood, California, and New York City and led to several meetings. Once in England, Barbalet paid some of his intellectual debt by soliciting Kemper's contribution for the collection *Emotions and Sociology,* which presented a selection of texts by mostly European researchers to the general sociological public.[49] It advanced the view that such alleged fortresses of rationality as the market, science, and bureaucracy are far from immune to emotions.

Why Import?

In some European countries, American sociology—which after World War II was in part brought back to Europe by scholars educated at American universities, in part imposed by the Allies, and in part voluntarily taken up[50]—was perceived as superior and attractive as a source of legitimate scientific knowledge. Stays in the United States and ties to American sociologists were valued because they buttressed careers. Just like pizza, spaghetti westerns, and Max Weber's sociology, everything tasted much better when it came to Europe via the United States: US pedigrees legitimated them, even in the face of domestic opposition. The intellectual questions that moved specific individuals to seek answers in the American sociology of emotions were of course very important, but they were interwoven with the intranational dynamics that, with some exceptions (esp. France), acknowledged the superpower position of the United States even within the social sciences. Taken together, personal intellectual pursuits and the standing that American sociology still enjoyed at the time account for the imports that I sketch out next. The sketch shows that a few sociologists based in Germany and the United Kingdom imported the sociology of emotions in the late 1980s and early 1990s but that only the British sociologists, mainly in the sociology of organizations, had scored some successes. In the Scandinavian countries, the sociology of emotions took off first in the early years of the twenty-first century. It is now spreading to Southern and Eastern Europe.

The British sociology of emotions became visible by the mid-1980s on the initiative of Rom Harré, a philosopher by education and an anthropologist at heart. His collection *The Social Construction of Emotions*, which included contributions by anthropologists, sociologists, and psychologists, pursued a wide range of questions.[51] Shared was the assumption that there is a single universal human physiology that becomes shaped by various cultures—that is, by their cultural codes or norms.

The major thesis was that, although as human beings we all have the same capacity to feel a wide range of emotions, cultural spheres and national cultures shape their own emotional repertoires, while specific cultural codes, scenarios, and rituals further limit the range of emotional expression. In 1989, a section on emotions was established within the British Sociological Association, but apparently it offered little that an organizational sociologist could latch onto: Hochschild's exploration of the role of emotional labor in organizations inspired Fineman, and his enthusiasm was shared from the very beginning by his colleague Yiannis Gabriel.[52] The organizational sociology of emotions took off largely on Fineman's initiative—a development that was endorsed by a major organizational sociologist, Steward Clegg.[53] Martin Albrow's early independent interest in emotions and organizations helped quite a bit too.[54]

In German-speaking countries, Georg Simmel, who defined emotions as the cement of society and devoted many essays to a subtle exploration of various emotions, has until recently been dismissed as too associative and impressionistic, just as he was in his own time. Despite this and other barriers,[55] Birgitte Nedelmann and Jürgen Gerhards published some texts on Simmel and emotions in the 1980s, encouraged by Friedhelm Neidhardt, a coeditor of *Kölner Zeitschrift für Soziologie und Sozialpsychologie*, the oldest and most prestigious German journal of sociology.[56]

The sociology of Norbert Elias stressed the role of the historically intensified capacity to feel shame and embarrassment as a motor of the civilizing process, but as of 1990 Elias was still not very well-known in Germany.[57] According to Hans Joas, Max Scheler merits recognition as a classic German thinker interested in emotions, but the fascist period had cut off postwar West German sociology from its roots.[58] And, not to be forgotten, many German professors of sociology even today either insist that all action is dictated by rational choice or simply stay away from emotions for fear of compromising their discipline. Although newer studies, even those on Weber, have proved them wrong, they can always justify their position by referring to Max Weber's quick aside to the effect that expressive action does not lend itself to a sociological study. Ignorance is bliss.

In Germany, the time around 1990 brought some important texts on emotions. Campus—a semiacademic publisher—issued Hochschild's *The Managed Heart* in German in 1990. On Joas's suggestion, the translation was included in a series edited by major (left-liberal) German sociologists: Claus Offe, Axel Honneth, and Joas. In 1985, Joas had spent some months at the University of Chicago together with Collins, who recommended *The Managed Heart* to him. In Joas's view, Hochschild's book would be interesting to the general reader; hence, compared to other books in the sociology of emotions, it merited special attention.

My "The Emotional 'Man,'" which appeared in *International Sociology* in 1990, made for some commotion. And Sighard Neckel's book *Status und Scham*, which drew on Simmel and on Elias, became a classic.[59] But these first texts did not suffice to fuel a true takeoff of the sociology of emotions. The true takeoff came with the next generation of scholars. My German-language introduction to the sociology of emotions, *Soziologie der Emotionen*, sold hundreds of copies in the last months of 2002—a lot for a publication in sociology and on emotions.

Although Starrin at Karlstad University in Sweden and Bloch at Copenhagen University in Denmark worked in the sociology of emotions, they were virtually alone with their special take on society. They cultivated a good relationship with Scheff, who was a frequent visitor, especially at Karlstad. It was the feminist professor of sociology Gerd Lindgren, along with her doctoral student Åsa Wettergren, however, who decided to build up Karlstad University as a center for the study of emotions in Scandinavia. Sweden decentralized its university system in the course of the 1980s and 1990s. The new universities have been under pressure to show that they have been worthy of reform. A high profile in the sociology of emotions could help achieve this goal.[60] Karlstad jump-started the Scandinavian sociology of emotions, paying special attention to emotions in everyday life.

In 2003, Barbalet accepted an invitation to organize the very first sessions on emotions at the congress of the European Sociological Association (ESA). Soon thereafter, the Research Network 11, devoted to the sociology of emotions, was established within the ESA. (It later became known as the European Emotions Network.) Its first midterm conference in Augsburg took place in 2004. It was followed by a midterm conference in Karlstad in 2006.[61] The Googlemail list of the European Emotions Network showed two hundred members by 2012. At the biannual ESA congresses, about seventy presentations are usually given.

Value-Added

I do not mean to give the impression that European sociologists have simply adopted American sociology in the form in which it was imported, so I turn now to a discussion of the ways in which the European sociology of emotions differs in its emphases from its American counterpart. There is much interest in the intellectual history of emotions, tracing what European philosophers and the classics of sociology had to say about emotions as well as the intellectual or contextual history of specific emotions. This is in part due to the fact that a couple of prominent Elias scholars—Helmut Kuzmics and Cas Wouters—have been in the European network on emotions from the outset. Kuzmics's work focuses on military history, Wouters's on sex, manners, and emotions. Since historians these days are enthusiastically embracing the turn to emotions, it has become possible to engage in a dialogue with them about emotions in the course of history. In fact, a shared midterm conference took place in 2012.

In organizational sociology, the focus is not just on private enterprises but also on schools, the police, hospitals, ministries, and political parties. Whereas, earlier, employees' emotions were of interest, today's research is on the emotions in (i) the triad of managers, employees, and clients, (ii) discrimination against women and migrants in public and private work settings, (iii) silencing reports of harassment or (sexual) violence on the job, and (iv) further theorizing about emotions in and around organizations.

The European Emotions Network has a very active publishing program, having so far brought out collections focusing on theory, organizations, finance, and the Internet and a special issue of the online journal *Sociological Research Online* on friendship.[62] A collection focusing on the methods of unearthing and exploring emotions is in preparation. Cutting across all these intellectual pursuits is a shared interest in a critical take on contemporary society and the massive problems that enterprise managers, state bureaucracies, the financial elites, and anti-migrant and sexist discourses generate.

Final Observations

This admittedly impressionistic account of the emergence of the sociology of emotions suggests several nodal points in the development of the sociology of emotions. We find the first in the 1970s in the United States, in the wake of highly emotional movements and conflicts that shook the country starting in the 1960s. The second is in the late 1980s and the early 1990s in Europe and emerges under the influence of the US pioneers in the sociology of emotions. Arguably, even though European researchers turned to the sociology of emotions driven by their various intellectual concerns, the extremely dramatic, emotional context—beginning with Solidarność and extending through the student protests in Tiananmen Square, the Velvet Revolutions in Central Europe, the fall of the Soviet Empire, and the reunification of Germany in 1989–90—was conducive to an interest in emotions. The early years of the twenty-first century constitute another nodal point in the development of the sociology of emotions, but this one appears to have been an outcome of networking and organizational efforts in the context of rising interest in emotions in neuropsychology and management, along with the entrance of a new, interested generation of scholars in academe. The emotions associated with the attack on the World Trade Center and the War on Terror also had something to do with it. Durkheim would certainly think so.

What this chapter also suggests is that the sociology of emotions developed either at the margins or at the power centers of sociology. In the United States, it was at the margins. Although, as I indicated, the balance of power within American sociology was shifting, its initial center after World War II was on the East Coast. The California version of the sociology of emotions was at the sociological margins. In England, when Harré launched his offensive to make the emotions visible, it was from the center, but, when Fineman and Gabriel became interested in the sociology of emotions, they were both off center—at the School of Management at the University of Bath. In Germany, it was Neidhardt—the editor of *Kölner Zeitschrift für Soziologie und Sozialpsychologie*—who was open-minded enough to encourage Nedelmann's and Gerhards's publications on emotions. Similarly, I was at the prestigious Max Planck Institute when one of the institute's directors backed my project behind the back of the other. In Sweden, it was in contrast an off center university—Karlstad—that took on emotions. Today, research on emotions is greatly encouraged not only by the publication and recruitment efforts of the network members but also by excellence funds and regular research funding and awards—and not only in Sweden and Germany but also in Portugal, Spain, and France.

NOTES

I would like to thank Jack Barbalet, Charlotte Block, Jürgen Gerhards, Hans Joas, Theodore Kemper, and Bengt Starrin for answering my e-mails about their turn to emotions; Myra Marx Ferree for her e-mail and the article on the US academic feminists; and Mechthild Bereswill, Martina Löw, and Brigit Geissler for comments on Arlie Russell Hochschild's work. I also wish to thank the editors, Frank Biess and Daniel M. Gross, and Jochen Kleres and Åsa Wettergren for their helpful comments. Thank you too to Jan Plamper as a reviewer. Felipe Rubio big smacking thanks for correcting my English and Daniel M. Gross for the final corrections.

1. See David Riesman, Nathan Glazer, and Reuel Denney, *The Lonely Crowd* (Cambridge: Cambridge University Press, 1950); and Seymour M. Lipset, *Political Man: The Social Bases of Politics* (Garden City, NY: Doubleday, 1960).
2. See Jeffrey C. Alexander, "On the Social Construction of Moral Universals," in *Cultural Trauma and Collective Identity*, by Jeffrey C. Alexander, Ron Eyerman, Bernard Giese, Niel J. Smelser, and Piotr Sztompka (Berkeley and Los Angeles: University of California Press, 2002), 196–263.
3. Dennis H. Wrong, "The Oversocialized Conception of Man in Modern Sociology," *American Sociological Review* 16 (1961): 183–93.
4. Hans Gerth and C. Wright Mills, *Character and Social Structure: The Psychology of Social Institutions* (1953; New York: Harcourt, 1970).
5. C. Wright Mills, *White Collar: The American Middle Classes* (New York: Oxford University Press, 1951).

6. Peter Blau, *Exchange and Power in Social Life* (1964; New Brunswick, NJ: Transaction, 1992).
7. Neil J. Smelser, *Theory of Collective Behavoir* (New York: Free Press, 1962).
8. Harold Garfinkel, "Studies in the Routine Grounds of Everyday Activities," *Social Problems* 11.3 (1964): 225–50.
9. Robert Merton, *Sociological Ambivalence and Other Essays* (London: Free Press, 1976).
10. Erving Goffman, *The Presentation of Self in Everyday Life* (Garden City, NY: Doubleday, 1959).
11. Ibid., 15, 55. See also Helena Flam, "Extreme Feelings and Feelings at Extremes," in *Theorizing Emotions*, ed. Debra Hopkins, Jochen Kleres, Helena Flam, and Helmut Kuzmics (Frankfurt a.M.: Campus, 2009), 73–93.
12. Erving Goffman, *Interaction Ritual* (New York: Pantheon, 1967), 5–12, 22–23, and "Embarrassment and Social Organization," *American Journal of Sociology* 62.3 (1956): 264–71.
13. In a presidential address to the members of the ASA that he did not live to deliver, Goffman underscored that the modern nation-state guarantees this order in that it protects, if necessary with arms, our extremely vulnerable bodies against many forms of violation, secures peaceful, bourgeois-democratic space, upholds citizens' rights, and secures respect for the private sphere. See Erving Goffman, "The Interaction Order: American Sociological Association, 1982 Presidential Address," *American Sociological Review* 48.1 (1983): 1–17. See also Erving Goffman, *Asylums* (1961; London: Penguin, 1991).
14. Arlie Russell Hochschild, "The Sociology of Feelings and Emotions," in *Another Voice*, ed. Marcia Millman and Rosabeth M. Kanter (Garden City, NY: Doubleday, 1975), 280–307.
15. The work of these pioneers was inspired—albeit to different extents—not only by classical philosophy, Freud, and the European classics of sociology, such as Marx, Weber, and Durkheim, but also by Goffman, social psychology, Foucault, and even French structural Marxists such as Althusser. Randall Collins drew largely on Marx, Weber, Durkheim, and Goffman in his *Conflict Sociology* (1975), abridged and updated by Stephen K. Sanderson (Boulder: Paradigm, 2009), which I discuss later in this chapter. Hochschild took Marx's idea of class exploitation very seriously, arguing that capitalist enterprises in the growing service sector usurp not only manual and intellectual labor but also the new—emotional—labor. She discussed Freud's definition of emotions as signal but rejected the idea that sociology can study unconscious and suppressed emotions. From Goffman came the idea that most individuals are playacting in an effort to conform to social norms, but Hochschild focused on norms pertaining to emotions. See Arlie Russell Hochschild, *The Managed Heart: Commercialization of Human Feeling* (Berkeley and Los Angeles: University of California Press, 1983). Norman Denzin, a cultural sociologist, targeted

experienced emotions and emotional practices as his object of inquiry. He distinguished three types of experienced emotions: (i) feelings of the lived body, such as despair; (ii) value-derived emotions, such as resentment; and (iii) moral feelings that focus on the self, such as guilt. He believed that practices generate all experienced emotions. He understood all embodied and socially embedded actions as emotional practices: working, drinking, watching television, kissing, or fighting. See Norman K. Denzin, "On Understanding Emotion: The Interpretative-Cultural Agenda," in *Research Agendas in the Sociology of Emotions*, ed. Theodore D. Kemper (Albany: State University of New York Press, 1990), 85–116, 89. Similarly to Althusser and Foucault, Denzin wished to investigate how the relations of domination in postmodern, late capitalist societies, generating such values as freedom, democracy, rationality, efficiency, or economic growth, translate into the contents, action patterns, and emotional models mediated by cultural institutions, such as film or television, and become incorporated in the everyday practices and (emotional) experiences of individuals.

16. Norman K. Denzin, *On Understanding Emotion* (San Francisco: Jossey-Bass, 1984).
17. Susan Shott, "Emotion and Social Life: A Symbolic Interactionist Analysis," *American Journal of Sociology* 84 (1979): 1317–34; Arlie Russell Hochschild, "Emotion Work, Feeling Rules, and Social Structure," *American Journal of Sociology* 85 (1979): 551–75.
18. Theodore D. Kemper, *A Social Interactional Theory of Emotions* (New York: Wiley, 1978), and "Toward a Sociology of Emotions: Some Problems and Some Solutions," *American Sociologist* 13 (1978): 30–41.
19. Theodore D. Kemper, "Social Constructionist and Positivist Approaches to the Sociology of Emotions," *American Journal of Sociology* 87 (1981): 336–62.
20. Peggy Thoits, "Emotional Deviance: Research Agendas," in Kemper, ed., *Research Agendas in the Sociology of Emotions*, 180–203; Candace Clark, "Emotions and Micropolitics in Everyday Life: Some Patterns and Paradoxes of 'Place,'" in ibid., 305–33.
21. Peter M. Blau and Otis Dudley Duncan, *The American Occupational Structure* (New York: Free Press, 1967).
22. Davis Kinsley and Wilbert Moore, "Some Principles of Stratification," *American Sociological Review* 10 (1945): 242–49.
23. Thomas J. Scheff, *Catharsis in Healing, Ritual, and Drama* (Berkeley and Los Angeles: University of California Press, 1979), "Shame and Conformity: The Deference-Emotion System," *American Sociological Review* 53 (1988): 395–406, and *Microsociology: Discourse, Emotion, and Social Structure* (Chicago: University of Chicago Press, 1990).
24. Collins, *Conflict Sociology*, 21.
25. Ibid., 116.

26. See, e.g., ibid., 124, 131, 143–47, 159.
27. Nancy Whittier, *The Politics of Child Sexual Abuse: Emotion, Social Movements and the State* (Oxford: Oxford University Press, 2009).
28. The women's movement also addressed (iii) the basic asymmetrical power relation between women and men (the patriarchy) that permeated all of society and all its institutions, subjecting women and children not only to structural but also to domestic violence, including sexual harassment and rape, (iv) the insistence of men on the ethics of just or equal rights rather than on the much more important and all-encompassing ethics of care, and (v) the commodification and sexualization of women by means of pornography, etc.
29. Millman and Kanter, eds., *Another Voice.*
30. Whittier, *Politics of Child Sexual Abuse.*
31. Myra Marx Ferree, Shamus Rabman Kahn, and Shauna A. Morimoto, "Assessing the Feminist Revolution: The Presence and Absence of Gender in Theory and Practice," in *Sociology in America: A History,* ed. Craig Calhoun (Chicago: University of Chicago Press, 2007), 438–79.
32. Arlie Russell Hochschild, *The Second Shift* (New York: Avon, 1989).
33. In *The Time Bind: When Work Becomes Home and Home Becomes Work* (New York: Metropolitan Books Holt, 1997), finally, Arlie Russell Hochschild showed that work and home have traded places in speeded-up capitalism: whereas work offers quiet, respect, friendship, and exciting achievements, home has become a construction or management site, filled with noise, conflicts, and little appreciation. For her numerous other texts, see her homepage: http://sociology.berkeley.edu/professor-emeritus/arlie-r-hochschild.
34. See Helena Flam, *Soziologie der Emotionen* (Konstanz: Universitätsverlag Konstanz/UTB, 2002); and Jonathan H. Turner and Jan E. Stets, *The Sociology of Emotions* (Cambridge: Cambridge University Press, 2005).
35. Marx Ferree, Rabman Kahn, and Morimoto, "Assessing the Feminist Revolution," 444.
36. Jürgen Gerhards, "Emotionsarbeit: Zur Kommerzialisierung von Gefühlen," *Soziale Welt* 31 (1988): 47–65; Artur Bogner and Cas Wouters, "Kolonialisierung der Herzen? Zu Arlie Hochschilds Grundlegung der Emotionssoziologie," *Leviathan* 18.2 (1990): 255–79.
37. Helena Flam, "The Emotional 'Man': I. The Emotional 'Man' and the Problem of Collective Action," *International Sociology* 5.1 (1990): 39–56, and "The Emotional 'Man': II. Corporate Actors as Emotion-Motivated Emotion Managers," *International Sociology* 5.2 (1990): 225–34.
38. Disparate female sociologists doing research on labor, gender, and love relations or on domestics as well as the Women's Section (*Frauensektion*) of the German Sociological Society have shown great interest in Hochschild's work, inviting her to various national and local conferences as well as translating and introducing some of her work.

39. The statistics for the United States come from Marx Ferree, Rabman Kahn, and Morimoto, "Assessing the Feminist Revolution." The statistics for Germany come from Steffen Mau and Denis Huschka, "Die Sozialstruktur der Soziologie—Professorenschaft in Deutschland" (working paper, Wissenschaftszentrum Berlin für Sozialforschung, 2010), http://nbn-resolving.de/urn:nbn:de:0168-ssoar-258651.
40. See Steven Fineman, ed., *Emotion in Organizations* (London: Sage, 1993).
41. Charlotte Bloch, *Passion and Paranoia* (Aldershot: Ashgate, 2012), and "Moods and Quality of Life," *Journal of Happiness Studies* 3 (2002): 101–28.
42. Jack Barbalet, "A Macrosociology of Emotion: Class Resentment," *Sociological Theory* 10.2 (1998): 150–63.
43. Jack Barbalet, *Emotion, Social Theory and Social Structure: A Macrosociological Approach* (Cambridge: Cambridge University Press, 1998).
44. From 1992 to 1997, while still in Australia, Barbalet organized the emotions panel of the Australian Sociological Association.
45. Collins's senior colleague Stephen Sanderson has followed his career since about 1975. He initially recommended *Conflict Sociology* for publication and recently updated and abridged its new edition. He says that he had to rely in part on a joint article by Collins and Jörg Rössel to present what Collins has to say on emotions. Stephen K. Sanderson, afterword to *Conflict Sociology*, 258–80, 275.
46. Randall Collins, *Interaction Ritual Chains* (Princeton, NJ: Princeton University Press, 2004).
47. I could not locate Gerd Kahle, who translated and included Kemper's critique of Hochschild and Shott in what was the very first German-language collection on emotions. See Theodor D. Kemper, "Auf dem Wege zu einer Theorie der Emotionen: Einige Probleme und Lösungsmöglichkeiten," in *Logik des Herzens: Die soziale Dimension der Gefühle*, ed. Gerd Kahle (Frankfurt a.M.: Suhrkamp, 1981), 134–54.
48. Jürgen Gerhards, "Georg Simmel's Contribution to a Theory of Emotions," *Social Science Information* 25 (1986): 901–24, and *Soziologie der Emotionen: Fragestellungen, Systematik, Perspektiven* (Munich: Juventa, 1988).
49. Jack Barbalet, ed., *Emotions and Sociology* (Oxford: Blackwell, 2002).
50. Peter Wagner, Carol Hirschon Weiss, Björn Wittrock, and Hellmut Wollmann, eds., *Social Sciences and Modern States: National Experiences and Theoretical Crossroads* (Cambridge: Cambridge University Press, 1991).
51. Rom Harré, ed., *The Social Construction of Emotions* (Oxford: Blackwell, 1986).
52. Fineman, ed., *Emotion in Organizations*.
53. Steward Clegg, Cynthia Hardy, and Walter R. Nord, eds., *Handbook of Organization Studies* (London: Sage, 1996).
54. Martin Albrow, "Sine Ira et Studio; or, Do Organizations Have Feelings?" *Organization Studies* 13.3 (1992): 313–29, and *Do Organizations Have Feelings?* (London: Routledge, 1997). The American sociology of emotions

also indirectly influenced British sociologists. As an editor of the journal *International Sociology*, Albrow first published my "Emotional 'Man': I" and "Emotional 'Man': II" and then followed up himself with texts on emotions and organizations. Margaret Archer, although always supportive, criticized "Emotional 'Man': I" for not going far enough. See Margaret S. Archer, *Being Human: The Problem of Agency* (Cambridge: Cambridge University Press, 2000), "Homo Economicus, Homo Sociologicus and Homo Sentiens," in *Rational Choice Theory: Resisting Colonization*, ed. Margaret S. Archer and Jonathan Q. Tritter (London: Routledge, 2000), 35–56, and *Conversations about Reflexivity* (London: Routledge, 2010).

55. At the University of Bielefeld, Simmel experts worked to change his low status in German sociology, putting out one after another of his numerous volumes. But one could definitely gain more acclaim writing about his work on culture or on money than about his views on emotions.
56. In the 1990s, Gerhards and Nedelmann discontinued their work in the sociology of emotions.
57. A small and exclusive group of Elias scholars did little to change this state of affairs, but, for one such attempt, see Helmut Kuzmics and Ingo Mörth, eds., *Der unendliche Prozeß der Zivilisation: Zur Kultursoziologie der Moderne nach Norbert Elias* (Frankfurt a.M.: Campus, 1991). Gerhards found Elias's psychology unconvincing. From my point of view, only by turning into a full-time Elias scholar could one dare to engage with his thoughts on emotions. That was a high hurdle to overcome.
58. Hans Joas, *Entstehung der Werte* (Frankfurt a.M.: Suhrkamp, 1999), translated by Gregory Moore as *The Genesis of Values* (Chicago: University of Chicago Press, 2000).
59. Sighard Neckel, *Status und Scham* (Frankfurt a.M.: Campus, 1991).
60. The first Scandinavian workshop on the sociology of emotions was jointly organized in 2003 by the Universities of Karlstad and Örebro.
61. For more information, see http://www.socemot.com.
62. Hopkins et al., eds., *Theorizing Emotions*; Barbara Sieben and Åsa Wettergren, eds., *Emotionalizing Organizations and Organizing Emotions* (London: Palgrave, 2010); Mary Holmes and Silvana Greco, eds., "Friendship and Emotions," *Sociological Research Online* 16.1 (2011), http://www.socresonline.org.uk/; Jocelyn Pixley, ed., *New Perspectives on Emotions in Finance: Sociology on Confidence, Betrayal and Fear* (London: Routledge, 2012); Tova Benski and Eran Fisher, eds., *Internet and Emotions* (London: Routledge, 2013).

FOURTEEN

Feminist Theories and the Science of Emotion

CATHERINE LUTZ

Introduction

One of the earliest projects of second-wave feminists in the 1970s was to detail and rebel against the use of the power of science and its predecessor experts to construct and deform the self or personality of *the woman.* This authoritatively pieced-together woman clearly had something other than the capacities of the fully human as normatively construed. Broverman, Broverman, and Clarkson, for example, surveyed clinicians working in the 1960s and found that they associated emotionality with normal female functioning and with deficient human (sex unspecified) functioning.[1] Ehrenreich and English traced the history of expert thought positing great risks for women who pursued schooling—risks arising from competition between a woman's uterine and brain functions, expansion of the latter supposedly leading to contraction of the former.[2] When we ask whose emotions have been considered problematic, at least throughout the last century in the United States, we find that it is the sine qua non of many forms of deviance. Havelock Ellis was one of a long line of commentators who drew explicitly, far into the twentieth century, on the idea that among women, "as among children, savages, and nervous subjects," the emotions are dominant and inferior reflexes.[3]

With the comparative insight that anthropology began to afford into alternative self-systems and emotion discourses in societies around the world, the cultural construal of emotion in the twentieth-century West was to become clearer: it was considered an unfortunate block to rational thought, a link to bodily nature, and a route to certain kinds of social virtue. The power of this idea, then and still, is multifarious. One of the important aspects of this power is its ability to help position women in unwaged work in the household. This is a key accomplishment of both the positive and the negative valuation of emotion and its association with women: the empathetically emotional woman could be seen as the angel of the house, a view that quickly transformed into a workplace problem where she would react oversensitively to the rough and tumble of commerce and workforce discipline. In line with this, second-wave feminists began to home in on anger as the one emotion exempted from this gender association, using it as a key index of when or in what contexts women could make claims for respect.[4] Frye usefully drew attention to what she called the *uptake* of women's anger, that is, its recognition as having occurred and/or having occurred legitimately.[5] This feminist insight began in the 1970s and 1980s to place emotions within a fully social view of power.

Feminists also stepped back from the question of emotion per se to build a more fundamental critique of the "nonaccidental ideology" of abstract individualism on which psychologizing about women has been based, even including most feminist psychology.[6] The notion introduced by feminism in this period that "the personal is political" was critiqued from this perspective, having the sometime consequence of suggesting that emotional change (or personal life) can or should constitute the whole of feminist political work.[7] Therapeutic industries grew faster than feminist organizing in the West, however, as newly valorized or politicized emotions were not understood as connected or subsidiary to organized communities of feminist activism.

Black and Chicana feminists immediately followed up with the insight that those cultural and scientific definitions were often tacitly descriptions of white womanhood, of the privileged servitude of a class of white women in wealthy households.[8] The distinctive qualities that defined black women under slavery and into the late twentieth century in popular culture were framed less psychologically and more in terms of physical attributes (such as inflated sexuality and "deviant" morphology)[9] and of moral character (such as an emasculating tendency and a propensity to leisure and/or self-sacrifice). Feminist theorizing about emotions, this critique elaborated, has often exercised the power of white privilege to ignore the nonunitary nature of the category *woman*.[10] This is evident in the focus on women's love for other women and their orientation toward caring for others, to the near exclusion of attention to women's anger at or dismissive feelings toward other women, often on the basis of race, class, or sexual orientation.[11] It is also evident in the fuller development of nonindividualist frameworks within black feminism.[12]

I begin this chapter by contrasting feminist and normative approaches to the emotions in the second half of the twentieth century. Feminist approaches have been distinguished by their attention to the material, institutional, and cultural capillaries of power through which discourses of emotion operate. Normative approaches have restricted their questions to the limited power that emotion—as culturally and conventionally defined in Western academic circles—has to shape individual behavior. I go on to discuss the main varieties of feminist definitions and explorations of emotion. While any number of other sortings are possible, the following six types are examined: feminist rereadings of developmental emotion dynamics, emotion as authentic femininity, emotion as epistemic resource, emotion as cultural discourse on power, emotion as social labor, and emotion as life on the social margins. Finally, I speculate about the place of feminist approaches to emotion in the broader context of late twentieth-century knowledge production about emotions.

Feminist and Normative Approaches to Knowledge about Emotions

While the approaches taken to emotion by feminists have been varied, they share critical and pragmatic purposes. They are critical in the sense that they redefine what is worth knowing about emotions, asking new questions, and questioning the interests served by the old questions. They are pragmatic in the sense of aiming to apply the new questions and their answers to benefit in women's (and men's) lives.

Traditional philosophy of science and current normal science argue that all rational persons can imagine and ask any question of nature or social life. Nature, not social context, suggests the questions. In this framework, it matters not at all where a question comes from—only whether the answer is right. Scientific method begins not with how something is discovered but with how, once discovered, a proposition is tested for its truth value. Scientific questions usually purport to identify some of the most crucial problems requiring investigation. Even those who see themselves as doing basic, nonapplied science would usually claim to be working on questions that help define the essence or central features of a phenomenon like emotion. The work is meant to help define the object of study in a way that applied or practical science will then want to, even have to, use.

Feminism's challenge to this view has been this: which questions are asked is as constitutive of what we end up knowing as how we test any tentative answers. Harding has pointed out that the problems that prompt scientific questions do not occur in the abstract.[13] They occur *to* people, and they are distributed differentially across social groups. To ask a question, then, is often to identify what the problem looks like from a particular social position. And the questions that different social groups have wanted answered about emotion are often different. Oppressed groups, for example, want to know how to change the conditions they live with, how to alleviate the pain they feel on that account.

The questions that normative science has asked about emotion in the last several decades have included the following: How do children become "improperly attached" to their mothers (but not their fathers)? What have mothers done to produce this emotional complex in their children? Are women better than men at recognizing facial expressions of emotion? Where in or on the body can one identify anger, disgust, fear, etc.? How do college students' moods shift around the cycle of the academic year? Is disgust universal? What are the emotional symptoms of menopause, and what drug best treats them? What percentage of the female population has angry mood swings as a result of suffering from PMS? What emotional disturbances accompany or constitute late luteal phase dysphoric disorder (a newly defined disease that constitutes a kind of super-PMS)?

By the early 1970s, feminists were pointing out how many questions about the self had *not* been asked. They were the questions that women would have been asking, and some of them were the queries that specifically racial minority women would have been constructing had they been included in the academic debates (they certainly were asking questions privately, in fictional form, and in on-the-ground political and other practical activities). These previously unpublished questions included the following: Why have women been considered the emotional gender? Why has emotion generally been pathologized, and pathology emotionalized, and normality masculinized? Why have men's moods and hormone levels not been examined? How have black and working-class women felt about their assigned place in society? What are the emotional defenses that white and wealthy women have employed in their relations with less privileged women? Why is anger an emotion considered relatively more inappropriate for women than for men by most in the United States? How cross-culturally does a different gender division of labor influence women's prestige and self-esteem? How would different social structures and a different allocation of social respect between women and men influence the likelihood of women fearing rape or other violence against their persons throughout their lives?[14] Where does the emotional force for male sexual violence come from? Why have so many men found child care and housework so distasteful?[15] What are the consequences of a political discourse in the United States that is centered around questions of proper masculinity and emotional states in leaders? What are the emotional contexts of frequent male resistance to contraception?

The questions that feminists have asked are rarely inquiries into what most would consider basic or universal aspects of emotion. In emotion study, the call for basic research and the terms of its definition can seem orthogonal to women's concerns or even hostile to them. The fact is that for many scientists the most fundamental, basic, or important knowledge one can have about emotions concerns psychobiological, not psychosocial, processes. When the psychobiological both defines the emotions and sets a national research agenda as it does, the social world and a critique of it shrinks to insignificance or invisibility. Feminist analysis of emotion points out the power or interests served by normative work on emotions and demonstrates just how partial it is in the two senses of that term.

Varieties of Feminist Emotion Work

Feminist work is composed of a large and diverse set of voices. What I want to do here is summarize the varieties of feminist work on emotion I am most familiar with, which is mainly American academic work in or primarily drawing on psychology, anthropology, sociology, history, and philosophy.

Feminist Rereadings of Developmental Emotion Dynamics

One of the earliest kinds of feminist retheorizations of the self, this theme in feminism of rereadings of developmental emotion dynamics, is psychoanalytic and has worked to reanalyze the emotion dynamics of gender. These approaches have treated emotional life as a central feature of gender identity and of gender relations. Unlike traditional psychoanalytic approaches, however, they foreground the potentially variable social power of women and men, and they treat the parent-child relationship less as a timeless crucible of gender identity and more as a social and historical institution.[16] This work has been involved not just in rereading the family dynamic itself but in analyzing its correlates in the emotional aspects of popular and literary culture. It has included explorations of the psychodynamic underpinnings of the demonization of women in Hollywood films, explained gender ambivalence in popular culture figures, and traced the historical and cultural contexts of the associations between women, mass culture, and sentimentality.[17]

The key feminist work in emotional development has been by Jessica Benjamin, who takes as her central problem the question of how women can come to feel pleasure from being dominated by others or, in other words, how emotional life is deformed so as to allow women to participate in their own subordination. She treats cultural myths of women (as essentially masochistic, as natural, etc.) as important but insufficient sources of women's feelings about themselves, desire, and power. Social learning of feminine ideals of the self and affect cannot account for what she sees as a result, not the cause, of the propensity of women to experience "pleasurable fantasies of erotic submission."[18] In contrast to MacKinnon and Griffin, Benjamin sees the problem not as the imposition of a male pornographic, sadomasochistic imagination and practice but as the repression of women's sociability and social agency through the course of development in any family in which the mother does not assert her own separate selfhood. Like Chodorow, she sees defensive male fantasies of omnipotence or denial of the other as the outcome of the attempt of male children to break free of the mother. Girls' developmental "progress," however, is toward self-abnegation. The feelings associated with this system include female fear of independence, women's attempts to control anxiety about separation through service, and their longing for recognition in the midst of a gender-polarized world in which men are subjects and women objects.

The emotional life of women and men becomes in this paradigm central to understanding how feminist transformation will occur. Benjamin suggests marshaling the longings for interpersonal recognition in loving relationships as a device for instilling hope for both personal and social change. Her approach remains fundamentally a liberal one, however, one in which feelings remain the property of individuals even as she traces, often poetically, the dialectic of feelings between self and other. Her utopian notions about intersubjective recognition notwithstanding, she leaves to others the task of linking the self-other dualism—the splitting of feelings of mastery and submission—to historical and social contexts and specific changes in those that would accompany such interpersonal changes.

Emotion as Authentic Femininity

An early and still culturally very popular feminist idea about emotion claims it as one of the centers of a revalorized femininity. In these views, the dualism of emotion and rationality is not rejected, and the association of women with nature is extolled. Emotion qua natural capacity then becomes simultaneously something men fail to have and the sign of women's superiority. Women are advised to resist the repression of emotion that is seen as a form of male dominance.[19] The classic statement of this view is Griffin's: she sees the domination of women as related or equivalent to the repression of nature, rejects technocratic rationality, and lauds the identification of women with emotion and other aspects of what she associates with untamed nature. Emotion is seen as inherently transgressive.

Where Griffin makes a universal and essential argument, other feminists have focused on feeling as something that becomes female through social learning but that ought to be reclaimed as a virtue. This is particularly so when the focus is on women's capacity for empathy and other feelings that motivate their caretaking of others.[20] Although the idea of emotion as positive capacity is currently rejected by perhaps most academic feminists who have written on the subject, many still draw at least tacitly on it.[21] Feminist discussions of anger, in particular, seem to be prone to a naturalized and hydraulic view of emotion, as when feminist pedagogists ask how they can "help students [in feminist classrooms] channel their anger in healthy and productive ways."[22] Anthropological study of the cross-cultural variation in attributions of emotionality to women and men problematizes the cultural assumption of female emotionality that much Western feminist work draws on. Dalton gives the example of the Rawa of New Guinea, who generally expect emotional expressiveness of both women and men.[23] Those ethnographic cases also demonstrate how the confession of something like feminist anger might also be seen as a social process of simultaneous repression and hypersurveillance and production of nonangry states.

Emotion as Epistemic Resource

Many feminist philosophers came to an analysis of emotion through a critique of traditional philosophy of science, which heightened the cultural dualism of rationality and emotion, claiming that good science required its practitioner to be dispassionate. This claim's implicit gender and race politics were brought forward, its roots in the masculinist dreams of Bacon and others identified.[24] Jaggar states the case succinctly: "[The] function [of the myth of the dispassionate investigator], obviously, is to bolster the epistemic authority of the currently dominant groups, composed largely of white men, and to discredit the observations and claims of the currently subordinate groups. . . . The more forcefully and vehemently the latter groups express their observations and claims, the more emotional they appear and so the more easily they are discredited."[25]

An alternative feminist epistemology developed.[26] In doing a sociology of the science of primatology, for example, Haraway writes about "love" and "knowledge" as the two things that those studying primates want from their scientific practice.[27] She uses the notion of love in place of other possibilities such as desire or motive and does so to avoid at least two problems from this new feminist epistemological perspective. Her first attempt is to reorient the sociology of science away from presenting all "nonrationalistic" scientific motives in a negative light (hence love rather than projection). The second goal is to center analysis on a social relation (hence love rather than anxiety) between scientist and primate subject that helps construct the story told. This avoids an individualistic portrait in which male desire constitutes scientific practice.

Alison Jaggar has coined the phrase *epistemic resource* to characterize emotion's potential role in women's lives. Learning rather than being born to feel as they do, women most often do so in ways that support existing social arrangements (e.g., when women, including feminists, often feel disgust for their own bodies). Emotions can be outlaw, however (as when someone feels angry with a sexist joke), or feminist (as when they entail feminist perceptions and ideas). A dialectic between emotion and feminist theory is posited such that critical reflection on emotion becomes a necessary part of a developing feminist theory and feminist practice (not, as Jagger says, just a preliminary "clearing of the decks"): "Outlaw emotions . . . are necessary to the development of a critical perspective on the world, but they also presuppose at least the beginnings of such a perspective."[28]

Like Griffin, Jagger edges toward a view that women's emotionality constitutes their strength when she speaks of feelings as part of women's "epistemic advantage." Reflection on emotion is seen as a kind of political theory and practice that women are generally more adept at than men because of their social responsibility for caretaking (a connection that some socio- or psychobiology makes and reduces to genetic code).[29] That is, women, people of color, and other subordinate groups are more likely to experience what Jagger and others have called *outlaw emotions*. Women's culturally glorified emotional empathy is radically reconceptualized as a skill in political analysis rather than a sign of their intuitive and nurturant virtue.

While the possibilities for emotional self-deception in the subordinate are acknowledged, it is left to Benjamin and others to describe how this occurs.[30] Spelman shows how the language of moral emotions can and has been used self-deceptively when she analyzes how white feminists have used the language of guilt, shame, or regret to focus on their own feelings more than on the harm done to women of color via their race privilege. If, as she claims, emotions are "powerful clues of the ways in which we take ourselves to be implicated in the lives of others and they in ours,"[31] then assumptions about one's importance as a white and well-off woman will be reflected in those emotions as well. While theorists vary in the degree to which they focus on the misrecognition or ideological distortion of emotion, all make the claim that emotions can be remade through renaming and might constitute empowering forms of knowledge for feminist purposes.

Emotions as Cultural-Historical Discourses on Power

In all these last three perspectives (rereading emotional development, emotion as authentic femininity, and emotion as epistemic resource), emotions are viewed as tools for collective social change, but in only some is it seen as central to the reproduction of patriarchal social relations. Others have taken on questions of power in a more direct and/or socially and historically contextualized way.

Feminist anthropologists and historians have drawn on anthropology's standard notion of culture as well as on the Foucaultian notion of discourse to expand the questions asked about women and emotionality. The emphasis here shifts to describing and theorizing the connections between emotional life and relations of power (of gender but also and simultaneously of class and race) described in their historical and cultural variations.[32] Like the theorists just mentioned, they see emotion talk as political but are more attuned to social structures and political economies. They trace the place of emotion discourse in societies with different configurations of power and different kinds of gender politics emerging from such things as matrilineal descent, an ethos of honor in a pastoral, patrilineal segmentary lineage system, and child bearing and rearing under conditions of extreme privation in a class-stratified society. Mageo opens new directions in the study of emotion, power, and history by suggesting that the emotional suffering articulated by possessed women in Samoa and elsewhere be seen not as hysteric symptoms but as creative contributions that seek "the resolution of cultural-historical paradoxes suffered by the individual."[33] Such insight into emotional talk as history suffered and its meaning remade could be applied to other work that details the contexts of shifts in emotional norms for women and men in the United States over the last several centuries.[34] This includes important work on how global economic restructuring has entailed historical shifts in the affective requirements of work assigned on the basis of gender.[35]

Abu-Lughod, for example, has detailed the deterioration of Awlad Ali Bedouin patrilineal authority with sedentarism and with the advance of the Egyptian state into their community.[36] Key to the erosion of power has been the strategic and often rebellious deployment of love poetry by younger women and men. That emotion is a relatively direct affront to elders, for whom control of sexuality via codes of invulnerability and honor is key to maintaining strong bonds between the men of the lineage. Seremetakis takes a similar aesthetic phenomena—Greek women's funeral laments—and treats them as commentaries on and recoveries of cultural notice of women's labor.[37] The unrecognized work includes agricultural fieldwork as well as women's traditional labor of mourning at funerals. The dynamic of recognition and rebellion is evident in the screaming body of the woman during death rituals. The woman is literally made more socially visible and hence powerful, even as she metaphorically leaves society by ripping at her clothes and breaking through other constraints. Finally, Wiss notes that, though racist ideologies led to the scientific dissection of a !Kung woman infamously brought to Europe in the early nineteenth century, the close attention paid to her feelings did not entail interest in her voice.[38]

The cultural concerns of contemporary middle-class feminists are reflected not only in the question of how women can theoretically feel about their young children but also in a growing historical literature on love and particularly heterosexual love.[39] While much of this work focuses on norms of behavior, it also generally attends to the social contexts producing such shifts as the narrowing gap in expectations about the ideal emotional profiles of men and women. From a peak of emotional differentiation in the sexes during the Victorian period, increased contacts between women and men outside the family and the growing labor force participation rates of women, particularly in the 1920s and on, led to a decline in normative sex differences in anger, fear, and jealousy. There emerged concern over socializing both boys and girls for the workplace through such things as an equivalent control of anger. There has also been much work in this vein on depression. This emotional syndrome predominates in women both in the West and cross-culturally. It is associated with situational factors of powerlessness rather than constitutional factors in the most detailed, contextualized studies done.[40]

Some of the most challenging work on gender, affect, and politics is found in the study of German fascism. The problem is to explain human participation in institutionalized and quotidian evil on a grand scale, and what is relevant to us here is the fact that Nazism and the Holocaust were highly gendered in ways that require attention if we are to understand how gender identities and the emotional investment in them are malleable in changing social circumstances and how one resists. Koonz's work takes on the question of how women emotionally responded to Nazi demands that they both leave paid employment and collaborate in anti-Semitic and cultural change projects.[41] Theweleit examines the problem of the German fascist mentality and draws a picture of men whose fear of women was conflated with and fueled their anticommunism and their attempts to control those whom they considered the masses.[42]

The widespread use of gendered forms of torture in warfare has also been documented, in conjunction with the emotional discourses it violently shapes.[43] The horrors of counterinsurgencies, wedded to patriarchy in crisis, have made emotion unspeakable through the shame that attaches to rape and the retraumatizing character of memory itself.[44] Bringing the analysis of militarism and gender home, Krasniewicz's vivid study of American militarism gives a detailed ethnographic sense of the rage that greeted the feminist peace activists at the Seneca Falls Army Depot.[45] Krasniewicz convincingly suggests that it was their violations of expectations of proper femininity and (hetero)sexuality more than their opposition to nuclear weapons per se that so angered the surrounding community. In this ethnography, emotional response is indexical of the politics of militarist masculinity as well as the conflicts between women within the camp and between camp and village women.

In paying attention to relations of power enacted through emotion discourse, feminist ethnographers have used reflexive analysis to examine how their own fieldwork and writing can unwittingly reproduce, even as it resists, gender relations as they are. Morgan reexamined her work in a feminist health collective in New England for its tacit acceptance of the epistemology that dichotomizes rationality and emotion.[46] This led her to neglect to take seriously feelings expressed by clinic workers. Reflexive analysis has also been used to highlight differences between the culture under study and the researcher's cultural background of thought and feeling about gender and affect.[47] Lavie presents a particularly poignant example of the value of this kind of analysis in her description of such a contrast of feelings about circumcision among a Sinai Desert group of Bedouins. Describing the ceremony, her "participant observation," and the operation's aftermath, she tells of the young girls running "panic-stricken" into the sea to stop the bleeding and her own escape to the edge of the settled area. There she begins vomiting and crying, vomiting and crying, all the while wondering "whether I cry because of pleasures never known and already lost by the girls, or because of their mothers' firm belief in the power the circumcision gives a woman over a man by removing lust, or perhaps because my own sexuality seems diminished as I carefully walk the thin rope stretching between the worlds of Mzeina men and women."[48]

The rhetoric and force of Lavie's feelings in this excerpt provide a political analysis that is all the more nuanced and effective in the ambivalence it faces. The political analysis, in other words, is neither simple, reductive, nor Manichaean because it (emotionally/cognitively) recognizes the difficulty of feminist evaluation and action in this context, the multiple critical perspectives that merit attention.

Emotion as a Form of Social Labor

Connecting questions of a gendered division of labor with questions of emotional meaning, feminist historians of the West have elaborated on the correspondence between the separation of the workplace and the home under capitalism, the allocation of women to the domestic sphere of unpaid labor, and the ideological split between notions of emotion and interest, expressive and instrumental roles and personalities, and the association of women more intensively with the affective side of those dualisms. Over time, the family was reconceptualized as primarily an emotional unit. Women were thereby more firmly associated with both domesticity and affect as they came to stand as the heart of a heartless world. The effects of this include the preservation of women as a reserve pool of labor for business, the reproduction of labor power without cost to the corporate world, and so on. The cultural equation of woman, family, and affect is also reproduced in part through the pursuit of commercial interests, for example, the greeting card and floral industries and the therapeutic industries.

The idea that emotion is requisite for social life leads to seeing it as a form of labor required of women, at least in the modern industrial world. This formulation involves materialist rereadings of cultural feminism's (and to a lesser degree discourse approaches') tendency to focus on representation and to underplay connections between ideologies of gender and emotion and the allocation of resources. Hochschild has the first and most elaborate discussion of this in her work on airline stewardesses and bank repossession agents.[49] She takes the Marxist notion of exploitation into the psychological realm and allows connections to be made between women's and men's feelings and the division of labor, labor costs, and the reproduction of the labor force and of profit. Her focus is on stewardesses, whose emotional labor consists of smiling pleasantly throughout a flight and making each passenger feel that they (standing in for the airline) are happy to serve them. Most service occupations, dominated by women employees, have these emotional requirements. Hochschild also researched repossessors as an occupation group whose emotional requirements are for angry toughness. The recruitment practices of repossession—one of a set of predominantly male occupations—follow from the gender exceptionalism of anger.

Emotional labor is usually unrecognized as such, a mystification that is key to its commercial exploitation. The exploitation process occurs in the household as well, as di Leonardo demonstrates in describing "the work of kinship" done by Italian American women.[50] This work involves planning and executing the yearly cycle of sending greeting cards, buying gifts, and preparing family rituals, particularly as organized between households. These important practices are virtually defined by their evocation and reproduction of affective ties, but they are ignored or trivialized as a form of labor. If, moreover, emotion is defined as a natural expression, and if women's feelings toward kin are assumed to be naturally positive or maternal, then the ideological incentive to see women's kin work as pleasant or as leisure is even more powerful.

With this same emphasis on emotion as labor, Hochschild has provocatively described the complex "economy of gratitude" within married middle-class heterosexual couples and its role in sustaining an unequal division of household labor, as that is traditionally defined.[51] This research addresses the problem of some feminist work on women and nurturance that fails to note labor's strategic ends rather than its exploited nature; it shows that "'altruism' (read compassion) and 'self-interest' (read rationality) are cultural constructions that are not necessarily mutually exclusive."[52] Altruism's association with emotion (such as compassion, love, and fear for others' safety) and self-interest's with cold rationality is another reason to see these redefinitions of the nature of women's labor as key to unmasking the damage done by the dualism of affect and reason.

In the political economic frame, emotion is also seen as a resource that—like the commodity under capitalism—can be redistributed (out of the nuclear family, e.g.) or fetishized. Cvetkovich shows how this is so in her complex and telling reading of the historical development of sensational literature and its critique in the nineteenth century (and its corollary in mass culture criticism in the twentieth).[53] She shows that the emergence of the middle class in the 1860s and 1870s in Great Britain is linked to the growth of a large market of readers and writers of so-called sensational literature. What defined the latter for critics was the evocation of emotion in readers, an "unnatural excitement" elicited by reading about predominantly female characters engaged in unnatural (often criminal or sexual) acts. Critical focus at the time was less on the social relations indexed (between sexes and classes) than on the emotion itself, which was defined as a problem. The canon was established, then, in part through an opposition between base instinctual, emotional responses to reading and high aesthetic responses, although it was gender and class conflict rather than an autonomous cultural ideology of affect opposed to rationality that fired the critics' behavior and canon formation.

Whether they expressed it or restrained it, the middle classes, like women themselves, were being defined by their relationship to feeling in this way. And, like the increasingly domesticated and privatized middle-class woman's life, affect was defined as a hidden phenomenon. For both women and affect, liberation then seemed to require that the hidden pain or problem be made visible or even a spectacle (something not logically necessary); conservative critics clearly then had to respond negatively to sensational literature. Going beyond literature, Cvetkovich astutely points to the dilemma of drawing attention to concrete instances of emotional suffering in women's lives: doing so "can both call attention to and obscure complex social relations and can both inspire and displace social action."[54]

Feelings associated with motherhood in different societies and time periods have often been the center of feminist analysis tacitly using this definition of emotion as labor and resource.[55] Scheper-Hughes contested the normative view of a natural intense attachment between mother and child in her ethnography of Brazilian shantytown women.[56] She traces the emotional injuries of class and race for these women, noting the competition with and indifference toward children that sometimes develop under conditions of privation and racism. Her analysis struggles not to (but sometimes does) reproduce the injunction on women to have that surplus of emotional resource to give to their children, something these Brazilians demonstrably do not have.

Emotion as Life on the Social Margins

Women's life on the social margins can provoke emotional response or even constitute the idea of the emotional itself by establishing a contrast with those whose mainstream or central place is (mis)taken for rationality. Some theorists have generalized about the emotions/marginality of women as a class, while others have focused on the violence of women's emotions in madness and the truth or protest in those mad emotions.[57] Chesler gave a forceful early statement of this perspective, noting that the denial of full human status to women drives some mad. Such madness is essentially an intense experience of female biological, sexual, and cultural castration and a doomed search for potency. The search often involves delusions or displays of physical aggression, grandeur, sexuality, and emotionality. Such traits in women are feared and punished in patriarchal mental asylums.[58]

One case of the play between emotions and marginality has been taken up by Seremetakis, who eloquently speaks about Greek peasant women's lives as fragmented ones. Like Chesler, she explores the emotional contexts of the denial to them of the status of full person. She argues for the benefits or uses of the social margins, saying: "[The fragment] may be marginal, but it is not necessarily dependent, for it is capable of denying recognition to any center." These women's emotions are construed as "transformative," not merely expressive: as a materially powerful body practice, emotional pain or lament can reconstitute the fragmented self into at least "provisional, empowering wholes." The materiality of their emotions, bodies, and pollution is what gives women power. This is especially so given that they are on the margin of the modern Greek state and so have not experienced the split of the public from the private, affective exchange from economic exchange: they can deploy their tears as they have noted their urban counterparts cannot, as a sign of those connections between themselves and others. Fragments of "the self [i.e., one's tears] disappear with the absent other" at the funeral and assert that something has been sundered, including relations between the living and the dead and between women's agricultural work and their socially recognized work.[59]

Not surprisingly, these theorists have in several instances been people who have focused on questions of doubled marginality, as when Trawick describes an untouchable woman in South India and Butler the (Western) lesbian.[60] Both analysts consider the category of the abject, construed both as a socially excluded person and as a feeling. Both women also make major innovations in discussing the subversive pleasures as well as the pain of being denied full subject status. Butler's writing about the ambivalence or the impossibility of being "comfortably" identified as a lesbian is very suggestive of the problems with precise definitions and identifications of emotions and with standing under marginalized identity categories such as *a woman* or *the emotional.* Her argument could be used to contest the idea that a positive project for women and for feminist science should be emotional transparency, that is, a clear sense of and singular quality to one's feelings and politics. Her notion of gender trouble can be extended and applied to what can be called *emotion trouble.* This latter would highlight the performative aspect of emotion and its ontologically unstable character. Any emotion would be seen as a performance rather than an expression of a determinate underlying psychological or psychosocial phenomenon. Like gender itself in Butler's view, emotion would then be seen as shadowed by the notion of the biologically original feeling for which any instance of emotion becomes then simply a poor imitation.

Definitions and National Research Agendas: Normative Center and Feminist Margin

In the larger scheme of things, the vast majority of research on emotions has been conducted within the framework of psychobiological and nonfeminist definitions. Social definitions of emotion remain marginal in relationship to the centers of federally funded research, such as the National Institute of Mental Health, an institution ever more fundamentally organized around the brain and hence around the individual as the unit of analysis. Moreover, data on emotion produced in experimental work with (nationality-, race-, and class-specific) college students are still not treated as nongeneralizable, while data on emotion collected ethnographically among African American women are seen as particularistic and therefore as relatively unfundable. The upshot is that into the twenty-first century science continues asking some questions and not others about emotion, particularly those having to do with pressing problems facing the social groups most often associated with emotionality.

Even where social definitions of the emotions have been taken seriously, the role of feminism in their development and the role of gender in the social are marginalized. It can be no coincidence that intense interest in emotion per se in the academy developed in the 1980s with the maturation of the women's movement and the influx of women and, to a lesser extent, racial minorities into research positions. While these social changes have helped create the feminist literature just reviewed, normative science has also revalorized emotion as an object worthy of scientific study in this same period. This latter work may be at least partially motivated by the attempt to co-opt and evacuate the emotional of association with the female and irrationality. Feminist efforts are marginalized, if not in certain parts of the academy, then certainly in relation to the centers of emotion science in psychology labs. Per Modleski's analysis of the 1980s in popular culture, which is characterized sardonically as having deployed a "feminism without women," suggests the general cultural backlash might have its academic corollary in emotion study without women. It can also be said, however, that the emergence of much feminist literature on emotion from the white middle class has meant that it has often used definitions of emotion that tacitly erase race and class distinctions between women by allowing a conflation of affect, protest, and femininity.[61]

It is the normative science of emotions, however, that has been most influential beyond the academic community. It has been disseminated through women's magazines, science journalism, and other media. The effect funnels from the scientists out to the community and back again, something I was struck by in conversation with an academic psychologist. As we worked together on a panel reviewing federal grant proposals, she began to tell me about the differences between her two children, her son being much more aggressive than her daughter. She said that this initially made her quite anxious and that she made great efforts to redirect his behavior in more pacific ways. After she read a best seller a friend recommended to her entitled *Brain Sex*, however, she has been much less troubled.[62] The book argues from psychological research for the sex-specific hormonal and other biological factors in brain development and behavior. Boys are naturally more aggressive and angry, she and others can conclude, and social, emotional, and political change is not possible or possible only within stark limits.

A rich variety of feminist thinking about emotion, beginning in the 1970s and continuing to the present, has challenged normative notions of what emotions are and what they accomplish. The power of normative science rests in its continuing to be heavily funded, biologistic, and associated with social relational analysis that usually goes no further than the mother-infant relationship. Decades of feminist work in redefining emotions has allowed for some radically new questions, for the double negative insight that "there is not nothing to be done," and for the remaking of emotion definitions and research questions in service to new social relations.

NOTES

This chapter is a revised version of an essay originally published in *Power and the Self*, ed. Jeannette Marie Mageo (Cambridge: Cambridge University Press, 2002), 194–215. Copyright © 2002 Publication of the Society for Cultural Anthropology. Reprinted with the permission of Cambridge University Press.

1. Inge K. Broverman, Donald M. Broverman, Frank E. Clarkson, Paul S. Rosenkrantz, and Susan R. Vogel, "Sex Role Stereotypes and Clinical Judgments of Mental Health," *Journal of Consulting and Clinical Psychology* 34 (1970): 1–7.
2. Barbara Ehrenreich and Deidre English, *For Her Own Good: 150 Years of the Experts' Advice to Women* (Garden City, NY: Anchor/Doubleday, 1978).
3. Havelock Ellis, *Man and Woman* (Boston: Houghton Mifflin, 1929), 310.
4. Alma Gottlieb, "American Premenstrual Syndrome: A Mute Voice," *Anthropology Today* 4.6 (1988): 10–13; Elizabeth Spelman, "Anger and Insubordination," in *Women, Knowledge and Reality: Explorations in Feminist Philosophy*, ed. Ann Garry and Marilyn Pearsall (Boston: Unwin Hyman, 1979), 263–73.
5. Marilyn Frye, "A Note on Anger," in *The Politics of Reality: Essays in Feminist Theory* (Trumansburg, NY: Crossing, 1983), 84–94.
6. Naomi Scheman, "Anger and the Politics of Naming," in *Women and Language in Literature and Society*, ed. Sally McConnell-Ginet, Ruth Borker, and Nelly Furman (New York: Praeger, 1980), 174–86; Michelle Fine and S. M. Gordon, "Effacing the Centre and the Margins: Life at the Intersection of Psychology and Feminism," *Feminism and Psychology* 1 (1991): 19–27.

7. For a contrasting Chinese feminist perspective on this relationship, see Lin Chun, "Toward a Chinese Feminism," *Dissent*, Fall 1995, 477–85.
8. Patricia Hill Collins, *Black Feminist Thought: Knowledge, Consciousness, and the Politics of Empowerment* (New York: Routledge, 1991); Gloria Hull, Patricia Bell Scott, and Barbara Smith, *All the Women Are White, All the Blacks Are Men, but Some of Us Are Brave: Black Women's Studies* (Old Westbury, NY: Feminist, 1991).
9. Sander Gilman, "Black Bodies, White Bodies: Toward an Iconography of Female Sexuality in Late Nineteenth-Century Art, Medicine, and Literature," in *Race, Writing and Difference*, ed. Henry Louis Gates Jr. (Chicago: University of Chicago Press, 1985), 204–42.
10. Elizabeth Spelman, "The Virtue of Feeling and the Feeling of Virtue," in *Feminist Ethics*, ed. Claudia Card (Lawrence: University Press of Kansas, 1991), 214–370.
11. Carroll Smith-Rosenberg, "The Female World of Love and Ritual: Relations between Women in Nineteenth-Century America," *Signs: Journal of Women in Culture and Society* 1 (1975): 1–29; Nel Noddings, *Caring: A Feminine Approach to Ethics and Moral Education* (Berkeley and Los Angeles: University of California Press, 1984); Sara Ruddick, *Maternal Thinking: Towards a Politics of Peace* (Boston: Beacon, 1989); Spelman, "The Virtue of Feeling"; Cynthia Burack, *The Problem of the Passions: Feminism, Psychoanalysis, and Social Theory* (New York: New York University Press, 1994); Maiva Lam, "Feeling Foreign in Feminism," *Signs: Journal of Women in Culture and Society* 19 (1994): 865–93.
12. For example, Collins, *Black Feminist Thought.*
13. Sandra Harding, "Introduction: Is There a Feminist Method?" in *Feminism and Methodology*, ed. Sandra Harding (Bloomington: Indiana University Press, 1987), 1–14.
14. Peggy Sanday, *Fraternity Gang Rape: Sex, Brotherhood, and Privilege on Campus* (New York: New York University Press, 1990).
15. Constance Perin, *Belonging in America: Reading between the Lines* (Madison: University of Wisconsin Press, 1988).
16. Nancy Chodorow, *The Reproduction of Mothering: Psychoanalysis and the Sociology of Gender* (Berkeley and Los Angeles: University of California Press, 1978); Jessica Benjamin, *The Bonds of Love: Psychoanalysis, Feminism, and the Problem of Domination* (New York: Pantheon, 1988).
17. Elizabeth G. Traube, *Dreaming Identities: Class, Gender, and Generation in 1980s Hollywood Movies* (Boulder, CO: Westview, 1992); Tania Modleski, *Feminism without Women: Culture and Criticism in a "Postfeminist" Age* (New York: Routledge, 1991); Ann Cvetkovich, *Mixed Feelings: Feminism, Mass Culture, and Victorian Sensationalism* (New Brunswick, NJ: Rutgers University Press, 1992).
18. Benjamin, *Bonds of Love*, 81. This perspective runs parallel to Mageo's insight into how both (patriarchical) power and the self as culturally

constituted by domination "can each produce distortive knowledge about the world." Jeannette Marie Mageo and Bruce M. Knauft, "Introduction: Theorizing Power and the Self," in Mageo, ed., *Power and the Self,* 8–26, 8.

19. Susan Griffin, *Woman and Nature* (New York: Harper & Row, 1978); Elaine Showalter, *A Literature of Their Own* (Princeton, NJ: Princeton University Press, 1977).
20. Ruddick, *Maternal Thinking;* Noddings, *Caring.*
21. Alison M. Jaggar, "Love and Knowledge: Emotion in Feminist Epistemology," *Inquiry* 32.2 (1989): 151–76; Donna Haraway, *Primate Visions: Gender, Race, and Nature in the World of Modern Science* (New York: Routledge, 1989); Arlie Russell Hochschild, *The Managed Heart: Commercialization of Human Feeling* (Berkeley and Los Angeles: University of California Press, 1983); Scheman, "Anger and the Politics of Naming."
22. Janet Lee, "Teaching Feminism: Anger, Despair and Self Growth," *Feminist Teacher* 7.2 (1993): 15–18, 15.
23. Douglas Dalton, "Spirit, Self, and Power: The Making of Colonial Experience in Papua New Guinea," in Mageo, ed., *Power and the Self,* 117–40.
24. Susan Bordo, *The Flight to Objectivity: Essays on Cartesianism and Culture* (Albany: State University of New York Press, 1987).
25. Jaggar, "Love and Knowledge," 158.
26. Scheman, "Anger and the Politics of Naming"; Jaggar, "Love and Knowledge"; Martha Nussbaum, *Women, Culture, and Development: A Study of Human Capabilities* (Oxford: Oxford University Press, 1997); Sandra Morgan, "Towards a Politics of 'Feeling': Beyond the Dialectic of Thought and Action," *Women's Studies* 10 (1983): 203–23.
27. Haraway, *Primate Visions.*
28. Jaggar, "Love and Knowledge," 23.
29. See, e.g., Wayne Babchuck, Raymond Hames, and Ross Thompson, "Sex Differences in the Recognition of Infant Facial Expressions of Emotion: The Primary Caretaker Hypothesis," *Ethology and Sociobiology* 6 (1985): 89–101.
30. Although I treat her work in a separate section below, Arlie Hochschild takes a related perspective when she defines emotions as a straightforward, indisputable "signal from the self." The implication is that women might learn their true feelings through careful reflection and sorting through of the culturally induced noise of a sexist world and its learned secondary emotions. Hochschild, *The Managed Heart,* 316.
31. Spelman, "The Virtue of Feeling," 220.
32. Lila Abu-Lughod, *Veiled Sentiments: Honor and Poetry in a Bedouin Society* (Berkeley and Los Angeles: University of California Press, 1986), and "The Romance of Resistance," in *Beyond the Second Sex: New Directions in the Anthropology of Gender,* ed. P. Sanday and R. Goodenough (Philadelphia: University of Pennsylvania Press, 1990), 41–55; Dorinne Kondo, *Crafting Selves: Power, Gender, and Discourses of Identity in a Japanese Workplace*

(Chicago: University of Chicago Press, 1990); Catherine Lutz, *Unnatural Emotions: Everyday Sentiments on a Micronesian Atoll and Their Challenge to Western Theory* (Chicago: University of Chicago Press, 1988), and "Engendered Emotion: Gender, Power and the Rhetoric of Emotional Control in American Discourse," in *Language and the Politics of Emotion*, ed. C. Lutz and L. Abu-Lughod (Cambridge: Cambridge University Press, 1990), 151–70; Jeannette Marie Mageo, "Spirit Girls and Marines: Possession and Ethnopsychiatry as Historical Discourse in Samoa," *American Ethnologist* 23.1 (1996): 61–82; Virginia Maher, "Possession and Dispossession: Maternity and Mortality in Morocco," in *Interest and Emotion*, ed. Hans Medick and David W. Sabean (Cambridge: Cambridge University Press, 1984), 103–28; Emily Martin, *The Woman in the Body* (Boston: Beacon, 1987); Michele Rosaldo, "Toward an Anthropology of Self and Feeling," in *Culture Theory: Essays on Mind, Self and Emotion*, ed. Richard A. Shweder and Robert A. LeVine (Cambridge: Cambridge University Press, 1984), 134–57; Nancy Scheper-Hughes, *Death without Weeping: The Violence of Everyday Life in Brazil* (Berkeley and Los Angeles: University of California Press, 1992); C. Nadia Seremetakis, *The Last Word: Women, Death and Divination in Inner Mani* (Chicago: University of Chicago Press, 1991); Stephanie Shields, Pamela Steinke, and Beth Koster, "The Doublebind of Caregiving: Representation of Gendered Emotion in American Advice Literature," *Sex Roles: A Journal of Research* 33 (1999): 467–89.

33. Mageo, "Spirit Girls and Marines," 61.
34. Francesca Cancian, *Love in America* (Cambridge: Cambridge University Press, 1987); E. Anthony Rotondo, "Romantic Friendship: Male Intimacy and Middle Class Youth in the Northern United States, 1800–1900," *Journal of Social History* 23 (1989): 1–26; Peter Stearns, "Gender and Emotion: A Twentieth Century Transition," *Social Perspectives on Emotion* 1 (1992): 127–60.
35. Heather Ferguson Bulan, Rebecca J. Erickson, and Amy S. Wharton, "Doing for Others on the Job: The Affective Requirements of Service Work, Gender, and Emotional Well-Being," *Social Problems* 44.2 (1997): 235–56; Aihwa Ong, *Spirits of Resistance and Capitalist Discipline: Factory Women in Malaysia* (Albany: State University of New York Press, 1987); Hochschild, *The Managed Heart*.
36. Abu-Lughod, "The Romance of Resistance."
37. Seremetakis, *The Last Word*.
38. Rosemary Wiss, "Lipreading: Remembering Saartjie Baartman," *Australian Journal of Anthropology* 5.1–2 (1994): 11–40.
39. Cancian, *Love in America*; Barbara Ehrenreich, *The Hearts of Men: American Dreams and the Flight from Commitment* (New York: Doubleday, 1983); Jan Lewis, "Mother's Love: The Construction of an Emotion in Nineteenth Century America," in *Social History and Issues in Human Conscious-*

ness: Some Interdisciplinary Connections, ed. Andrew E. Barnes and Peter N. Stearns (New York: New York University Press, 1989), 209–29.

40. George Brown and Tirril Harris, *Social Origins of Depression: A Study of Psychiatric Disorder in Women* (New York: Free Press, 1978); Susan Nolen-Hoeksema, *Sex Differences in Depression* (Stanford, CA: Stanford University Press, 1990); Janis J. Jenkins, Arthur Kleinman, and Byron H. Good, "Cross-Cultural Aspects of Depression," in *Advances in Affective Disorders: Theory and Research*, ed. Joseph Becker and Arthur Kleinman, vol. 1, *Psychosocial Aspects* (Hillsdale, NJ: Erlbaum, 1991), 67–99; Bonnie Strickland, "Women and Depression," *Current Directions in Psychological Science* 1.4 (1992): 132–35.
41. Claudia Koonz, *Mothers in the Fatherland: Women, the Family and Nazi Politics* (New York: St. Martin's, 1987).
42. Klaus Theweleit, *Male Fantasies*, vol. 1, *Women, Floods, Bodies, History* (Minneapolis: University of Minnesota Press, 1987).
43. Ximena Bunster-Burotto, "Surviving beyond Fear: Women and Torture in Latin America," in *Women and Violence*, ed. Miranda Davies (London: Zed, 1994), 156–76.
44. See also Janis H. Jenkins, "Women's Experience of Trauma and Political Violence," in *Gender and Health: An International Perspective*, ed. Carolyn Sargent and Caroline Brettell (Upper Saddle River, NJ: Prentice-Hall, 1996), 178–291.
45. Louise Krasniewicz, *Nuclear Summer: The Clash of Communities at the Seneca Women's Peace Encampment* (Ithaca, NY: Cornell University Press, 1992).
46. Morgan, "Towards a Politics of "Feeling."
47. Kondo, *Crafting Selves*; Smadar Lavie, *The Poetics of Military Occupation: Mzeina Allegories of Bedouin Identity under Israeli and Egyptian Rule* (Berkeley and Los Angeles: University of California Press, 1990); Lutz, *Unnatural Emotions*.
48. Lavie, *Poetics of Military Occupation*, 146.
49. Hochschild, *The Managed Heart*.
50. Micaela di Leonardo, "The Female World of Cards and Holidays: Women, Families and the Work of Kinship," *Signs: Journal of Women in Culture and Society* 12.3 (1987): 440–53.
51. Arlie Hochschild, "The Economy of Gratitude," in *The Sociology of Emotions*, ed. David D. Franks and E. Doyle McCarthy (Greenwich, CT: JAI, 1989), 95–113.
52. Di Leonardo, "The Female World of Cards," 452. See also Medick and Sabean, eds., *Interest and Emotion*.
53. Cvetkovich, *Mixed Feelings*.
54. Ibid., 5.
55. For example, Maher, "Possession and Dispossession"; Scheper-Hughes, *Death without Weeping*; Shields, Steinke, and Koster, "The Doublebind of Caregiving."

56. Scheper-Hughes, *Death without Weeping.*
57. Phyllis Chesler, *Women and Madness* (New York: Avon, 1972); Lavie, *Poetics of Military Occupation*; Nancy Scheper-Hughes, "The Genocidal Continuum: Peace-Time Crimes," in Mageo, ed., *Power and the Self*, 29–47; and William S. Lachicotte, "Intimate Powers, Public Selves: Bakhtin's Space of Authoring," in ibid., 48–67.
58. Chesler, *Women and Madness*, 117.
59. Seremetakis, *The Last Word*, 1. 216.
60. Margaret Trawick, "Untouchability and the Fear of Death in a Tamil Song," in Lutz and Abu-Lughod, eds., *Language and the Politics of Emotion*, 186–206; Judith Butler, "Imitation and Gender Subordination," in *Inside/Out: Lesbian Theories, Gay Theories*, ed. Diana Fuss (New York: Routledge, 1991), 307–20.
61. Modleski, *Feminism without Women.*
62. Anne Moir and David Jessel, *Brain Sex: The Real Difference between Men and Women* (New York: Carol, 1991).

FIFTEEN

Affect, Trauma, and Daily Life: Transatlantic Legal and Medical Responses to Bullying and Intimidation

RODDEY REID

A deadly combination of economic rationalism, increasing competition, "downsizing," and the current fashion for tough, dynamic, "macho," management styles, have created a culture in which bullying can thrive, producing "toxic" workplaces. *BRITISH MEDICAL JOURNAL*

Have fear-producing mechanisms become so pervasive and invasive that we can no longer separate ourselves from our fear? If they have, is fear still fundamentally an emotion, a personal experience, or is it part of what constitutes the collective ground of possible experience? **BRIAN MASSUMI**

In this chapter, I wish to explore the differences between the United States and Europe in legal and medical responses to bullying and psychological intimidation, especially in the workplace. I want to argue that the different arenas of the school, the workplace, the media, and politics operate as multiple sites of bullying so as to produce a public culture of intimidation that links the most subjective, individual experiences with those of collective life itself. The example and achieved legitimacy of behaviors in one arena can have the effect of authorizing analogous conduct in other domains of civil society and thereby produce a reciprocal legitimation of like practices and actions. In this respect, the culture of intimidation or fear in its affective dimension has posed a particular challenge to contemporary social theory and to existing medical and legal frameworks on both sides of the Atlantic. It brings together the two levels of violence that Slavoj Žižek prefers to view as epistemologically opposed. For him, the question of harassment is simply a symptom of the decadence of liberal tolerance and stands as the personalized, displaced form of objective social forces such as globalization that have ruined the protective walls of civility.[1] Žižek's position replicates something of an economist view still common in some Marxist cultural criticism and sociology that discounts the power of the affective dimension of the minutiae of daily life in these various arenas and how it may contribute to producing a collective experience that in turn shapes a personal sense of self.

In what follows I adopt a working definition of *affect* that understands it as fundamentally dynamic in William Reddy's early sense with respect to the transformative power of emotional gestures and utterances.[2] However, it is inflected by Brian Massumi's stress on the autonomy of affect in terms of its virtual character and its disregard of boundaries between personal and collective experience. In Massumi's view, affect operates as an event whose intensity is not fully assimilatable by existing linguistic and social structures and thus displays an openness to future possibilities whose ideological or political valence cannot be determined in advance.[3] In other words, affect is fundamentally volatile and labile. And, as we shall see, it has come to be understood as powerfully transformative of even one's identity and personhood. These future possibilities play out in *daily life,* which Lawrence Grossberg defines as an endless dynamic of "structured mobility."[4] In turn, this play of possibilities that derive from affects may give rise to a cultural formation of what Daniel M. Gross calls *social passions* with its own subjective *structure of feeling* (as Raymond Williams would say) or sensibility for particular groups within the social hierarchy.[5] In this fashion the concept of affect seems to offer the advantage of grasping the fluid and intensive dimensions of subjective daily life in societies undergoing rapid change within the structures of globalization and free market capitalism rather than viewing them as merely symptoms of the latter.

Bullying as Transnational Event

In recent years, bullying and intimidation have become widely reported phenomena and much-discussed topics in public and private discourse in North America, Europe, and East Asia. Commentators in transnational media especially note that new communication technologies (e-mail, anonymous listservs, and social network media) have multiplied the opportunities for verbal and psychological bullying and have become the means of choice for intimidating others. In the United States, one has only to think of the rash of suicide cases in 2010 attributed to cyberbullying, including those of Phoebe Prince, an Irish high school student living in Massachusetts, and of Tyler Clementi, a gay first-year student at Rutgers University in New Jersey. What attracted international attention was that for the first time in US legal history criminal charges were filed against the alleged perpetrators of bullying.[6] The previous year, media reports focused on the sentencing by a British court of an eighteen-year-old woman to three months' detention in a young offenders institute for posting death threats on Facebook.[7] In Japan, where bullying-related suicides of schoolchildren have been debated since the 1990s, the sensational three-year-long trial of a young Japanese man who claimed to be a victim of cyberbullying and went on a killing rampage with his car in the Akihabara District of Tokyo, concluded in March 2011 with a death sentence.[8]

Alongside and prior to the recent wave of cyberbullying, incidents of workplace intimidation have also dominated journalists' reporting in Europe, North America, and Asia for some time. With the notable exception of the United States, such cases had begun to be subject to public policy and legal action in many countries. For example, in Germany, where there is no special law against bullying but where there exists well-established recourse against such behaviors in the federal labor court, in January 2010 a case in which Siemens was being sued by a former employee for bullying as well as racial and sexual discrimination came to trial in Nuremberg.[9] In France, where a strong antiworkplace bullying law has been in force since 2002, there has been a flurry of court cases involving major French companies and institutions, including the automobile manufacturers Renault and Peugeot, the supermarket chain Carrefour, the French army, and, most spectacularly, France Télécom after a rash of suicides following its restructuring and downsizing.[10] The success of the 2002 French law laid the foundation for a second piece of legislation, approved unanimously in 2010 by the French National Assembly, to extend protections against bullying and intimidation to the domestic sphere.[11] Meanwhile, in China in May and June 2010, the topic of workplace bullying erupted in public discourse in reaction to a wave of suicides in the huge factories of the Taiwanese electronics firm Foxconn that manufacture leading brands of smartphones and electronic tablets, including Apple's iPhone and iPad.[12] It is at the affective level of distress and even life-changing trauma that personal and collective struggles and revendications are being played out.

Naming It

It would appear then that bullying as a phenomenon of daily life has begun to acquire a transnational character. It goes by many names: *ijime* in Japan, *qiwu* in Mandarin, *mobbing* in Scandinavia, *bullying* in the United Kingdom and Commonwealth countries, *psychological intimidation* or *harassment* in French- and Spanish-speaking countries (*le harcèlement moral* or *psychologique; el acoso moral* or *sicológico*), *mobbing* or *psychological terror* (*Psychoterror*) in Germany, and *harassment, emotional abuse*, and *bullying* in the United States. The renewed attention to bullying in the media, political discourse, and legal and medical circles arguably dwarfs past discussions, which focused mainly on the rough and tumble play of the schoolyard and confined it to a developmental stage that is superseded by adulthood. The new discourse troubles these standard accounts by framing bullying and intimidation as the eruption of life-changing affective trauma into the routines of daily life at any age.

My discussion of bullying as a transnational phenomenon will focus mainly on the United States, which has been the starting point of my reflections, and to a lesser degree Europe. As I have argued elsewhere, in the aftermath of the civil rights and women's movements, defeat in Vietnam, and early globalization's massive deindustrialization of the US economy, a new macho white populism arose in the 1970s in the United States that forged strong links between deep-seated feelings of individual insecurity, collective economic uncertainty, and distress at the diminishing fortunes of a superpower on the verge of decline. As a result, over the last thirty years there has been a surge in public intimidation from the workplace (with corporate restructuring and privatization and now the financial and economic crisis) to the media (the rise of talk radio and television shows, reality television series, and social media) and politics (the War on Terror and recent political campaigns). For example, the public physical and psychological intimidation to which Barack Obama has been subjected since assuming the office of president has been unprecedented in recent US history.[13] Meanwhile, across the Atlantic, the persistent recession, the sovereign debt crisis, and the extreme fiscal austerity policies adopted by EU countries in 2011–12 beginning with the Cameron government in the United Kingdom have exacerbated already tense relations in civil society and the workplace stemming from market liberalization, privatization, and downsizing of companies that have been marked by outbreaks of violence and psychological intimidation targeting immigrants, guest workers, Muslims, Roma, and organizations associated with socialist parties.[14] These forms of aggression could be considered the return of violent sovereignty in people's everyday existence, and they would appear to have come to shape private and public life in Europe and the United States.

The Responses of Human and Psychiatric Sciences

The subjective experience of contemporary suffering in its connections to collective life in terms of its social causes and the emergence of a new sensibility has been the subject of research by social scientists, psychologists, psychiatrists, and social and political philosophers in Europe and the United States from Pierre Bourdieu, Richard Sennett, Christian Dejours, and Marie-France Hirigoyen to Alain Ehrenberg, Axel Honneth, Nancy Fraser, and Judith Butler. Their work constitutes a key milestone in the "emotional turn" in the humanities, social sciences, and therapeutic disciplines. Their studies highlight how the interplay of economic and noneconomic humiliations and indignities encountered in daily life under market economies and the War on Terror undermines personhood and identity. This dynamic constitutes a set of repeated experiences that cut most deeply and elicit the most consequential affective responses. The workplace, where adults spend the majority of their waking hours, is central to most of these analyses. Facing an uncertain job market, cuts in wages, and threats of downsizing, workers must live the grueling contradiction between management's arbitrary incitements to ever greater personal investment in the workplace and teamwork, on the one hand, and demands by these same employers for flexibility and continuous retooling, on the other. This is a form of daily injustice that Bourdieu and Dejours call *ordinary* or *social suffering,* a term that has enjoyed greater currency in Europe than in the United States.[15] Those who are targeted in the workplace and in other spheres of daily life can lose not only control over their self-representation but even their capacity to pursue their studies or training, work productively, enjoy effective voice in public discourse, or participate in political life. Arguably, it is within this culture of daily life that an aggressive affective politics of degraded subjecthood has emerged consisting in the practice of demeaning or dishonoring others, often in terms of stigmatizing deficient social identities that at times are assigned the status of the viscerally repulsive abject other whose life—to echo Judith Butler—is not deemed livable or even mournable.[16] Just what protections the different legal and medical systems in the United States and Europe offer against this form of psychological violence are explored below.

In their conceptualizations of this type of injury, philosophers such as Honneth, Fraser, and Sennett refer to older notions of human dignity and recognition that go back to the writings of Kant and Hegel and Habermas's theory of distorted communication.[17] Dejours not only appropriates Kant but also incorporates Arendt's concept of everyday evil and, with Hirigoyen and Ehrenberg, recent reformulations of the clinical and psychiatric concept of psychic trauma (including posttraumatic stress disorder [PTSD], understood as an incapacitating individual and collective wound) that have emerged in the late twentieth century (more about these below).[18] In response to these pervasive forms of everyday violence exacerbated by the War on Terror, the US feminist philosopher Judith Butler, drawing on the psychoanalytic concept of narcissism and Emmanuel Levinas's concept of the neighbor, has even gone so far as to propose vulnerability and violent responses to it as the fundamental ethical condition of human existence.[19]

Workplace Bullying Studies

Even as social scientists, philosophers, and psychiatrists have investigated the affective dimension of contemporary suffering, the history of the transatlantic emergence and circulation of more encompassing concepts of bullying and emotional abuse is one of uneven developments. First of all, with respect to the workplace, in the United States bullying and intimidation are not the subjects of occupational health policies and legal statutes, unlike in Canada and a number of EU countries, and until lately have been little studied by academic and health science researchers.[20] Paradoxically, one of the earliest studies was *The Harassed Worker* (1976) by the University of California, Berkeley, psychiatrist and anthropologist Caroll Brodsky. Based on research in the archives of workers' compensation claims, its discovery of large numbers of patients incapacitated by "cruel" interactions on the factory floor enlarged the scope of injury requiring medical attention and compensation to include crippling psychological harm that exceeds the wear and tear of occupational stress.[21] However, this new departure in workplace literature enjoyed little influence in Europe and even in the United States remained stillborn, for no further sustained research on workplace dynamics (besides those of sexual and racial harassment) was conducted in the United States until the 1990s, and even then academic studies were few in number and focused on extreme single acts of physical violence such as felonious assaults and shootings. It was left to business journalists, a few activists, and employees writing on the Web to draw public attention to the depredations of aggressive, if less spectacular, workplace behavior, especially on the part of management.[22]

In contrast, a sustained wave of research on mobbing began first in Germany and Scandinavia in the 1970s and 1980s (and somewhat later in other countries, especially the United Kingdom and France), presumably due to their stronger trade union traditions and more extensive labor laws and welfare systems and to the influence of new social movements (propelled by younger workers, feminists, antinuclear activists, and environmentalists) with their emphasis on quality-of-life issues over simple economic revendications. In 1972 in Germany, the child psychologist Peter-Paul Heinemann applied the ethologist Konrad Lorenz's term *mobbing* for aggressive animal group behavior to that of children and included human behaviors such as insults, ostracism, and occasional acts of violence that threaten targets' "personal dignity" and "respect." Then the German industrial psychologist Heinz Leymann, working in Sweden in the 1980s, transferred the term to the adult world of work and extended it to describe both individual and group harassment of employees. Out of these studies and subsequent workplace policies emerged the standard international definition of bullying: "Bullying at work is about repeated actions and practices that are directed against one or more workers, that are unwanted by the victim, that may be carried out deliberately or unconsciously, but clearly cause humiliation, offence, and distress, and that may interfere with job performance and/or cause an unpleasant working environment."[23] These aggressive workplace behaviors may target subordinates, coworkers, or superiors and are "primarily psychological in nature." A sense of the complex array of behaviors that have affective consequences and that fall under the definition of bullying or mobbing can be found in an early 1993 Swedish occupational health and safety regulation that defines acts of mobbing as follows:

- Slandering or maligning an employee or his/her family.
- Deliberately withholding work-related information or supplying incorrect information of this kind.

- Deliberately sabotaging or impeding the performance of work.
- Obviously insulting ostracism, boycott or disregard of the employee.
- Persecution in various forms, threats and the inspiration of fear, degradation, e.g. sexual harassment.
- Deliberate insults, hypercritical or negative response or attitudes (ridicule, unfriendliness, etc.).
- Supervision of the employee without his/her knowledge and with harmful intent.
- Offensive "administrative penal sanctions" which are suddenly directed against an individual employee without any objective cause, explanations or efforts at jointly solving any underlying problems.[24]

In the United Kingdom, academic studies appeared in professional business and management journals and human resources publications such as *Personnel Review* and *Journal of Business Ethics* and were part of public discussions that led to the dignity at work bill that since the 1990s has nonetheless repeatedly failed to pass in Parliament. In France, the work of two psychiatrists—Christian Dejours and Marie-France Hirigoyen—in the 1990s focused on affective relationships between men and women both in the workplace and in the domestic sphere. Their studies tended to highlight men as the main perpetrators of bullying and emotional abuse in these settings. Hirigoyen, trained in the emerging field of victimology in the United States,[25] built on Leymann's early work on mobbing and Dejours's concept of social suffering and adapted the English term *sexual harassment* to formulate a theory of psychological or "moral" harassment and abuse that can lead to the destruction of one's personhood both in the workplace and in the life of families and couples: her work became the basis of major revisions of the French Penal Code and workplace and domestic regulations in 2002 and 2010. Of workplace bullying Hirigoyen writes: "In all other forms of workplace suffering, and especially in cases of excessive occupational pressure, if the stimulus ceases, then so too does the suffering, and the sufferer can recover her or his normal state. Moral harassment, on the contrary, leaves indelible traces that can go from posttraumatic stress to a recurring experience of shame or even to permanent personality changes."[26] By contrast, it would appear that the American Psychiatric Association's diagnostic category *PTSD* plays a very limited role in US court cases and workplace and education policies, on the one hand, and in US-based academic studies and policies on bullying, on the other. It is featured mostly in journalistic and popular accounts.[27] The limited legal scope of PTSD as a diagnostic category in relation to psychological intimidation may have something to do with the US juridical traditions of privacy and freedom of expression (more about this below).

In European and US learned and popular literature on bullying and intimidation, much time is spent outlining the difficulties posed by the arbitrary and invisible nature of psychological aggression, especially when it is verbal and undocumented. Either the acts of violence take place in the privacy of a conversation out of earshot of possible witnesses, or, if others are present, the destabilizing nature of verbal insults, derision, and disrespect is not apparent, for the witnesses are unaware of the preceding conversations or relevant context. Moreover, the cumulative effect of many small acts of senseless aggression that take place in the workplace, school, or social circles—silence, unfounded rumor, unexplained withdrawal of support, isolation, exclusion, and so forth—seems disproportionate in the eyes of outside observers. And, if the affective harm appears, it often does so with little or no legitimacy. So for the target to name or even describe her or his distress to outside parties seems to be an impossible task. More organized or institutional forms of bullying and intimidation, such as those found in media organizations or private companies and public bureaucracies undergoing downsizing, are not much easier to name and oppose, for they often enjoy the tolerance of coworkers and the indifference or even unofficial sanction of owners, editors, supervisors, human resources departments, and senior managers.[28]

The Evolving Medical Status of Psychological Trauma

When Leymann, Dejours, and Hirigoyen introduced psychic trauma as a framework for understanding harassment and emotional abuse, it already had a long, conflicted history as a clinical and diagnostic category. Their work drew on the changing medical and cultural status of trauma, whose historical sufferers on both sides of the Atlantic have been bedeviled by obstacles similar to those faced by targets of bullying: credibility, proof, etc. Up to the Second World War, there appears to be something of a shared history between North America and Europe. Medical historians have documented the genealogy of the concept of psychic trauma and its sufferers that goes back to the nineteenth century and industrial and railway accidents and the experience of war. Past physical injury or shock under spectacular circumstances was understood to persist in the present in the form of psychological symptoms that disrupted the otherwise unremarkable routines and events of daily life. This was viewed more sympathetically by psychoanalysts than by psychiatrists. The questions of care, compensation, testimony, traumatic memory, and proof quickly came to the fore, and for many decades victims and their alleged condition were often treated with suspicion by the authorities and the general population: workers and soldiers in particular suffering from "trauma neurosis" were commonly classed as malingerers, cowards, weak, phony, or dishonest and never enjoyed the respect of those who simply suffered corporal harm with no noticeable psychological aftereffects. As the psychiatrists and anthropologists Daniel Fassin and Richard Rechtman put it, their conditions did not fit within the heroic national narrative.[29]

The advent of the two world wars introduced important changes, though with some variation between countries. With the Second World War, even as sympathy for psychic suffering deepened on both sides of the Atlantic, the history of psychic trauma as a medical and cultural category underwent a growing divergence between Europe and the United States. A debate among US military psychiatrists and government officials over the nature of the affective traces of the horrors of war and "combat fatigue" in soldiers signaled a shift in how psychic trauma was regarded,[30] but, according to Fassin and Rechtman, the soldiers' experience and clinical records did not win the uniform respect of European and American authorities. Rather, in the end it was the mental struggles of survivors of the Nazi death camps and atomic bombings of Hiroshima and Nagasaki that eventually did, through US-based studies of traumatic memory conducted in the 1950s and 1960s. And later the work of US anti–Vietnam War psychiatrists such as Robert Lifton extended the early research on concentration camp and atomic bombing survivors to US soldiers traumatized by their involvement in combat and atrocities in Vietnam. Here, the antiwar movement and its desire to disculpate US soldiers, many of whom were drafted against their will into the armed services, together with the feminist movement intersected with a medical field—psychiatry—that was in the throes of a legitimization crisis set off by persistent attacks by progressive critics. What emerged from this historical encounter was a new edition of the Diagnostic and Statistical Manual of Mental Disorders in 1980 (i.e., the *DSM-III*) that moved away from the diagnostic framework of psychoanalysis, depathologized homosexuality, broke with the clinical category of trauma neurosis dear to US and European psychiatry, and introduced the new category of posttraumatic stress disorder.[31] In Fassin and Rechtman's view, psychic trauma was now understood in the US-dominated English-language psychiatric world to be a normal response by an ordinary person to an extraordinary event. The unbearable nature of the event became paramount and determining. This imparted a new sense of legitimacy to the victims of psychic trauma, whose experience was moreover understood to be on the order of the unknowable. As a consequence, the victim's voice and narrative—and no longer clinical accounts—became decisive in establishing the veracity of the experience.[32] However, in the early 1990s, the deployment of traumatic recovered or repressed memory in spectacular court cases involving childhood sexual abuse that sent parents, guardians, teachers, and child-care workers to long prison terms on the basis of unsubstantiated child testimony undermined the rising legitimacy of psychic trauma as an operative category in the United States.[33] "Psychic trauma" as a name for individual and collective experience would return in popular discourse with the spectacular terrorist attacks on New York and Washington in September 2001.[34]

The destigmatization of psychological trauma would take a somewhat different path in Europe and in France in particular, where, still under the sway of psychoanalysis, psychiatry did not enter into an alliance with a social movement but rather became involved in the emerging fields of emergency and humanitarian psychiatry in which the category of psychic trauma (if not strictly speaking PTSD, an American, nonpsychoanalytic concept) became operative. It was during the 1990s that this diagnosis was extended by European and international government and nongovernment aid organizations to the victims of oppression, torture, terror, and exile and to entire collectivities and groups. This universalization of trauma sometimes came at the cost of erasing the socially determined forms of suffering of specific groups, whereas in the United States it had been a means for precisely articulating the grievances of groups such as women and Vietnam War veterans.[35] Meanwhile, as we have seen, in Europe psychological trauma and even PTSD began to emerge in the clinical frameworks developed by Leymann, Hirigoyen, and Dejours for the everyday victims of workplace and even nonworkplace bullying and harassment.[36] In this growing clinical literature, trauma as extraordinary event enters daily life not only in the spectacular form of a rare immediate experience of a natural disaster or a terrorist attack but also and more often in the very fabric of relationships in school, the workplace, political life, and the public media sphere.

Thus, it can be said that, on both sides of the Atlantic, there is a common picture in contemporary popular and scholarly literature on emotional abuse and bullying: the traumatic event is locatable not only in the past but also potentially in every workday, walk down the school hall, dinner conversation, or newscast: the past trauma returns not simply as a disabling memory but also as an unpredictable and ungraspable future, as a terrifying possible yet unknowable act of aggression. In this view, traumatized targets enter unprotected into the infinite, fearful regress and fact of future threat. However, the discourse is unevenly distributed and consequently enjoys different degrees of legitimacy. In the United States through the psychiatric diagnostic category of PTSD, emotional or psychic trauma stemming from physical violence associated with war, terrorist attacks, and sexual abuse enjoys increasing recognition and legitimacy in medical circles. However, it is mainly within the popular culture of media reports, business literature, and Internet-based discussions that there is public recognition of psychological or verbal violence in the form of bullying as a medical and social issue of daily life in educational establishments and the workplace. In Europe, by contrast, the general concept of psychic trauma (if not PTSD) has greater currency across all public arenas, and psychologically degrading behaviors in the workplace fall under the purview of workplace regulations. This transatlantic difference emerges sharply in existing penal and civil law.

Bullying and Intimidation in Juridical Frameworks

I now want to return to the question of the different legal status of bullying and intimidation in their affective dimension in the United States and Europe. It would appear that, as a consequence of recent trends in American jurisprudence, nondiscriminatory acts of psychological violence in and of themselves in the form of verbal aggression, harassment, and repeated acts of humiliation in public arenas from school and the workplace to media and political life remain beyond the reach of US tort law. As I stated earlier, there are few occupational health and legal protections against workplace bullying as such in the United States, and only since 2010 has there been any meaningful attempt to establish legal regulation of bullying in US schools (see below). Legal scholars have offered several explanations. To begin with, in the United States there is the powerful legacy of civil rights protections, going back to the 1964 Civil Rights Act, that target discriminatory acts against classes of persons. These protections regard primarily the spheres of the workplace and commercial and public services, not political life or the media.[37] Thus, in the United States, sexual harassment in the workplace falls under the logic of legislation banning discrimination but not so bullying, which is not covered by existing law unless the victim is targeted in terms of his or her belonging to a protected group (in terms of race, color, religion, sex, national origin, disability, age, or marital status).

Moreover, according to the US comparative legal scholars Gabriela Friedman and James Whitman, who look at the fate of US sexual harassment law when it was adopted in Europe, although American sexual harassment law does consider the affective issues of stress and abuse in its hostile working environment clauses, actual court rulings have focused on the measurable loss of tangible opportunities for employment, advancement, and financial rewards, and they have not extended protection against nondiscriminatory sorts of harassment. In Friedman and Whitman's view, this legal interpretation stems from two assumptions: (1) that of a fluid job market in which workers do not remain in the same job for very long and (2) the federal legal doctrine of "at-will employment" whereby, in many states, employers have the right to fire employees without cause. By way of contrast, the quality of working conditions matters more in European laws and codes, where the reigning assumption is presumably the opposite: that workers wish to and do stay in the same jobs and, as a consequence, expect and demand more humane working conditions. As Friedman and Whitman put it, reprising the words of an unnamed German feminist: "European sexual harassment law became the law of 'dignity' in a stable job, rather than the law of 'equality' in a fluid job market."[38]

The operative word and concept here is of course *dignity*. As stated earlier, it has deep roots in modern European philosophy going back to Kant. It is also a key concept in twentieth-century constitutional and legal texts. For example, it can be found in many documents from the UN Universal Declaration of Human Rights (1948) and the French Penal Code (1992) and Social Law (2002) to the post-Franco Spanish Constitution (sec. 10) and the German Basic Law or Constitution (1949), where it appears in the form of the "right to freely develop one's personality" (art. 2, sec. 1). Article 26 of the European Social Charter reads: "All workers have the right to dignity at work."[39] Dignity also figures in allied concepts such as "social dignity," "social standing," "personal integrity," "personal honor," "right to reputation," and "public image." None of these concepts is explicitly recognized by the US Constitution, and, although they had some purchase in early court cases involving slander and privacy, they are less and less operative in current American tort law. There are many reasons for this. First of all, dignity in its various acceptations in European texts has very little to do with the American idea of *privacy*, the term by which it is commonly translated into English, as in the expression *invasion of privacy*. In US law, this latter expression has a very distinct meaning, namely, the violation of the household and private life by a third party.[40] In the US legal tradition, the liable third party has most often been municipal, state, or federal agencies that have placed wiretaps or cameras in victims' dwellings or hotel rooms. Thus, outside the physical confines of the home, in the United States citizens and residents enjoy few protections against what others may say, film, or record concerning one's person, image, or reputation as long it is deemed to be political or simply expressive speech.[41] This places a strong restriction on the regulation of verbal and psychological aggression in the United States.

This brings us to another legal tradition in the United States, that of freedom of expression, which enjoys exceptional status in the arena of international law. The focus of court rulings has been on the potential consequences of utterances rather than on their content.[42] Moreover, in the wake of the interpretive tradition of the First Amendment going back to Justice Oliver Wendell Holmes's opinions in the early twentieth century, all public forms of speech, however hateful and injurious they may be, are protected by US law unless they lead immediately to acts of *physical* violence.[43]

This tradition interprets the First Amendment to mean the protection of robust, uninhibited, and wide-open political debate deemed crucial for effective self-government. Comparative legal scholars term this an *absolutist* doctrine of free speech; that is to say, it is one of no "proportionality," whereby freedom of expression is not balanced with other values such as privacy, dignity, safety, equality, respect, and so forth.[44] This doctrine reflects both a long-standing American concern with freedom of the press and, more recently, a movement within the legal world dating from the 1960s that has militated in favor of an expanding concept of the public's right to know and, as a consequence, of what is deemed newsworthy.[45] First applied to speech acts directed at public officials, it was then extended to utterances targeting public celebrities and finally to almost anyone, effectively gutting defamation law.[46] So what began as the protection of political speech has been extended to public expression *tout court*. As a result, hate speech codes regulating racist utterances commonly found in Europe are nonexistent in the United States, and all attempts to establish them there have come to grief in the courts. The courts have explicitly steered clear of judging the content of public statements, which they consider to be a subjective exercise, and their tendency has been to leave it up to editors and the public to decide.[47] Indeed, in recent rulings, the Supreme Court has offered little in the way of protection against defamatory public speech so long as the speaker claims to be engaging in expression on issues of public concern.[48] As the Court wrote in 2011: "Speech is powerful. It can stir people to action, move them to tears of both joy and sorrow, and—as it did here—inflict great pain. On the facts before us, we cannot react to that pain by punishing the speaker. As a Nation we have chosen a different course—to protect even hurtful speech on public issues to ensure that we do not stifle public debate."[49] Things have reached such a pass in the United States that legal scholars now claim that, were the *contents* of the secret video of Tyler Clementi's tryst with another man that provoked his suicide to be the subject of public debate, nothing in American law would prevent their circulation to the general American public. Such are the weaknesses of protection against emotional harm due to unwanted publicity.[50]

In the Offing

However, in the United States things may be changing. The building emotional turn in public opinion over the last decade has begun to reach the halls of state and federal government. Several months after Phoebe Prince's death in January 2010, the Massachusetts state legislature passed a law on school bullying targeting the harassment of students by their peers—be it on the school grounds or elsewhere, via the Internet, or over the phone—that results in the creation of a hostile environment that effectively prevents students from the successful pursuit of their studies. What is striking in the new law is, on the one hand, the total absence of any reference to the discriminatory nature of the crime as its defining element and, on the other, its emphasis on the psychological harm suffered by victims. Moreover, in October 2010, the US Department of Education published a circular warning school and college authorities of the dangers posed by bullying. While in this text the federal authorities continue to insist on the discriminatory character of the acts of aggression, they now speak of entire groups—not just individuals belonging to groups—who are victims of a hostile environment resulting from the accumulation of acts that up until then had been considered as isolated, unrelated incidents. It remains to be seen how the courts will view these new regulations and laws targeting populations of minors under the paternalistic care of school authorities.

However, with respect to the workplace, the public media sphere, and the world of politics in the United States, it would appear that little has evolved. One possible agent of future change is the growing legitimacy of PTSD as a medical diagnosis and condition requiring care. Already since the 1980s it has served as a defense in civilian courts primarily in death penalty cases involving male veterans and battered women. More recently, yet other advances have been made with respect to veterans suffering from psychological trauma, to whom, in a sense, PTSD owes its beginnings. Faced with a tidal wave of cases from the wars in Iraq and Afghanistan, military authorities have begrudgingly begun to recognize the incapacitating effects of PTSD, especially on combat troops, and a diagnosis of PTSD can now serve as the basis for release from combat duty, an honorable discharge, and therapeutic care.[51] That said, for other types of cases, the advent of the diagnosis of PTSD has yet to serve as grounds for legal action against bullying and psychological intimidation. It is interesting to note that in Europe, where there is already an existing legal and political vocabulary recognizing related concepts of subjective life and where regulations and even laws exist targeting workplace bullying, diagnoses of PTSD have yet to be taken in account in cases of bullying that have come before the courts where indeed other grounds already exist for filing a legal complaint.[52] The expansive application of PTSD and emotional trauma to the injustices of daily life is one attempt in the United States to address the absence of a psychological safety net in contemporary civil society. In a similar manner in the realm of US academic research, the emotional turn in the humanities and social sciences generally and the strong interest in affect theory in particular (in its validation of experience that presumably is not readily accounted for in long-standing intellectual frameworks) may perhaps be understood not only as a new approach to understanding societies undergoing rapid change owing to globalization and free market capitalism but perhaps also as an attempt to come to grips with dimensions of subjective and collective existence that do not yet enjoy public recognition in the domain of social policy and law.

NOTES

Epigraphs: "Workplace Bullying: The Silent Epidemic," *British Medical Journal* 326 (April 12, 2003): 776–77; Brian Massumi, preface to *The Politics of Everyday Fear* (Minneapolis: University of Minnesota Press, 1993), ix.

1. Slavoj Žižek, *Violence* (London: Verso, 2008), 1–6, 41–42, 56–58, 197–205, 207.
2. William Reddy, "Against Constructionism: The Historical Ethnography of Emotions," *Current Anthropology* 38 (1997): 327.
3. Brian Massumi, *Parables for the Virtual* (Durham, NC: Duke University Press, 2002), 27–28. For Massumi, emotion has a more conservative, less dynamic character: "It is intensity owned and recognized." Ibid., 28. For this same reason William Reddy's notion of emotional regimes, which focuses on ritualized, normative emotions, does not fully capture the sheer dynamism of affect that is at play in the current culture of bullying and psychological intimidation. See William M. Reddy, *The Navigation of Feeling: A Framework for the History of Emotions* (Cambridge: Cambridge University Press, 2001), 129. This dynamism can, by virtue of its tendency to forge unforeseen connections and changes, trouble identities, unmoor significations, bypass cognitive processes, and unsettle intentional structures. This is what Massumi means by *the autonomy of affect*: not its lack of relation to these dimensions but rather "its openness," "its participation in the virtual." Ibid., 35. The relationship with these other dimensions is at once one of "disconnection" and "resonance," of "irreducible alterity and infinite connection." Ibid., 39. For example, affective intensities may upset delineations based on largely stable ideological differences between subjects who may end up participating in a shared sense of vulnerability (and possibly fear) when otherwise they would stand opposed. The affective dynamics are such that they can even blithely ignore facts and established data as, e.g., in the case of the Bush administration's campaign to invade Iraq, which was based on the conditional certainty of what Saddam Hussein would have done had he possessed weapons of mass destruction. For Massumi, this strange temporality of a projected future based on a conjectural past underwrites the new fear-based regime of virtual facts and citizens' vulnerability to it that has emerged over the last several decades. See Brian Massumi, "Potential Politics and the Primacy of Preemption," *Theory and Event* 10.2 (2007), https://muse.jhu.edu/journals/theory_and_event/. This "frighteningly reactive body politic" in the United States is what motivates his philosophical inquiry because of the political quandaries it poses to traditional political analysis. Massumi, *Parables for the Virtual*, 41. Judith Butler has also focused on the affective politics of the temporality of the endless War on Terror and the doctrine of indefinite detention. See Judith Butler, *Precarious Life: The Powers of Mourning and Violence* (London: Verso, 2004), 68, 130. I thus see Massumi's autonomy of affect in less absolutist terms than does Ruth Leys in her analysis of his position, which, according to her, posits no relation whatsoever to intentionality and signification however broadly defined. See Ruth Leys, "The Turn to Affect: A Critique," *Critical Inquiry* 37 (Spring 2011): 434–72. For a discussion of Leys's essay and her concept of inten-

tionality, see Murray Schwartz, "Locating Trauma: A Commentary on Ruth Leys's *Trauma: A Genealogy*," *American Imago* 59.3 (2002): 367–84; and Charles Altieri, "Affect, Intentionality, and Cognition: A Response to Ruth Leys," *Critical Inquiry* 38 (Summer 2012): 878–81.

4. Lawrence Grossberg, *We Gotta Get Out of This Place: Popular Conservatism and Postmodern Culture* (New York: Routledge, 1992), 107–9.
5. Daniel M. Gross, *The Secret History of Emotion: From Aristotle's "Rhetoric" to Modern Brain Science* (Chicago: University of Chicago Press, 2006), 6; Grossberg, *We Gotta Get Out of This Place*, 72.
6. Jesse McKinley, "Suicides Put Light on Pressures of Gay Teenagers," *New York Times*, October 3, 2010.
7. Luke Salkeld, "Facebook Bully Jailed: Death Threat Girl, 18, Is First Person Put behind Bars for Vicious Internet Campaign," *Daily Mail*, August 21, 2009, http://www.dailymail.co.uk/news/article-1208147/First-cyberbully-jailed-Facebook-death-threats.html#ixzz11kBWFvlc.
8. "Tomohiro Kato Sentenced to Death over Tokyo Stabbings," *BBC News*, March 24, 2011, http://www.bbc.co.uk/news/world-asia-pacific-12844673.
9. Philipp S. Fischinger, "'Mobbing': The German Law of Bullying," *Comparative Labor Law and Policy Journal* 32.1 (2010): 153–84; Vanessa Johnston, "Workplace Bullying Trial Begins," *Deutsche Welle*, January 20, 2010, http://www.dw.de/dw/article/0,,5150735,00.html.
10. François Krug, "Chez France Télécom, le nombre de suicides a chuté," *Le monde*, September 15, 2011.
11. Anne Chemin, "Débat sur la violence psychologique au sein du couple," *Le monde*, February 25, 2010.
12. Olivia Chung, "Foxconn Suicide Toll Mounts," *Asia Times Online*, May 22, 2010, http://www.atimes.com/atimes/China_Business/LE22Cb01.html.
13. Roddey Reid, "The American Culture of Public Bullying," *Black Renaissance Noire* 9.2–3 (Fall–Winter 2009–10): 174–87; "La culture d'intimidation aux Etats-Unis," *Esprit*, August–September 2009, 50–68.
14. Stephen Erlanger, "Norway Suspect Denies Guilt and Suggests He Did Not Act Alone," *New York Times*, July 25, 2011.
15. Pierre Bourdieu et al., *The Weight of the World: Social Suffering in Contemporary Society*, trans. Priscilla Parkhurst Ferguson (Stanford, CA: Stanford University Press, 1999); Christian Dejours, *Souffrance en France: La banalisation de l'injustice sociale* (Paris: Seuil, 1998).
16. Butler, *Precarious Life*, 19–49.
17. Axel Honneth, *Disrespect: The Normative Foundations of Critical Theory* (Cambridge: Polity, 2007); Nancy Fraser, "From Redistribution to Recognition? Dilemmas of Justice in a 'Post-Socialist' Age," in *Redistribution or Recognition? A Political-Philosophical Exchange*, by Nancy Fraser and Axel Honneth, trans. Joel Golb, James Ingram, and Christiane Wilke (London: Verso, 2003), 68–93; Richard Sennett, *The Corrosion of Character: The Personal Consequences of Work in the New Capitalism* (New York: Norton, 1998).

18. Marie-France Hirigoyen, *Le harcèlement moral: La violence perverse au quotidien* (Paris: Syros, 1996), *Le harcèlement moral dans la vie professionelle* (Paris: Découverte, 2001), and *Stalking the Soul: Emotional Abuse and the Erosion of Identity*, trans. Helen Marx (New York: Helen Marx, 2000); Alain Ehrenberg, *La société du malaise* (Paris: Odile Jacob, 2010).
19. Butler, *Precarious Life*, 128–51. For a critique of the ideological limits of injury and victimhood as the basis for a feminist redress against contemporary violence for replicating nineteenth-century traditional middle-class concepts of womanhood and liberal subjectivity, see Wendy Brown, *States of Injury: Power and Freedom in Modern Society* (Princeton, NJ: Princeton University Press, 1995).
20. Lorealeigh Keashly and Karen Jagatic, "By Any Other Name: American Perspectives on Workplace Bullying," in *Bullying and Emotional Abuse in the Workplace: International Perspectives in Research and Practice*, ed. Ståle Einarsen, Helge Hoel, Dieter Zapf, and Cary L. Cooper (London: Taylor & Francis, 2003), 31–61; Peter M. Glendinning, "Workplace Bullying: Curing the Cancer of the American Workplace," *Public Personnel Management* 30.3 (2001): 269–86.
21. Caroll M. Brodsky, *The Harassed Worker* (Lexington, MA: Heath, 1976).
22. See, e.g., Stanley Bing, *Crazy Bosses: Spotting Them, Serving Them, Surviving Them* (New York: William Morrow, 1992); Harvey A. Hornstein, *Brutal Bosses and Their Prey* (New York: Riverhead, 1996); Robert A. Baron and Joel H. Neuman, "Workplace Aggression: The Iceberg beneath the Tip of Workplace Violence: Evidence on Its Forms, Frequency, and Targets," *Public Affairs Quarterly* (Winter 1998): 446–64; and Gary Namie and Ruth Namie, *The Bully at Work* (Maperville, IL: Sourcebooks, 2000).
23. Ståle Einarsen, Helga Hoel, Dieter Zapf, and Cary L. Cooper, "The Concept of Bullying at Work: The European Tradition," in *Bullying and Emotional Abuse in the Workplace: International Perspectives in Research and Practice*, ed. Ståle Einarsen, Helga Hoel, Dieter Zapf, and Cary Cooper (London: Taylor & Francis, 2003), 3–30, 6.
24. Gabrielle S. Friedman and James Q. Whitman, "The European Transformation of Harassment Law: Discrimination versus Dignity," *Columbia Journal of European Law* 9.2 (2003): 247–53.
25. Victimology began in the 1970s as a subfield of criminology that focused on the victims of felonious crime and their experience with the criminal justice system, with a view toward obtaining greater acknowledgment of their suffering and losses in the noneconomic sense. See William Parsonage, introduction to *Perspectives on Victimology* (Beverly Hills, CA: Sage, 1979), 9–11.
26. Heinrich Leymann, "Mobbing and Psychological Terror at Work," *Violence and Victims* 5.2 (1990): 119–26; Heinrich Leymann and Annelie Gustafsson, "Mobbing at Work and the Development of Post-Traumatic Stress Disorders," *European Journal of Work and Organizational Psychology* 5.2 (1996):

251–75; Hirigoyen, *Le harcèlement moral dans la vie professionnelle*, 202 (my translation).

27. Tara Parker-Pope, "Meet the Work Bully," *New York Times*, March 11, 2008.
28. Bing, *Crazy Bosses*; Hirigoyen, *Le harcèlement moral dans la vie professionnelle*.
29. Mark S. Micale and Paul Lerner, "Trauma, Psychiatry, and History: A Conceptual and Historiographical Introduction," in *Traumatic Pasts: History, Psychiatry, and Trauma in the Modern Age, 1870–1930* (Cambridge: Cambridge University Press, 2001), 1–27; Ruth Leys, *Trauma: A Genealogy* (Chicago: University of Chicago Press, 2000); Didier Fassin and Richard Rechtman, *The Empire of Trauma: An Inquiry into the Condition of Victimhood*, trans. Rachel Gomme (Princeton, NJ: Princeton University Press, 2009), 77–99.
30. See Plant, chapter 8 in this volume.
31. Allan Young, *The Harmony of Illusions: Inventing Post-Traumatic Stress Disorder* (Princeton, NJ: Princeton University Press, 1995), 109–11; Fassin and Rechtman, *The Empire of Trauma*, 70–76, 90–92.
32. Fassin and Rechtman, *The Empire of Trauma*, 94–96.
33. Lawrence Wright, *Remembering Satan* (New York: Knopf, 1994).
34. Fassin and Rechtman, *The Empire of Trauma*, 1–4.
35. Ibid., 119–20, 124–27.
36. Leymann and Gustafsson, "Mobbing at Work," passim; Hirigoyen, *Le harcèlement moral dans la vie professionnelle*, passim; Christian Dejours, ed., *Conjurer la violence: Travail, violence et santé* (Paris: Payot & Rivages, 2007), 74–78. This application of the concept of trauma to the more routine domains of daily life such as the workplace and school is unaccountably overlooked by Fassin and Rechtman, whose stated goal was nonetheless to perform a study of "the traumatization of daily life" and an "anthropology of common sense." Fassin and Rechtman, *The Empire of Trauma*, 22, 76.
37. David C. Yamada, "The Phenomenon of 'Workplace Bullying' and the Need for Status-Blind Hostile Work Environment Protection," *Georgetown Law Journal* 88.3 (2000): 475–546.
38. Friedman and Whitman, "The European Transformation of Harassment Law," 245.
39. Council of Europe, "Art. 26: The Right to Dignity at Work," *European Social Charter*, rev. (Strasbourg, 1996), 15.
40. James Q. Whitman, "The Two Western Cultures of Privacy: Dignity versus Liberty," Yale Law School Faculty Scholarship Series, Paper no. 649 (2004), http://digitalcommons.law.yale.edu/fss_papers/649.
41. Neil M. Richards, "The Limits of Torts Privacy," Washington University School of Law Legal Studies Research Paper Series, Paper no. 11-06-06 (June 2011), 369–70.

42. Stephen Brooks, "Hate Speech and the Rights Cultures of Canada and the United States," *49th Parallel: An Interdisciplinary Journal of North America Studies* 13 (Spring 2004), http://www.49thparallel.bham.ac.uk/back/issue13/brooks.htm.
43. American jurisprudence surely accounts in part for the dearth of sustained academic research, mentioned above, in the United States into the strictly emotional harm of bullying in the workplace as opposed to studies of outbreaks of felonious assaults and shootings.
44. Frederick Schauer, "The Exceptional First Amendment," Faculty Working Papers, John F. Kennedy School of Government, Harvard University (February 2005), 18–19.
45. Richards, "The Limits of Torts Privacy," 369–70, 378–79.
46. Schauer, "The Exceptional First Amendment," 15, 19, 24.
47. Richards, "The Limits of Torts Privacy," 377–78.
48. Jeffrey Schulman, "Free Speech at What Cost? *Snyder v. Phelps* and Speech-Based Tort Liability," *Cardozo Law Review De Novo* 313 (2010): 314–15.
49. *Snyder v. Phelps*, 131 S.Ct. 1225 (2011).
50. Richards, "The Limits of Torts Privacy," 382.
51. Ralph Slovenko, "The DSM in Litigation and Legislation," *Journal of the American Academy of Psychiatry and Law* 39 (2011): 6–11.
52. Claire Bonafons, Louis Jehel, and Alain Coroller-Béquet, "Specificity of the Links between Workplace Harassment and PTSD: Primary Results using Court Decisions, a Pilot Study in France," *International Archives of Occupational and Environmental Health* 82 (2009): 663–68.

CODA

Erasures: Writing History about Holocaust Trauma

CAROLYN J. DEAN

It is certain, that the same object of distress, which pleases in a tragedy, were it really set before us, would give the most unfeigned uneasiness; though it be the most effectual cure to languor and indolence. DAVID HUME (1757)

In her important book on emotions in the early Middle Ages, Barbara H. Rosenwein has underscored how histories of emotions have long assumed a trajectory from the so-called barbarism of the Middle Ages to the internalized self-restraint characteristic of civilized culture. She argues that such a paradigm should no longer hold sway.[1] Her call to examine emotions in the context of thicker descriptions of the contexts in which their parameters and normative frameworks are defined as well as to analyze the impact of various social and cultural influences shaping emotions is extremely important. At the same time, such efforts to use contextualization may falter in relation to certain questions, and it is one of those questions that I would like to pose: How do we assess the emotions invested in the reception of Holocaust survivors' testimony? How do historians' own emotions intrude into the interpretation and use of such memoirs? Obviously, context matters. Indeed, I seek to contextualize recent approaches to survivors' experiences, which have changed over time and in relation to different kinds of affective responses to them. Yet, as many critics have argued,[2] contextualization has also proved inadequate to describing extraordinary events because reconstructing and interpreting context is itself fraught with difficulty: the reduction of emotions to context, however creatively wrought, is compromised by the way in which the imaginative work of interpretation inevitably supplements texts and relies on multiple contexts (which are often hypostatized by reference to *the* context).[3] Context must still be interpreted in the present, and judgments must be made about which contexts are more significant than others.

Perhaps to avoid the dramatic way in which survivor memoirs raise these problems about how to interpret traumatic emotional experiences and their relationship to events, historians and other scholars often bracket emotions in their approaches to memoirs by creating a conceptual division between history and memory. History is a reliable source of factual data, and memory is an unreliable source tainted by emotions. This division, which most historians understand is never entirely stable, obscures the problem of relying on a purely contextual and empirical reconstruction of the past when analyzing enormous emotional events such as the Holocaust and in particular the traumatic memory of victims. The divide between history and memory in the vast historical work on the Holocaust not only brackets emotional response but also often frames it as an impediment to veracity. Here, contextualization simply reduces memoirs to a binary between the emotional experience of an event and that which historians can verify really happened. Many historians still treat emotion as a surplus of meaning that cannot be analyzed or should be controlled for in sample data. In what follows, I argue that victims of traumatic events pose a particular challenge for historians who want to reconstruct victims' emotional experiences in all their complexity. I argue that victims of the Holocaust of European Jewry, and victims of other catastrophic or genocidal events, elicit negative emotions when they do not conform to a concept of ideal victim behavior that has progressively asserted itself. Thus, analyzing historians' and others' scholarly efforts to conceive the experiences of victims of persecution is an important means of approaching the challenges we face in writing the history of emotions, especially if we are to move beyond contextualization.

In discussions of victims of genocide, *emotions* is usually a reference to trauma or traumatic memory. Trauma is now a common reference in psychiatric studies and even in popular cultural idiom (trauma centers, e.g., are set up for people in the aftermath of a natural or man-made catastrophe). Yet the vast array of discussions about trauma by public intellectuals and literary theorists as well as historians embeds victims' testimony within a set of contested rhetorical and often implicitly negative claims about victims. The historian Annette Wieviorka dubs ours *the era of the witness*,[4] in which Holocaust survivors and other victims become subjects of media spectacles, the greater their suffering, the better. Moreover, this quest to appeal to the popular voyeurism attached to violent and extreme events encourages some people to make up stories about past persecution in attention-seeking ploys that undermine the credibility of those who have really suffered. Thus, the famous child memoir by Binjamin Wilkomirski won its author prizes and acclaim until investigation revealed that he had made it all up.[5]

Many critics blame all the attention paid to Auschwitz as an icon of "evil in our time" for developing a public appetite for more and more suffering that affects our views of victims. In any event, there is little question that a dominant strain of public debate holds that media transform survivors' experiences into melodrama. In this context, some scholars like Michael A. Bernstein and Christopher Browning seek to return to an empirical history of injury uncontaminated by memory and favor a documentary style that avoids the potential conversion of suffering into kitsch or hyperbole. For the most part, critics neglect questions about how broader cultural views of media exploitation and the skepticism or adoration of victims these views have encouraged shape how victims' emotional experiences are perceived and expressed. But how views of victims are shaped by media exploitation is one context of many we might explore. Moreover, we might ask more broadly too how affective relations to victims are mobilized, constituted, and institutionalized and thus how victims' experiences are shaped not only by contexts but also by unconscious emotional responses to them.

In order to identify real victims and focus on empirical truths about their experiences, some historians often inadvertently erase victims' experiences and identities in ways great and small.[6] Other discussions, mostly by literary theorists, attend more closely to victims' emotions but also often dehistoricize them. Victims' memoirs say a great deal about how their authors are made to feel that they impose on others, become objects of contempt or pity by virtue of their demands, or have experiences whose painfulness is blotted out by those who care the most for them. These experiences are often distorted by a usually inadvertent refusal to acknowledge fully victims' past or present suffering. Moreover, this erasure of real experience is repeated, ironically, in diverse disciplinary discourses, all of which seek to recover the survivor's experience.

When survivors write about their postwar experiences, they tend to report how well-intentioned people refused in various ways to acknowledge victims' disempowerment. But scholars do not usually address this kind of testimony or discuss it as part of a narrative of what happened. The Austrian survivor Ruth Klüger recounts the contempt with which she was treated by American Jews when she settled in the United States, as if having been in a camp were a source of contamination and degradation. Eva Hoffman tells of the overt condescension with which her Canadian benefactors treated her family when they emigrated from Poland after having survived the war.[7] Primo Levi's anxiety about others' indifference and his own isolation is perhaps the most famous expression of the fear that one's experience may not be acknowledged. In Auschwitz, Levi had a recurring dream in which he tries to explain the camps to his sister, her friend, and many other people, but they remain uninterested, and his sister finally walks away.[8] Charlotte Delbo's memoir is a long testimony not only to her time in Auschwitz but also to the difficulty of being heard: to the awkwardness she provoked in others and even to the self-protective narcissism of a friend's husband, who could not bear what his wife had suffered. In the guise of feeling for her, he takes her place and narrates her own experiences, "remembering," for example, the geography of Birkenau better than she does.[9]

In Anna Langfus's fictionalized memoirs, the Polish social world in which she lives is not interested in what she endured during the war. She becomes, in a quest for normalcy, the companion of a man who speaks about his own experience of the war as if it were comparable to hers, about which he never asks. He merely pities her for being so alone.[10] And, though her memoir hardly exhausts a long list, Hélène Berr most vividly expresses the experience of having had her persecution entirely erased even before she herself was murdered at Bergen-Belsen. A gentile woman is horrified when she realizes that the Nazis are deporting Jewish children from France and on a visit to the Jewish Berr family rushes in to share her feelings:

> She asked Maman: "You mean to say they are deporting children?" She was horrified. It's impossible to express the pain that I felt on seeing that she had taken all this time to *understand*, and that she had only understood because it concerned someone she knew. Maman presumably felt the same thing I did and replied: "We have been telling you so for a whole year, but you would not believe us."
>
> Not knowing, not understanding even when you do know, because you have a closed door inside you, and you only can *realize* what you merely know if you

open it. That is the enormous drama of our age. Everyone is blind to those being tortured.[11]

An interesting set of essays by Christopher Browning about how historians might use both survivors' and perpetrators' testimony confirms my observations thus far. Browning does not address how victims' experiences are always embedded in broader cultural discourses, nor does he provide a way of recognizing these kinds of erasures, though he seeks to capture victims' experiences. He argues that there are essentially two ways of approaching testimony: from the critical perspective of the compassionate historian or from the perspective of those who refuse to sit in judgment on survivors and thus transform history into a form of commemoration. He establishes the risks involved in the latter by reference to the fraud perpetrated by Binjamin Wilkomirski, a Swiss gentile who passed himself off as a child Holocaust survivor in a best-selling memoir, an incident to which I have already alluded. Browning also refers to an Israeli court's conviction and then admission of error in the case of John Demjanjuk, a Ukrainian living in the United States who had been identified at trial by survivors as the sadistic camp guard Ivan the Terrible. Later evidence proved survivors' memories wrong, though Demjanjuk was deported from the United States to Germany in 2009, this time to face charges of war crimes in Sobibor, a Nazi extermination camp.[12]

Browning insists that he wishes to examine not the content of testimony but the form, by which he does not mean its rhetorical style. He is less interested in "retelling and narrative construction," in "how survivors live their experience," or in trauma than in how to construct a history from testimonies that are often contradictory and even erroneous.[13] His aim is to establish patterns of evidence whose preponderance points primarily to one conclusion rather than another. He does not dismiss testimony. To the contrary, he uses it in a gentle and compassionate manner and seeks to demonstrate its value. But he reiterates the necessity of returning to a critical history that would do away with the idealizations and projections typical of commemoration.

Browning is too sophisticated a historian to draw a bold line between history and memory, as his effort to constitute patterns of evidence suggests. But, if memory is always part of the reconstitution of the past, such patterns necessarily privilege some forms of memory over others because they are verifiable and neglect other, messier memory traces that might be extremely revealing. In the end, Browning insists finally on a neopositivist use of as empirically verifiable as possible memories and reintroduces the divide between history and memory through the back door.

In contrast, literary theorists understand that traumatic memory in particular compromises history conceived as a straightforward narrative account of the unfolding of events that the force of trauma by definition defers, represses, renders opaque or all too vivid but difficult to recount. Sidra DeKoven Ezrahi and Sem Dresden have both discussed how testimonies and memoirs complicate the divide between history and memory because of the very nature not only of trauma but also of language and the representation of atrocity itself.[14] More recently, Zoë Waxman has argued that testimony that is documentary evidence must be conceived not only in relation to its empirical utility but also as a mode of understanding how the constraints placed on victims may have altered the ways in which victims express their feelings. "The nature of the camps rendered obsolete" the "model of witnessing prescribed by Emmanuel Ringelblum and the staff of Oneg Shabat [in the Warsaw Ghetto]." Waxman notes that "concentration camp life militated against" the kind of documentation of Jewish life still possible in the ghettos and that physical and psychological suffering may well mean that testimonies from the camps were not necessarily written, as she puts it, with "objectivity" and "comprehensibility."[15]

When the literary critic Thomas Trezise questions the psychoanalyst Dori Laub's manner of interviewing survivors, he argues that no understanding of victims can be divided neatly into what victims have told us and what historians have established happened (the former emphasizing a performative self-making before an audience and the latter aimed at developing a constative narrative of events).[16] He would seem then to offer an alternative to narrow contextualization that echoes the concerns of other literary critics about the importance of the fluid boundary between history and memory. These different levels of meaning-conferring language, Trezise argues, require consistent attention to how different aims of inquiry limit our full grasp of testimony's hybrid nature and in particular its multidimensional relation to various registers of language. If our aim is to witness an individual experience *or* to understand testimony as evidence of a collective experience of persecution, we will fail to grasp that "not all our memory-institutions, not even all the genres [such as history or literary theory] in which testimony might be housed, have proven or are likely to prove adequate to 'transmit the dreadful experience.'" He argues that, "since the experience itself pertained to the very destruction of community and hence could only leave testimony to seek a temporary home in its cultural ruins, to haunt the remnants of genre just as more generally the Holocaust continues to haunt its own historical aftermath."[17] In other words, no genre is ever adequate to represent the testimony of a community's obliteration. Memory will forever haunt contemporary consciousness, but its meaning can never be fully recovered.

In Trezise's interesting but problematic view, history is not a set of empirical referents but made up of fragments of a world that cannot be recovered. But if this is true—and we cannot deny that it is partially so—we cannot explain how memory is figured by and embedded within a particular historical and social organization or location. We cannot identify more specifically how the meaning of victimization is constituted at the nexus of victims' testimonies and a generalized set of cultural assumptions about what it means to be or have been a victim. That is, the incommensurable relationship between testimony and documentary evidence asserted by Trezise fails to address how memory that does not correlate with a stable set of empirical references is nonetheless embedded in specific disciplinary and even broader cultural construction of victims' suffering. When Trezise, for example, argues that the testimony of the victims and historians' efforts to document their stories are in and of themselves inadequate to "transmit the dreadful experience," he seeks to make a larger point about the necessity of interdisciplinary inquiry in ensuring how the legacy of the Holocaust will be taught in the future. Selective listening to one kind of inquiry or another at the risk of missing the larger experience of victimization derives from an "anxiety of transmission," the power of which intensifies as those who lived through the events pass away.[18] Trezise is concerned primarily with addressing the consequences of selective listening rather than dwelling on this anxiety, and the reference to anxiety remains purely descriptive and suggestive. But, however casual, his invocation of this anxiety is a reference to some organization of affective response channeled through disciplinary inquiry, in which questions and assertions about victims are already embedded. We might, for example, refer again to the bracketing of emotions in much historical inquiry on the topic and understand it now as derived from this anxiety about getting it right before eyewitnesses have perished. Thus, Laub, himself a survivor, seeks to support the victim at all cost, while historians, whose living witnesses are disappearing, are more than ever eager to establish facts and often disdain the victim's testimony when it may not accord with them. Hence the irony: historians' own emotional and ethical investment in capturing the past leads them to attend to some victims and not others. And we might ask further, Is not the anxiety of transmission also about who future audiences will be and what they will be capable of learning? In what rhetorical contexts will this anxiety of transmission be embedded? Such audiences remain abstract external referents in Trezise's account.[19]

Indeed, with no conceptualization of any cultural referents organizing affective response, it is impossible to grasp fully the meaning and impact of an alleged anxiety of transmission. Surely a reference to the anxiety about the disappearance of Holocaust survivors is one of the most interesting collective emotional responses to the passing of history in recent memory and more than simply a defense against forgetting. But Trezise provides no way of thinking about this emotional response historically. Neither his concept of history in ruins nor a concept of history as metaphoric brick building that always requires revision can account for constructions of victimization and their consequences. That is, without a concept of history conceived in terms of the social and rhetorical organization of affect, it is impossible to understand not only why we remember victims the way we do but also why so many victims were and continue to be erased by contempt, overidentification, condescension, and other forms of projection or aversion attested to by Klüger, Hoffman, and others.

The extent to which historical and other forms of understanding not only reiterate but also depend on various erasures of victims' injuries remains underacknowledged, in spite of all the nuanced commentary on Holocaust memory, on testimony, and the many moving efforts in such work to grasp victims' emotional experiences. Though we may possess extraordinary readings of victims' testimony, questions about what happens culturally when we know who is speaking, how he or she speaks, and various investments in how we listen all shape the reception of what is said. The more innocent the speaker, for example, the more dramatic the injury inflicted will seem to be, yet presumptions of victims' innocence and powerlessness may also be defenses against or anxieties about their rage.

The French historian and essayist Georges Bensoussan notes that victims are told not to hate in the name of being reintegrated into the community. "May your suffering be discreet," the victim is counseled. "May your memory be calm and your desire for revenge muted, for it is a matter of assuring the goodwill of humanity." Society, Bensoussan continues, "has little interest in taking charge of victims, and is only interested in the peace between its component parts."[20] Andrés Nader describes how the final stanzas of some of Ruth Klüger's poetry were removed from German-language anthologies of poems from the camps because their desire for vengeance undercut the depiction of an "innocent victim worthy of pity."[21] The preference among many critics for the ostensibly gentle Primo Levi over the angry memoirist and survivor Jean Améry also implies greater comfort with familiar noble constructions of enduring suffering discreetly rather than giving vent to rage. Améry argues that Nazi crimes will be rendered indistinct by those who will want to move on and writes: "Everything will be submerged in a general 'Century of Barbarism.' *We*, the victims, will appear as the truly incorrigible, irreconcilable ones, as the antihistorical reactionaries in the exact sense of the word, and in the end it will seem like a technical mishap that some of us still survived."[22]

As has been implicit thus far, testimony cannot be assigned either to history or to memory except as it has already been interpreted by the criteria that define and legitimate these two categories of meaning. The cultural imperative that opposes one to the other and the literary theoretical insistence, among some critics, on their infinite incommensurability reduce the complexity of testimony either to contextualization or to incongruence, rendering efforts to explain how their relationship might be affectively and rhetorically constituted difficult to grasp. Without attending to these sorts of questions about the cultural and psychic constraints that shape interpretation, we cannot understand how so much compassion for and knowledge of victims can coexist alongside a profound indifference or aversion to claims to injury and thus cannot acknowledge powerful defenses against recognizing how and why the reality of victimization itself is inextricable from the rhetorical fashioning of the exemplary victim.

Among an increasing number of critics, the power of particular self-protective, normative rhetorical constructions of victims as modest, reticent (they proclaim their suffering quietly and never angrily), life affirming, or innocent stifles other possible representations, including those prominent in a great deal of victim testimony.[23] In particular, this construction of the ideal victim renders all but opaque the depth of cultural ambivalence toward victims underlying so many debates about Jewish victimization that have endured to the present: suspicion of victims, numbness or impatience with them, compassion tinged with revulsion, and revulsion yoked to attachment, some of which emerged in various forms in heated discussions about Jewish passivity during the Holocaust, to which I will turn shortly.

In a 1980 preface to a memoir by a Dutch survivor reminiscent of many critics' preference for minimalism in testimony, the psychiatrist Albert J. Solnit writes: "Dr. Micheels is low-key in personality, in writing, and in remembering. Yet his unhistrionic, matter-of-fact style of telling is itself a reassuring demonstration of the patience and indomitability of those who survived and found a way to bring order out of chaos, coherence out of horror, and an affirmation of life out of mass murder."[24] Twenty-six years later, writing of his concerns about the exploitation of atrocity by false victims, the literary critic Charles Bigsby asserts:

Nor is it so unusual for people to present themselves as victims, itself a familiar tactic of those who wish to deny responsibility for their own lives or seek compensation for their supposed sufferings. In the twenty-first century, this is a commonplace of an increasingly litigious culture. But something else is in play. Money does not seem to have been primary. They share something with the assassin whose marginal existence is transformed by attaching himself to history. They live vicariously. They live publicly. They lay claims to that twenty-first century grail—fame. They emerge out of the shadows into the bright light of attention. . . . And so they synthesise memories, often extracting what seems to them to be the active ingredient of the concentration camp experience. Not for them the extraordinary reticence shown by so many survivors. They steal memories and when those memories are insufficiently dramatic they tend to show a preference for gothic horrors.[25]

Thus, the preference for minimalist style in victim testimonies is but one component of a much broader discourse about the exemplary victim. Over and over, the presumptive cultural demands of false victims (here defined by the dramatic contrast between them and the putatively "extraordinary reticence" shown by many survivors) trump any inquiry into how some victims are deemed more credible than others since we are presumed to know that "real" survivors would by definition be reluctant to seek attention. Reticence thus distinguishes real and false victims. Bigsby made this pronouncement before the headlines screamed a few years later about the very real Holocaust survivor who wrote a fantastic love story about a girl who threw him apples across the prison barrier. He meets her by pure chance after the war, and they marry. The survivor's undoing was not that he had made up having survived the camps but that he had allowed this love story to be passed off as nonfiction, bringing him interviews, television appearances, and thus a tremendous amount of attention.[26]

Jewish Innocence and Jewish Passivity

If we ask how victims are rhetorically constituted in these terms, it becomes clearer how certain cultural truisms are shaped to conform to redemptive or at least to familiar moral narratives. The truism, for example, that all victims of genocide are fundamentally innocent obviously does not exclude passing judgment on them and distinguishes between some victims and others—and not only men in contrast to women and children. We should thus analyze under what conditions assertions of innocence appear to be more or less credible, of what innocence really consists, and what purposes it fulfills.

Some of the judgments imposed on Jewish victims are by now a relic of earlier, self-conscious efforts to address a presumptively anti-Semitic audience by survivors or heirs of victims worried about the narrative representation of Jewish death. One important example is the ongoing, though now more muted, discussions about the Jewish councils in European cities and ghettos. The councils were constituted normally by Jewish community elders and responsible for negotiating with the Nazis, including making decisions about who would be deported and who would not. Though Hannah Arendt's famous denunciation in *Eichmann in Jerusalem* of the councils as fundamentally complicit with the Nazis may have been derided, the harsh judgment of or ambivalence about those Jews deemed to have been complicit cannot be denied.[27] Was the leader of the Jewish council of the Warsaw ghetto who committed suicide, Adam Czerniaków, guilty for having done so instead of having told his community what he knew to be true, or was he a martyr because he refused to sign the "resettlement" edict, knowing that children would be sent to Treblinka? His innocence hardly spares him judgment.

One might easily respond that the simple fact that participants were themselves victims does not and should not relieve them of all agency and that those victims who decided, for example, to became Jewish kapos should indeed be judged for having chosen (for the most part illusorily) to save themselves by hurting others. Such judgments avoid asking difficult questions about the parameters of action in a radically redefined moral economy: the point is not to dismiss the idea that victims had agency (some victims had more than others) but to ask how agency might be evaluated in different contexts.[28] The debate about Jewish passivity in the face of annihilation provides more fodder for discussion because it exposes to what extent assertions of innocence appear to be more credible when victims have resisted their tormentors against the odds, especially when those victims behaved heroically and nobly rather than vengefully or in anger. Arendt was much denounced for having raised the biblical cry about Jews going like sheep to the slaughter, though she had actually noted that Jews were no different from any other victimized group in this respect.[29] Nonetheless, the controversy over her apparent insensitivity raged because the question about Jewish passivity clearly touched a very raw nerve. Samuel Moyn has demonstrated to what extent the issue also resonated in the context of the French controversy over Jean-François Steiner's book on Treblinka some five years later. One of the central concerns of the controversy was whether Jews had been passive in the camps or had organized resistance and how. Denunciations and defenses of Steiner were rampant.[30] Similarly, when André Schwarz-Bart won the Prix Goncourt for *The Last of the Just* in 1959, several prominent Jewish intellectuals believed not only that he had distorted Jewish religious tradition but also that he had presented the Holocaust in terms of Christian martyrdom, charges he bitterly rejected.[31]

Lucy S. Dawidowicz was still troubled in the early 1980s by the struggle among Jewish survivors and historians to prove that Jews had resisted the Nazi onslaught. She speculated, however controversially, that the cultural tendency to blame the victim derived from a "modern sensibility which values activism and misunderstands the heroism of martyrdom" (meaning, as I interpret it, that heroism applies to the mass murder of innocents who could not fight back but went knowingly, and therefore heroically, to their deaths).[32] Her reference was to the ongoing ambivalence toward Jews who did not evidently resist when rounded up to be shot or deported, but it confirms the necessity of turning disempowerment itself into a form of heroism in order to cope with the shame associated with victimization, even when shame is related to industrialized murder.

In his 1992 preface to a book on the history of persecution under the Vichy regime, the historian François Bédarida demonstrates just how vibrant such questions remain, especially in Europe, even as they have been discredited. Bédarida notes that all the questions about Jewish passivity make no sense. "Without any doubt," he writes, "the heroes of the resistance saved Jewish honor, even if one may still ask what dishonor might reside in being the victims of a massive and systematic operation of extermination. Can one think of reproaching the hostages killed in reprisal for armed resistance to the Germans for having been shot? Or the victims of Oradour-sur-Glane, or of the Ardeatine

Caves, for letting themselves be massacred?"[33] And Berel Lang has recently sought to dismantle questions about why Jews did not resist by proving them to be logical fallacies, and by engaging such questions he demonstrates their continuing relevance.[34] One might also point to the preface to the first English version of Filip Müller's *Eyewitness Auschwitz*, in which the historian Yehuda Bauer leads the reader casually through a series of possible responses to the testimony that stand in stark contrast to established truisms. Some readers, he notes, "will notice the lack of successful resistance," a phrase suggesting that efforts to resist are prevalent in the memoir. He goes on gratuitously: "Some will notice the fact that the author does not mention any case of Jews begging for their lives; some will emphasize the fact that most of the Jews going to the gas chambers had no inkling of what was happening to them; others will analyze the behavior of the SS murderers."[35] In short, Bauer anticipates and then "corrects" any misapprehensions about Jewish passivity, in the process directing readers to envision Jews as unknowing and courageous in the face of death. Evidently, Israel's military victory in 1967 did not put ideas about passive Jewry to rest. It underscored a new fantasy that Jews, now embodied by the state of Israel, were all-powerful, and generated the "tough Jew" phenomenon dissected by Paul Breines, but did not decrease the contempt for Holocaust survivors in Israel itself because of their alleged passivity.[36]

In intra-Jewish debate, post-Holocaust discussions no longer engage with these questions. They nonetheless reiterate skepticism about victims, as if victim blaming were no longer a cultural problem in reference to Jews. There appears to be confidence that skepticism about victims provides a welcome counterweight to victim sanctifiers. The questions we have mostly left behind (Were Jews too passive? Why didn't they resist? Didn't some of them collude with the Nazis?) focused primarily on degrees of agency—projecting too much or too little of it onto victims rather than asking what kind of agency they had and how it might have been exercised. Some historians' sober assessments have had little impact on these questions, which are symptoms of a more profound cultural problem with and skepticism of constrained agency, constrained intention, and constrained will, even or perhaps especially in the context of the terrorized disempowerment that confronts so many victims.

Surely these questions were given their particular force by efforts both to fashion and to counter images of Jews as a group deserving of collective punishment, weak and parasitical: a group thought not unlikely to have sought to survive by parasitism or by selling out their own.[37] Thus, many questions about Jewish passivity seem quaint or irrelevant now (though some critics still take time to respond to them) because, from the perspective of many contemporary critics, we are for the most part no longer in thrall to the stereotypes that motivated them: indeed, they say, we have sanctified the survivors and transformed the Holocaust into an icon of human-made catastrophe about which we want to speak endlessly. Most of these critics have not been too interested in exploring dimensions of indifference, repulsion, and ambivalence toward Jewish victims, except in terms of irrational fantasies that can be demonstrated empirically to be irrational. They then determine Jewish attitudes toward gentiles to be either overblown and paranoid or reasonable. That is why the historian Peter Novick dismisses Jewish claims about anti-Semitism in the United States as fueled by Jewish paranoia and the allegedly high status that Western culture accords to victims.[38]

I do not wish to insist that victims should embrace their victimization. I wish simply to note that questions about Jewish passivity reveal as much about profound discomfort with or contempt for constrained agency and abjection as they do about anti-Semitic beliefs that Jews occupied a different moral universe than others. Attitudes toward victims are hard to grasp not simply because of various prejudices against them that can be clearly identified and challenged. They are also hard to grasp because scholars' and other critics' affective relationships with victims, especially with those already marginalized psychically and structurally, complicate the ability to do so. The rejection or sacralization of victims of catastrophe are rarely examined as modes by which nonvictims cope with how some victims' behavior disrupts a secure moral universe in which people are presumed to struggle against their humiliation or in some fashion retain their honor (including their equanimity once liberated). Critics rarely address unconscious contempt, fear, and envy of victims that cannot be designated as a self-consciously irrational attitude and controlled for as bias.

All Too Human

The sadly condescending responses to survivors by those who deemed themselves superior to victims before much was known or fears (like Primo Levi's) that one's own family will be indifferent merely confirm the pervasiveness of these and other, equally insidious forms of nonrecognition, especially those most dramatic because they took place when knowledge about the Holocaust was already widespread. Ruth Klüger recounts an evening with her friend the famous German writer Martin Walser, who "forgets" in spite of their friendship that she had survived Auschwitz and, in reference to the tragedy that befell German Jews, suggests that, after all, the fear of foreigners is deep-rooted and universal. "Martin," she writes, with a combination of weariness and pain, "says hatred of Jews was one of those variants of xenophobia which comes naturally to all men. No one wants to deal with differences if you haven't been brought up to tolerate them. I wonder, however: do I really act and look so different from him and his family, who have invited me into their house and at whose table I am sitting?" And she asks now more contemptuously: "Didn't our friendship come easily? Or did it really take virtue, courage, and deep insights on your part?"[39]

Walser's oscillation between knowing and not seeming really to know, between forgetting Klüger's past and remembering the general tragedy that befell German Jews, is not (or does not seem to be) driven by awkwardness or guilt (it is Walser who declared that the Holocaust had become a moral cudgel in postwar Germany that had outlived its purpose). If anything, Walser's own friendship with Klüger becomes, as she notes, evidence of his own particular virtue. Indeed, Walser explains prejudice as natural unless "men" are educated otherwise. His assertion not only naturalizes and thus reduces the impact of prejudice and difference and their effects but also diminishes the violence that was done and is now being replayed again, albeit in the form of a less lethal refusal really to acknowledge the experience of his friend Ruth. As questions about Jewish resistance and passivity denied the impact of terror under certain conditions in favor of motifs of courage and heroism, assertions that hatred of difference is tied to a universal human fear of foreigners also diminish the impact of this fear on victims, leaving them out of consideration. From the victim's point of view, this logic renders it impossible to understand the impact of having been a victim, of having been rendered abject, on recovery.

It is the absence of discussions of the abjection to which victims were reduced as well as its effects (in psychological rather than descriptive terms) that is most striking in discourses about victims: victims speak of contempt, of silence, of arrogance, and, of course, of their own fear (born of having survived something they knew would be difficult to believe). But those who surround and might even be sympathetic to them speak primarily in Walser's vocabulary. Thus, Walser's ideas about the naturalness of attitudes toward foreigners, however painful their effects on his friend, are extremely familiar: certain social attitudes, including that which involves inflicting pain on others, are naturalized so that from an ideological point of view they can be rationalized and explained. Prejudices can coexist with human decency and a commitment to universal rights because they have been transformed into a human attribute without cause: in this case, Walser refers to a hollowed-out tribalism possessed by everyone and thus by no one (though more prominent, he claims, in the uneducated, thus rendering this particular trait even more profoundly human and universal because more primitive). Moreover, since so many murdered Jews were assimilated, Klüger's implicit question about why *foreign* is even applicable in this context reveals that Walser's logic is even more insidious because it renders anti-Semitic hatred and its effects benign.

The naturalization of discrimination and its effects, including the abjection of the victim, occurs in a wide variety of contexts, and its flattening of the social world most dramatically obscures injury. The transposition from the historical to the transhistorical, the slippage from a conception of history as a brick-building-like narrative that attends only to the facts to a narrative about human history that is utterly dehistoricized in its assertions about how human beings are "like that," makes even traumatic injury part of the natural course of things. In Tzvetan Todorov's long essay about how Bulgarians saved Jews, there is little about the absence or presence of anti-Semitism and a lot about human frailty, decency, and historical accident.[40] According to Todorov, Bulgarians are not particularly anti-Semitic and tend to be tolerant of minorities: the Orthodox Church opposed deportation (though particularly of converted Jews), and other Bulgarians did so because the Germans had not occupied the country and the regime still permitted opposition. When things heated up, some men emerged as heroes because they were decent, courageous, and unselfish, in spite of having once supported anti-Semitic laws. Todorov distinguishes Boris Peshev, a member of the Bulgarian National Assembly who publicly opposed the government's effort to deport Jews, "from other courageous and unselfish men" not by virtue of his conscience "but [by] his strategy. He did what had to be done under these particular circumstances in order to achieve his goal." Peshev voted for anti-Semitic laws and thus presumably believed in some of their premises and goals. But, when he realized that the Nazis wanted to murder the Jews, he rose to the occasion, behaving virtuously, courageously, and exceptionally. Todorov's essay is finally about the virtues of decent men rather than about what happened to Jews, how, and why. Anti-Semitism has little to do with it: decent people, not necessarily destined for greatness, triumph or fail. Indeed, Jews appear in the narrative only as people who plead for help or who are already dead: "The sacrifice [*sic*] borne by the Thracian Jews would thus serve to save their brothers, because it would rouse the conscience of Bulgaria's legislators and high clergy." "Under other circumstances," Todorov adds, open dissent might have been "futile, even suicidal."[41]

In such accounts, ordinary people do extraordinary things when natural and profoundly human attributes emerge under extraordinary circumstances. In other, more sophisticated historical and sociological accounts, this naturalization tends to be the result not of regression to some deep recess of what constitutes the human, for better or worse, but, paradoxically, of complex social processes that favor the development of distance between victims and perpetrators, turning presumably feeling people into numb ones. This post-Weberian argument in its mostly functionalist forms about the relationship between Nazism and modernity and extermination and bureaucracy is so well-known (or perhaps by now so seemingly self-evident) that I will summarize it but briefly to demonstrate the impact of its cognitive and unsatisfying approach to emotions. Emotions are not only dehistoricized because rendered human but also naturalized in the sense that they are cognitively universal responses to external stimuli.[42] Christopher Browning argues in his work on Nazi perpetrators: "This approach emphasizes the degree to which modern bureaucratic life fosters a functional and physical distancing in the same way that war and negative racial stereotyping promote a psychological distancing between perpetrator and victim."[43] The sociologist Zygmunt Bauman also argues that this enabling distance is socially produced by the institutions of modernity, in particular bureaucracy and its corollary, instrumental rationality.[44] Others have criticized the lack of any attention to anti-Semitic ideology in such arguments.[45]

Here, I want less to emphasize ideology than to note that for the purposes of explaining bystander or perpetrator behavior these functionalist arguments have more in common with the radically different but ideologically commonsensical approaches of Walser and Todorov than meets the eye. Of course, the reference to negative racial stereotyping and the modern bureaucracies and technologies that enabled its diffusion and implementation cannot explain why these institutions and tools were effective, as those who wish to focus on anti-Semitic ideology have long claimed. But it is also true that, in spite of their efforts not to revert to ahistorical categories like *human nature* and to put Nazi action into a specific historical context, these accounts finally explain how modernity turned Jews into things by reference to what human beings (in these cases, non-Jews) do under extraordinary circumstances—meaning that these are somehow natural emotional responses. That is, the brutalization of bystanders effected by modernity and its institutions may mean simply that under brutal conditions human beings behave brutally. Thus, as Browning and others have noted, when circumstances generate an insuperable divide between communities, people do not wish to be excluded from the group to which they are designated as belonging.[46] It is perhaps not surprising that responses by social scientists to the question of how the vast majority in Nazi Germany or societies occupied by the Nazis allowed the Jews to be exterminated also use references to how humans behave universally under extraordinary circumstances: not wanting to know, indifference, tacit consent under pressure, and a focus on one's own deprivation and fear.[47]

It is not simply that none of these accounts provides explanations about how anti-Semitism took the extreme forms that it did—a particularly difficult question and one that has not been answered by those who focus on ideology—but that all these accounts naturalize the responses of bystanders (and sometimes perpetrators) as quintessentially human behavior under pressure.[48] They do so by universalizing human emotion, with the result that anti-Semitism disappears into the vast and self-evident history of how human beings behave, shorn of any historical specificity, aim, or object.[49] Thus, many social scientists as well as humanist writers do not address the sometimes hallucinatory quality of victims' experiences, especially those that cannot be assimilated into normative understandings of everyday life.

The recasting of anti-Semitism into human experience under pressure also fails to acknowledge the affective relations between Germans and Jews, for example, or between the writers we have discussed and the victims themselves. It is difficult to understand how these writers think about victims or whether they do so at all. Walser re-creates his and Klüger's past in his surely unconscious effort to exculpate Germans, and one has to presume that Todorov's desire to redeem Bulgarians comes at the expense, again unconsciously, of any attention to Jewish victims (other than those who were already dead). Other historians do not address Jewish victims at all except as already other, already abject.

Historians of Jewish experience have decried in other terms the scholarly reiteration of Martin Walser's erasure of Ruth Klüger's experience of victimization (at least as I have characterized it here). They have noted that functionalist and cognitive psychological accounts not only neglect anti-Semitic ideology (and thus neglect the force of prejudice understood as the projection of fears about one's self onto alien others) but also transform Jews into figures in the background that are hard to make out and focus solely on Nazi perpetrators (and bystanders to a lesser extent). Saul Friedländer has suggested that historians may even deflect their anxiety about the abjection of the victims by rendering them invisible or mass-like.[50] Most historians of anti-Semitism counter these tendencies primarily, if not exclusively, by documenting the victim's point of view. But they often still engage testimony as exemplary of the verifiable consequences of discrimination. Though literary theorists tend to focus on texts and the representation of atrocity,[51] with only moderate interest in those texts' relationship to historical context, historians, as noted previously, often bracket the quality of testimony that exceeds contextualization (its often hallucinatory quality, its gaps, its performative potential to heal) because it is insufficiently related to the context that generated it.

To the extent that prejudice is defined as natural, if regrettable, discrimination and its impact are evacuated of power to injure. To the extent that most, if not all, historians who emphasize anti-Semitism acknowledge that injury disturbs the putatively natural order of things (meaning that injury can really be recognized as catastrophic), it is reduced for the most part to a condition that must be documented rather than an ongoing struggle to live and recover. This tendency to bracket those aspects of testimony that exceed such documentation and to identify them with memory goes far in explaining how scholars as well as other critical commentators interpret this ongoing struggle to live or to recover in the form of a return to the ordinary, that is to say, to a recognizable human condition characterized by all that history is and memory is not: reticence, caution, modesty, and the absence of histrionics. Michael A. Bernstein argues that these experiences (or the event *the Holocaust*) should be rendered ordinary and thus remain within the realm of the recognizably human and historical not because they are ordinary but because the Holocaust proves that human beings react in extreme fashion to extreme circumstances.[52]

This logic still begs the question not only of how ordinary or extreme human response is constituted ideologically rather than as a natural reaction to events as they occur but also of what counts as ordinary and extraordinary and thus how victims' experiences are evaluated and constituted as real injuries. It also begs the question of what these experiences tell us about the making and unmaking of social identity (a question that mostly literary theorists have pondered primarily in psychological or psychoanalytic terms, as has Trezise) as well as of how people might be made to disappear psychically and even literally. Most important, perhaps, we can see now that *human, history,* and *ordinary* are tautologies whose impact is to cause the presumption that even extraordinary responses are ultimately ordinary because human and historical, leaving those experiences that cannot be so assimilated no conceptual space other than that of the thoroughly suspect realm of memory, now cast as a traumatic injury that has not been mastered and must be bracketed. The tendency to address attitudes toward these experiences as human responses to crises embedded or not in modes of distantiation leaves theoretical room for a tendency to treat these experiences as things that are sacred and defy any sort of human understanding other than religious or in terms of secularized forms of abjection that evoke fascination and pleasure. In short, under pressure of the so-called surfeit of Holocaust memory, the imperative to return to the ordinary and to history against memory has intensified at the expense of understanding how victims must be constituted to be credible and at what cost and tends, as do all forms of neopositivism, to deepen the abyss in which mysticism and things sacred reside.

Toward an Ethics of Writing about Victims

The idea that we have embraced victims in this era of the witness (at least Jewish victims of the Holocaust) evades to what extent this new moral economy is a part of the ongoing historical refashioning of cultural attitudes to victims. These attitudes can be understood only in relational and affective terms in which suspicion, envy, attachment, and aversion are all in the process of being reformulated in reference to a new historical context in which the Holocaust of European Jewry has become problematically the paradigmatic catastrophe. It is crucial to recognize the power of a moral economy in which victims who demonstrated no agency under pressure of circumstances were least respected, least dignified, and perhaps even suspect. The narrative of the Holocaust is one of human willfulness against the odds, of human willfulness and its power to make or unmake the world, that is always implicated in any discussion about victims and victimization, including a profound cultural investment in the moral soundness of those who suffer. The introduction of the traumatized victim of catastrophe in his or her extreme disempowerment and traumatic, deferred, and thus often empirically indecipherable grief appears to have generated as much aversion and suspicion as sympathy. We might recall that Dr. Louis Micheel's memoir, to which I alluded earlier, was extolled because it was not histrionic and was thus reassuring and life affirming. His words not only suggest that we cannot identify with victims who do not reassure and affirm life but also beg the question of why we need victims to do so. This praise of Micheels, like the more pervasive praise of a far more famous memoirist like Primo Levi, avoids focusing on the aversion to their plight that victims themselves so often discuss.[53]

One might also note that those theorists who are not skeptical of victims, like Wendy Brown, still fail to diagnose why victims' recovery is so difficult except in terms that reiterate the importance of insight, will, and agency in seeing through and resisting oppression. According to her, minorities believe that pluralistic political systems will better hear their grievances if they work within the system by defining themselves as having injuries they believe it might respond to. In Brown's work, the appropriation of traumatized identity (or "identity politics") by victims themselves constitutes a misguided concession to dominant culture: the victim's embrace of his wounds reveals an attachment to the recognition accorded his suffering. If neoliberalism is the source not simply of wounds created by inequity but also of an attachment to the wounds it creates, then only a willful struggle for what you want against the state, the refusal to accept its handouts in the form of acknowledgment that you suffer, can make a difference. This argument, whether we agree with its substance or not, does not address the existence of the deep cultural aversion to having been wounded, to the victim's debilitation (as opposed to the perpetrator's self-lacerations) and its impact on self-making and the constitution of exemplary and less desirable forms of personhood. The history of injury may be constituted by the patterns of survivors' testimony that correlate so that memories are verifiable and trustworthy and thus immune to the vagaries of victims' efforts to remember. But it is only criticism that embeds injury in the social relations that shape responses to it (and thus criticism that refuses to avert its eyes from injury) that will allow us to conceptualize impediments to recovery. In so doing, it might enable us to address from victims' points of view how a cultural aversion to or discomfort with their pain might generate the need for narratives that in one way or another reiterate that pain by denying, repressing, or erasing it, thereby compromising recovery.

Cultural discomfort both with constrained agency and with fantasies about others' abjection is present even when victims are unmistakably survivors. Indeed, there is perhaps no better current statement of the discomfort generated by the state of having been victimized than the historian Annette Wieviorka's claim that, merely because she was a *historian* of Auschwitz, her colleagues behaved as if the subject matter would contaminate her, transform her "into a mournful being, living in between evil and death, forbidden the pleasures of life."[54] Surely Wieviorka's own denunciation of our era of the witness is informed by anxiety that media representations of survivors invoke and encourage such a response.

In conclusion, the emotions implicit in Holocaust survivor memoirs appear to be commensurate with a culture in which critics deem emotion to dominate over a rational quest for the truth about victims. Does the proliferation of melodrama pressure victims to appropriate minimalism and emotional neutrality to sustain credibility? Does the rejection of happy endings in nonredemptive works of some memoirists contribute to their marginality (or at least a preference for those with another style)? Surely the postwar emphasis on national redemption in different countries led to the sidelining of victims' stories and to the deemphasis of Jewish suffering in favor of the universal category *victims of fascism*.[55] That is, narratives of redemption—from *Schindler's List* to De Gaulle's touting of the French Resistance against Nazism—risk stigmatizing victims' anger. But one might argue tentatively that a general cultural emphasis on the comportment of victims is a legacy of dignified bearing under pressure so much a part of humanist constructions of dignity. At the very least, we learn by contextualizing views about such memoirs that they offer a surplus of meaning that exceeds efforts to make them conform to an emotional regime.

It is less than clear how we interpret that surplus, which for the most part critics have deemed overwrought, inessential, important but impossible to discuss, and, even in the worst cases, to distort the truth about events. For the victims we have discussed, there is as yet no way of writing the history of the traumatic emotions they have witnessed that at once places them in historical context while respecting the particular truths that memoirs, even those whose truths are not empirically verifiable, can help us understand about events. Gabrielle Speigel has recently remarked: "To incorporate 'terror' into historical representation will mean acknowledging and accepting as historiographically legitimate the differing status of analytically recuperated 'facts' and victim testimony; to find a way to theorize (as has yet to be done) the materiality and reality of 'voices' from the past, without assuming the necessary 'truth' of what they convey, at least in terms of the factuality of its content."[56] Her words caution us against writing a history of the emotions of persecuted people in whatever context that cannot account for different kinds of truths. Her words expand Rosenwein's warning about avoiding reductive normative parameters that define how the scope for feeling changes over time (the modern paradigm of internalized self-restraint). When they write the history of emotional traumas, it is crucial that historians become conscious of their own affective responses to victims and recognize that pursuing one kind of truth may well lead to the neglect of the very voices they seek to recuperate.

NOTES

Portions of the coda have been previously published as "Erasures," in *Aversion and Erasure: The Fate of the Victim after the Holocaust*, by Carolyn J. Dean. Copyright © 2010 by Cornell University. Used by permission of the publisher, Cornell University Press.

1. Barbara Rosenwein, *Emotional Communities in the Early Middle Ages* (Ithaca, NY: Cornell University Press, 2006), 5–15.
2. For a recent and judicious synopsis of the challenges to contextualization, see Martin Jay, "Historical Explanation and the Event: Reflections on the Limits of Contextualization," *New Literary History* 42.4 (2011): 557–71.
3. For the now standard treatment of this aspect of Derridean thought and its impact on intellectual history, see Dominick LaCapra, "Rethinking Intellectual History and Reading Texts," in *Modern European Intellectual History: Reappraisals and New Perspectives*, ed. Dominick LaCapra and Steven L. Caplan (Ithaca, NY: Cornell University Press, 1982), 47–85.
4. Annette Wiewiorka, *The Era of the Witness*, trans. Jared Stark (Ithaca, NY: Cornell University Press, 2006).
5. See Stefan Maechler, *The Wilkomirski Affair: A Study in Biographical Truth* (New York: Schocken, 2001).
6. Raul Hilberg was famously accused of blaming Jews for their own slaughter (see Friedländer's comments on the difficulties of working with traumatic material, cited in the text at n. 50). In his memoir, he ascribes this misinterpretation to the politicized reception of his scholarship. See in particular the anecdote recounted in Raul Hilberg, *The Politics of Memory* (Chicago: Ivan R. Dee, 1996), 153–54. Exceptions to this kind of historical literature are Saul Friedländer's magisterial *Nazi Germany and the Jews*, vol. 1, *The Years of Persecution, 1933–1939*, and vol. 2, *The Years of Extermination* (New York: HarperCollins, 1997, 2007); and Jan T. Gross, *Neighbors: The Destruction of the Jewish Community in Jedwabne, Poland* (Princeton, NJ: Princeton University Press, 2001). I cannot comment at length on these works, but have done so in Dean, *Aversion and Erasure*, 115–28.

7. Ruth Klüger, *Von hoher und niedriger Literatur* (Göttingen: Wallstein, 1996), 35; Eva Hoffman, *Lost in Translation: A Life in a New Language* (New York: Penguin, 1989), 102–4.
8. Primo Levi, *Survival in Auschwitz: The Nazi Assault on Humanity*, trans. Stuart Woolf (New York: Collier, 1961), 256.
9. Charlotte Delbo, *Auschwitz et après*, vol. 3, *Mesure de nos jours* (Paris: Minuit, 1971), 95–98.
10. Anna Langfus, *The Lost Shore*, trans. Peter Wiles (New York: Pantheon, 1964), esp. 86.
11. Hélène Berr, *The Journal of Hélène Berr*, trans. David Bellos (New York: Weinstein, 2008), 204.
12. Christopher R. Browning, *Collected Memories: Holocaust History and Postwar Testimony* (Madison: University of Wisconsin Press, 2003), 42.
13. Ibid., 37–38.
14. Sidra DeKoven Ezrahi, *By Words Alone: The Holocaust in Literature* (Chicago: University of Chicago Press, 1980); Sem Dresden, *Persecution, Extermination, Literature*, trans. Henry G. Schogt (Toronto: University of Toronto Press, 1995). For a summary of this material, see Carolyn J. Dean, *Aversion and Erasure: The Fate of the Victim After the Holocaust* (Ithaca, NY: Cornell University Press, 2010), 147n10.
15. Zoë Waxman, *Writing the Holocaust: Identity, Testimony, Representation* (Oxford: Oxford University Press, 2006), 86. On poetry writing in the camps, see Andrés Nader, *Traumatic Verses: On Poetry in German from the Concentration Camps, 1933–1945* (Rochester, NY: Camden House, 2007).
16. See Thomas Trezise, "Between History and Psychoanalysis: A Case Study in the Reception of Holocaust Survivor Testimony," *History and Memory* 20 (2008): 7–47; and Dori Laub's bitter response, "On Holocaust Testimony and Its Reception within Its Own Frame, as a Process in Its Own Right: A Response to 'Between History and Psychoanalysis' by Thomas Trezise," *History and Memory* 21 (2009): 127–50.
17. Trezise, "Between History and Psychoanalysis," 35–36.
18. Ibid., 28.
19. For a rigorous and accurate treatment and extension of my argument in this chapter, see Gabrielle Spiegel, "The Final Phase?" *History and Theory* 51 (2012): 1–13, a review of Dean, *Aversion and Erasure*.
20. Georges Bensoussan, *L'Auschwitz en héritage: D'un bon usage de la mémoire* (Paris: Mille et une nuits, 2003), 206.
21. Nader, *Traumatic Verses*, 59.
22. Jean Améry, *At the Mind's Limits: Contemplations by a Survivor on Auschwitz and Its Realities*, trans. Sidney Rosenfeld and Stella P. Rosenfeld (Bloomington: Indiana University Press, 1980), 80. The contrast between Levi and Améry is well-known in the literature and discussed in Nader, *Traumatic Verses*, 40–43.

23. *Reticence* is a useful way to describe some survivors' feelings about recounting their past, for fear, as Michael Pollak puts it, that they might render the extraordinary banal in the process of trying to represent their experiences. See Michael Pollak, *L'expérience concentrationaire: Essai sur le maintien de l'identité sociale* (Paris: Suites Sciences Humaines, 2000), 20. For a somewhat different view, see Aharon Appelfeld's memoir *The Story of a Life*, trans. Aloma Halter (New York: Schocken, 2004), 181.
24. Louis J. Micheels, *Doctor #117641: A Holocaust Memoir* (New Haven, CT: Yale University Press, 1980), viii.
25. Charles Bigsby, *Remembering and Imagining the Holocaust: The Chain of Memory* (Cambridge: Cambridge University Press, 2006), 374.
26. The survivor is Herman Rosenblat. See "False Memoir of Holocaust Survivor Cancelled," *New York Times*, December 29, 2008.
27. Hannah Arendt, *Eichmann in Jerusalem: A Report on the Banality of Evil* (1963; New York: Penguin, 1994).
28. Calel Perechodnik, *Am I a Murderer? Testament of a Jewish Ghetto Policeman*, ed. and trans. Frank Fox (Boulder, CO: Westview, 1996). See also Ruth Leys's effort to attribute agency to survivors by emphasizing their guilt. Ruth Leys, *From Guilt to Shame: Auschwitz and After* (Princeton, NJ: Princeton University Press, 2007), 1–16.
29. Arendt, *Eichmann in Jerusalem*, 11–12.
30. Samuel Moyn, *A Holocaust Controversy: The Treblinka Affair in Postwar France* (Waltham, MA: Brandeis University Press, 2005), 117–18.
31. André Schwartz-Bart, *The Last of the Just*, trans. Stephen Becker (New York: Atheneum, 1960).
32. Lucy S. Dawidowicz, *The Holocaust and the Historians* (Cambridge, MA: Harvard University Press, 1981), 134.
33. François Bédarida, foreword to Adam Rayski, *The Choice of the Jews under Vichy: Between Submission and Resistance*, trans. Will Sayers (Notre Dame: University of Notre Dame Press, 2005), ix–xiv, xiv.
34. Berel Lang, *Post-Holocaust: Interpretation, Misinterpretation, and the Claims of History* (Bloomington: Indiana University Press, 2005), 86–99.
35. Yehuda Bauer, foreword to Filip Müller, *Eyewitness Auschwitz: Three Years in the Gas Chambers*, ed. and trans. Susanne Flatauer (New York: Ivan R. Dee, 1979), ix–x, x.
36. Paul Breines, *Tough Jews: Political Fantasies and the Moral Dilemma of American Jewry* (New York: Basic, 1990); Tom Segev, *The Seventh Million: The Israelis and the Holocaust*, trans. Haim Watzman (New York: Henry Holt, 1991).
37. Perechodnik, *Am I a Murderer?* xiii.
38. Peter Novick, *The Holocaust in American Life* (New York: Houghton Mifflin, 2000), 189.
39. Ruth Klüger, *Still Alive: A Holocaust Girlhood Remembered* (New York: Feminist, 2001), 167–69.

40. Tzvetan Todorov, *The Fragility of Goodness: Why Bulgaria's Jews Survived the Holocaust*, trans. Arthur Denner (Princeton, NJ: Princeton University Press, 1999).
41. Todorov, *The Fragility of Goodness*, 30, 39.
42. For a criticism of the historian William Reddy, who sought to merge historical and cognitive psychological approaches, see Rosenwein, *Emotional Communities*, 16–23.
43. Christopher Browning, *Ordinary Men: Reserve Battalion 101 and the Final Solution in Poland* (New York: Harper Collins, 1992), 162. Browning's work leaves room for ideological mobilization regarding how the eventual extermination of European Jewry developed in the Nazi bureaucracy. But he does not integrate these insights into his work on how the killing was rendered possible.
44. Zygmunt Bauman, *Modernity and the Holocaust* (Ithaca, NY: Cornell University Press, 1989). The not entirely outdated functionalist concept of Nazism, in which killing Jews was a by-product of other military and state demands and had little to do with anti-Semitic ideology, simply mirrors this approach in which the distantiation generated by modernity and its demands (or war and its imperatives) renders victims less than human.
45. The most famous example of an ideological approach is Lucy S. Dawidowicz, *The War against the Jews, 1933–45* (New York: Holt, Rinehart & Winston, 1975).
46. Browning, *Ordinary Men*, 188.
47. For a detailed discussion of how what historians referred to as *indifference* to German Jewry was really a form of violence rendered invisible or oddly agentless, see Carolyn J. Dean, *The Fragility of Empathy After the Holocaust* (Ithaca, NY: Cornell University Press, 2004), 76–105.
48. For sophisticated newer accounts, see Dan Stone, *Constructing the Holocaust: A Study in Historiography* (London: Vallentine Mitchell, 2003); and Alon Confino, "Fantasies about the Jews: Cultural Reflections on the Holocaust," *History and Memory* 2 (2005): 296–322.
49. Though Timothy Snyder, an exception, argues that the brutality of human behavior in the "bloodlands" where most of the killings during the Second World War derived from the "double occupation" of Germany and the Soviet Union, he also writes as if Polish anti-Semitism did not exist. Timothy Snyder, *Bloodlands: Europe between Hitler and Stalin* (New York: Basic, 2010).
50. Saul Friedländer, *Memory, History, and the Extermination of the Jews of Europe* (Bloomington: Indiana University Press, 1993), 130–31.
51. Aujke Kluge and Benn E. Williams, introduction to *Re-Examining the Holocaust through Literature*, ed. Aujke Kluge and Benn E. Williams (Newcastle upon Tyne: Cambridge Scholars, 2009), 9.
52. Michael A. Bernstein, "Homage to the Extreme: The Shoah and the Rhetoric of Catastrophe," *Times Literary Supplement*, March 6, 1998, 6–9.

53. "I saw one of the soldiers bend over and vomit, and then another one. And then I understood. It disgusted them to look at us. They found us repulsive." Clara Greenbaum, liberated from Bergen-Belsen, quoted in François Bédarida and Laurent Gervereau, eds., *La déportation, le système concentrationnaire Nazi* (Paris: Collections des Publications de la BDIC, 1995), 201.
54. Annette Wieviorka, *Auschwitz: La mémoire d'un lieu* (Paris: Plurielle, 2005), 11.
55. Pieter Lagrou, *The Legacy of Nazi Occupation: Patriotic Memory and National Recovery in Western Europe, 1945–1965* (Cambridge: Cambridge University Press, 2000). See in particular Lagrou's own discussion of how some victims were defined. Ibid., 202.
56. Gabrielle Speigel, "Writing Terror: The Representation of Violence in Contemporary Historical Writing" (unpublished manuscript cited with permission of the author), to appear in English in Flocel Sabaté, ed., *Por Politica, Terror Social"* (Lledia: Pagès Editors, in press). See also Speigel, "The Final Phase?" 12.

Contributors

JORDANNA BAILKIN is the Giovanni and Amne Costigan Professor of European History at the University of Washington. She is the author of *The Culture of Property* (Chicago, 2004) and *The Afterlife of Empire* (University of California Press, 2012).

FRANK BIESS is professor of history at the University of California, San Diego. He is the author of *Homecomings: Returning POWs and the Legacies of Defeat in Postwar Germany* (Princeton, 2006) and is currently working on a history of fear and anxiety in postwar Germany.

CAROLYN J. DEAN is professor of History at Yale University. Several of her books focus on the relationship between emotions and culture in modern France, Germany, and the United States, including *The Fragility of Empathy after the Holocaust* (Cornell, 2004) and *Aversion and Erasure: The Fate of the Victim after the Holocaust* (Cornell, 2010).

OTNIEL E. DROR is the Joel Wilbush Chair in Medical Anthropology and Head of the Section for the History of Medicine in the Medical Faculty at the Hebrew University of Jerusalem. His coedited *Knowledge and Pain* is forthcoming with Rodopi Press, his coedited Osiris volume *History of Science and the Emotions* is forthcoming with the University of Chicago Press, and his *Blush, Flush, Adrenaline: Science, Modernity and Paradigms of Emotions, 1850–1930* is under revision for the University of Chicago Press. He is currently working on the history of the study of pleasure during the post–World War II period.

HELENA FLAM is professor of sociology at the University of Leipzig and author of one of the key syntheses on the sociology of emotions: *Soziologie der Emotionen—Eine Einführung* (UVK/UTB, 2002; reprinted 2007).

UTE FREVERT is Director of the Center for the History of Emotions at the Max Planck Institute in Berlin. Recent publications include *Emotions in History: Lost and Found*, Natalie Zemon Davis Annual Lecture Series at Central European University, Budapest (Central European University Press, 2011).

CATHY GERE is associate professor of history at the University of California, San Diego. She is the author of *Knosses and the Prophets of Modernism* (Chicago, 2009) and currently works on the history of neurology.

DANIEL M. GROSS is associate professor of English at the University of California, Irvine. His numerous publications on the history and theory of emotion include *The Secret History of Emotion: From Aristotle's "Rhetoric" to Modern Brain Science* (Chicago, 2006) and "Defending the Humanities with Charles Darwin's *The Expression of the Emotions in Man and Animals* (1872)," *Critical Inquiry* (2010).

BETTINA HITZER is a research fellow at the Center for the History of Emotions in Berlin. She works on the history of body-related emotions, especially with regard to cancer in twentieth-century Germany. Publications in the history of emotion include *Im Netz der Liebe: Die protestantische Kirche und ihre Zuwanderer in der Metropole Berlin, 1849–1914* (Böhlau, 2006).

UFFA JENSEN is a research fellow at the Center for the History of Emotions in Berlin. He is working on a study of the reception of psychoanalysis in Europe and in India, and he is also the coeditor of *The Rationalization of Feeling: Science and Emotions during the Turn of the Century, 1880–1930* (in German).

RUTH LEYS is the Henry Wiesenfeld Professor of Humanities, with a secondary appointment in the Department of History, at Johns Hopkins University. She works on the postwar history of experimental and theoretical approaches to the study of the emotions. Publications in the history of emotion include *From Guilt to Shame: Auschwitz and After* (Princeton, 2007).

CATHERINE LUTZ is the Thomas J. Watson Jr. Family Professor of Anthropology and International Studies at Brown University, where she holds a joint appointment in the Department of Anthropology. She is also codirector of the Costs of War research project based at the Watson Institute. Her numerous books and articles include a foundation of emotion studies, *Unnatural Emotions: Everyday Sentiments on a Micronesian Atoll and Their Challenge to Western Theory* (Chicago, 1988).

REBECCA JO PLANT is associate professor of history at the University of California, San Diego. She is the author of *Mom: The Transformation of Motherhood in Modern America* (Chicago, 2010) and is working on a study of combat fatigue during and after World War II.

STEPHANIE D. PRESTON is associate professor of neuroscience at the University of Michigan, Ann Arbor, where she is Director of the Ecological Neuroscience Laboratory, which uses an interdisciplinary approach to study the interface between emotion and decisionmaking. She is coauthor with Frans de Waal of the influential "Empathy: Its Ultimate and Proximate Bases," *Behavioral and Brain Sciences* (2006), along with numerous other publications.

RODDEY REID is professor of French studies and cultural studies at the University of California, San Diego. His work focuses on modern cultures and societies in France, the United States, and Japan as well as on the interdisciplinary study of science and medicine.

WILLIAM M. REDDY is the William T. Laprade Professor of History and professor of cultural anthropology at Duke University. He is the author of a seminal work in the field, *The Navigation of Feelings: A Framework for the History of Emotions* (Cambridge, 2003).

NAYAN B. SHAH is professor of ethnic studies at the University of Southern California. He works at the intersection of the history of race, sexuality, and emotions.

Index